FPL
3.—

Additional Praise for *Widen the Window*

"Dr. Stanley writes with a clarity and intelligence that results in a truly accessible scholarly work on understanding stress, trauma, and a path to healing. This is one of the most important books on meditation since Jon Kabat-Zinn's *Full Catastrophe Living* brought meditation into the mainstream. *Widen the Window* is about healing and recovery. It is about a pathway beyond self-improvement to self-understanding."

—Gary Kaplan, D.O., author of *Total Recovery* and founder of the
Kaplan Center for Integrative Medicine

"This book holds a template for enhancing performance—attention span, focus, rapid recovery from shock, and stress. Although I was initially skeptical, I became convinced when I was able to see, review, and understand the supporting science, in particular the data that shows physiological changes. These outcomes hold value in all environments and conditions, and can help people gain more mastery of their bodies to improve their daily performance, and their overall lives."

—Major General Melvin G. Spiese, U.S. Marine Corps (Ret.), former commander of the
U.S.M.C. Training and Education Command

"In *Widen the Window*, Elizabeth Stanley takes us on her profound journey into experiencing, understanding, and treating the devastating impact of trauma. Through her personal journey, she shares her transformation from numbness to an awareness of the language of her body with an understanding of the important role that the autonomic nervous system plays in our mental and physical health. These experiences have enabled her to develop an innovative treatment model to provide the heroic survivors of trauma with tools to enhance regulation and resilience."

—Stephen W. Porges, Ph.D., author of *The Polyvagal Theory* and Distinguished
University Scientist and Founding Director of the Traumatic Stress
Research Consortium in the Kinsey Institute, Indiana University

"Stanley offers a brave and skillful journey through her personal experiences and the science of stress and trauma. This is a stark look at how society defines strength and success, and how achieving these at the highest levels undermines their very foundation. Stanley gives us an opportunity to rethink and shift our approach to strength and resilience. A must-read for the many driven individuals who find themselves victims of their own drive and success."

—Sarah Bowen, Ph.D., author of
Mindfulness-Based Relapse Prevention: A Clinician's Guide and
Associate Professor of Psychology, Pacific University

"Elizabeth Stanley's new book, *Widen the Window*, is a well-researched, thorough exploration of the causes and treatments of stress and trauma. This book offers hope to those of us who have lived within a narrow window of our potential by offering a composite of proven principles and practices that can free us from a lifetime of conditioning."

—Rodney Smith, author of *Touching the Infinite* and
founding teacher of the Seattle Insight Meditation Society

"Dr. Stanley's book offers invaluable insight on how to handle the stress of everyday life as well as more severe emotional trauma. Her strategies should be a lifeline for military veterans dealing with the wrenching memories of combat, but also for all of the rest of us trying to deal with the mental strain of our lives. Most important, this book offers hope. It shows a path to better mental health."

—Congressman Adam Smith, Chair of the House Armed Services Committee

"Full of stories mixed with science in a friendly, stimulating, and hopeful read, this pioneering book offers an exciting new perspective on stress and trauma. Explaining that we often fail to recognize, and thus neglect, the effects of such ordeals, Liz Stanley teaches the reader how to exchange maladaptive conditioned responses for new, adaptive strategies that increase concentration, enhance performance, and heal mind and body. By integrating top-down with bottom-up approaches, our windows of tolerance can be widened, both individually and collectively, bringing out the best in humanity."

—Pat Ogden, Ph.D., founder of the Sensorimotor Psychotherapy Institute

"At a time when our culture is catching up with the physical, mental, and social costs of stress and trauma, Elizabeth Stanley, in her groundbreaking book, brings a new and much-needed understanding of how these phenomena are intimately linked, and how we can heal them. A must-read for anyone who is in the helping professions."

—Richard Strozzi-Heckler, Ph.D., author of *In Search of the Warrior Spirit* and *The Leadership Dojo*

"Drawing from her personal experience as a longtime meditator and trauma survivor, teacher, and researcher of mindfulness-based programs for high stress, high trauma populations, and training in trauma-oriented therapies, Dr. Stanley's *Widen the Window* provides an accessible and valuable contribution to the burgeoning field of trauma-informed mindfulness."

—Willoughby Britton, Ph.D., Director of the Clinical and Affective Neuroscience Laboratory at Brown University Medical School

"With the excellent books on trauma, Dr. Stanley's book on resilience stands out. It adds a vital link in understanding how we regulate stress and soften the corrosive effects of trauma. Her book offers an evidence- and theory-based understanding of the intimate influences of bonding and attachment, processes that solidify the early roots of resilience. But then, it goes further by illustrating how, at every age, and in any situation, we can learn powerful skills that let us tap into the healing power of resilience. It does this by solidifying the intrinsic and welded unity of mind, brain, and body. A must-read for all therapists and for all of us seeking healing and wholeness."

—Peter A. Levine Ph.D., author of *In an Unspoken Voice and Trauma and Memory*

"In *Widen the Window*, Liz Stanley offers an in-depth understanding of the physiological and psychological impact of survival responses to stress and trauma. Drawing from cutting-edge research, she offers both professional and lay readers practices and strategies for dealing effectively with the challenging and rapid-fire responses that arise when we feel overwhelmed or threatened. I cannot recommend this book highly enough."

—Nancy J. Napier, MA, LMFT, SEP, Faculty, Somatic Experiencing® Trauma Institute, and author of *Getting Through the Day: Strategies for Adults Hurt as Children*

"Liz Stanley has opened a window onto a promising new vista for human healing and thriving. Her powerful insights transform the narrative of how societies and organizations approach (and often ignore, dismiss, or deny) trauma. She offers a state-of-the-art understanding of how traumas are formed as well as effectively repaired. For people whose work routinely puts their lives on the line—and pay a terrible price for it—she offers a ray of hope."

—Jeremy Hunter, Ph.D., Founding Director and Associate Professor of Practice at the Executive Mind Leadership Institute, Peter F. Drucker and Masatoshi Ito Graduate School of Management

WIDEN

the

WINDOW

TRAINING YOUR BRAIN AND BODY TO THRIVE DURING STRESS AND RECOVER FROM TRAUMA

ELIZABETH A. STANLEY, PH.D.

AVERY

an imprint of Penguin Random House

New York

AVERY

AN IMPRINT OF PENGUIN RANDOM HOUSE LLC
penguinrandomhouse.com

Most Avery books are available at special quantity discounts for bulk purchase
for sales promotions, premiums, fund-raising, and educational needs. Special
books or book excerpts also can be created to fit specific needs. For details, write
SpecialMarkets@penguinrandomhouse.com.

LIBRARY OF CONGRESS CATALOGING-IN-PUBLICATION DATA

Names: Stanley, Elizabeth A., 1970–author.
Title: Widen the window : training your brain and body to thrive during
stress and recover from trauma / Elizabeth A. Stanley.
Description: New York : Avery, [2019] | Includes bibliographical references and index. |
Summary: "A pioneering researcher gives us a new understanding of stress
and trauma, as well as the tools to heal and thrive"—Provided by publisher.
Identifiers: LCCN 2019015841 | ISBN 9780735216594 (hardcover) |
ISBN 9780735216617 (ebook)
Subjects: LCSH: Stress (Psychology) | Post-traumatic stress disorder. | Stress management.
Classification: LCC BF575.S75 S726 2019 | DDC 155.9/3—dc23
LC record available at https://lccn.loc.gov/2019015841
p.   cm.

Printed in the United States of America
1 3 5 7 9 10 8 6 4 2

Book design by Lorie Pagnozzi

For my family, with gratitude and love

For those who suffer,
that they may widen their windows

and

For the warriors, from all walks of life,
who bring wisdom and courage to help widen the
collective window

CONTENTS

Foreword

A famous researcher friend of mine once told me that "all research is me-search." Most of us study issues that are immediately relevant to our own well-being. I doubt whether anybody spends his or her life studying the impact of horrendous life experiences and tries to find solutions, unless they themselves have come face to face with their own personal horror. Over the past thirty years there have been numerous excellent books about traumatic stress. These generally fall into roughly two categories: biographies of survivors who tell the story of their journey, and academic books that explain mechanisms, research studies, and prescriptions for treatment. *Widen the Window* embodies the very best of both categories.

I don't think I have ever read a book that paints such a complex and accurate landscape of what it is like to live with the legacy of trauma as this book does, while offering a comprehensive approach to healing that is simultaneously based on both the author's own personal experiences and journey into health, as well as on a thorough scientific understanding of the underlying issues about the ways that the mind, brain, and body are affected by traumatic stress. This insightful and innovative synthesis is all the more remarkable, given that Dr. Stanley is not a trauma clinician—but was trained as a political scientist and teaches about international security.

This book has a solid foundation in Dr. Stanley's own unflinchingly told experiences with multi-generational trauma, childhood abuse, and family alcoholism. Her military experience involved both deployment trauma and, as is all too common for women in the military, harassment within the command structure. One of the things I loved about reading this book was the precision with which Dr. Stanley describes both her post-traumatic symptoms and her systematic way of dealing with them.

The imperative for Dr. Stanley to deal with a plethora of post-traumatic sequelae—problems that every trauma survivor will recognize in very personal way—led to her development of Mindfulness-based Mind Fitness Training (MMFT). But those problems don't begin to define the human being. Like so many other survivors I've known, she is also brilliant, courageous, persistent, competent, self-reliant, focused, and tenacious. This book is the product of her superb intellect, terrific organizational skills, and her deep and courageous self-exploration.

The first step in that process was to learn to courageously meet herself by cultivating nonjudgmental curiosity, a capacity without which true healing seems to be impossible. She, like all survivors, had to meet and befriend the parts of herself that she most despised, avoided, and tried to ignore. That is probably the single most important element in recovering from trauma: finding ways to allow yourself to feel what you feel, and to know what you know.

That simply is the nature of trauma: It's not just something very unpleasant that will fade with time. It is horrendous, too awful to face head on. The memory of traumatic events is fragmented into little pieces: intense emotions, bizarre behaviors, unbearable physical sensations, images, and fragmented thoughts. They are stored outside of our conscious awareness in bodily symptoms and self-destructive actions. These post-traumatic reactions help traumatized individuals to cope and survive. Trauma changes how we perceive the world we live in. The traumatic event itself may have been in the past, but these post-traumatic reactions rob us of the capacity to feel fully alive in the present. Nobody in his or her mind would want to deal with all that, unless their life depended on it. And since you have picked up this book, it is quite possible that your life actually does depend on it.

Most of us need support in this process: a coach or therapist who can help us to widen our window of tolerance and guide us into places that are not contaminated by the horror and shame of the past. But aside from such a guide, we also need to dedicate ourselves to the practice of learning to take care of ourselves—to take care of our bodies as if our lives depend on them, because they do.

This book can help you develop such a practice. It has a solid foundation in our current scientific understanding of how the mind, brain, physiology, and immunology are impacted by trauma. Post-traumatic reactions—unwanted feelings, sensations, and behaviors—do not emanate from the thinking brain but from deep inside the survival brain and the autonomic nervous system that function outside of conscious awareness, yet essentially run the show whether we like it or not. Liz Stanley, being the courageous warrior that she is, has put together a rich collection of personal experiences and scientific studies, and has synthesized them into a systematic treatment approach that she has tested and applied to thousands of traumatized soldiers, veterans, and people in a range of other high-stress environments. That is the way to grow, not just individually but collectively: by taking the deepest personal lessons that we have learned, organizing them into an understandable format, and sharing them with others on the road to recovery.

—Bessel van der Kolk, M.D., author of *The Body Keeps the Score:*
Mind, Brain, and Body in the Healing of Trauma

Part I

LIFE ON
THE GERBIL
WHEEL

CHAPTER ONE

Stone Age Physiology in a Digital World

In the summer of 2002, I worked incessantly to complete my Ph.D. dissertation on deadline. My faculty advisors at Harvard had already set my defense date so I could begin a prestigious fellowship starting in September. Everything seemed on track for a successful start to my academic career. Well, everything except for that *one minor detail* I'd neglected to share with my committee: Of the ten chapters and appendixes in my dissertation, I still needed to write seven of them.

In mid-June, I finally quit my full-time job to finish it. Early one August morning, after weeks of pushing myself to write sixteen hours a day without any days off, I carried my coffee mug into my study and turned on the computer. I opened my draft, reread the paragraph I'd finished late the night before, and started writing.

I was halfway through my first sentence when I puked all over the keyboard.

After running for paper towels to clean up my mess, it quickly became apparent that my vomit was permanently lodged under some of the keys. (The space bar was especially hard hit.) No amount of wiping it up could rectify the situation.

I brushed my teeth, washed my spew-speckled arms, and found my shoes and my wallet. Outside, I threw the keyboard into the trash can and climbed into the car. I drove to a shopping center and parked. It was seven fifty in the morning. When Staples opened at eight, I was the first one in the door.

New keyboard in hand, I was back at my computer finishing that first sentence of the morning by eight thirty.

SUCK IT UP AND DRIVE ON

To be clear, I didn't have a stomach bug or food poisoning. Rather, I'd been living for years with relentless bouts of nausea and lack of appetite.

Here's a snapshot of me—and my overscheduled, extremely compartmentalized, and rigorously well-organized life—circa 2002: I was compulsively driven to achieve. I was addicted to demanding workouts, to maintain my body's physical prowess. I was incessantly cheerful at work, while experiencing radical mood swings and crying jags at home. My mind raced with thoughts about my never-ending to-do list and "what-if" worst-case scenarios. My body was hypervigilant and tense from projecting an external aura of self-confidence while internally bracing against when the other shoe would drop. I was severely claustrophobic and hypersensitive to crowds, traffic, loud noises, and bright lights. Between insomnia and terrible nightmares, I rarely slept.

In retrospect, I see that the message that my body transmitted to me that morning was clever, dramatic, and spot on: *At that moment, I was literally sick of this (expletive here) project and I desperately needed a break.*

However, I didn't have the time to think about that right then. I had a dissertation to finish, and I was running out of time.

And so I overrode this rather extreme signal from my body and just kept writing.

I delivered my completed manuscript by deadline. I successfully defended my Ph.D. dissertation and started my fellowship on schedule that fall.

I was also an anxious, workaholic wreck.

So how did I get here? How did I end up literally puking out a Harvard Ph.D. dissertation? Why did my body present me with such an extreme signal that morning? And why was my (mostly unconscious) default response simply to ignore and override that signal and keep pushing?

In many ways, finding answers to these questions has motivated my work over the last fifteen years. Perhaps not surprising, since I'm a political scientist who teaches about international security, in 2002 I made sense of the Keyboard Incident as my body waging an insurgency against my mind's drive to perform and succeed. Of course, inherent in this explanation is its own recommended cure: counterinsurgency. In other words, just dig in, access deep wells of willpower and determination, and power through. Otherwise, it's just mental weakness and laziness, right?

For many decades, I considered my capacity to ignore and override my body and my emotions in this way to be a *good thing*—a sign of strength, self-discipline, and determination. And from one perspective, it was. But as I'll explain in this book, from another perspective, this default strategy was actually undermining my performance and well-being.

Of course, I'm not alone in this conditioning. It's a common way of relating to experience that many people call "suck it up and drive on" or "powering through." Contemporary American culture in general—and warrior culture in particular—prizes this approach to life. We've all heard and perhaps even admire stories of people overcoming extreme adversity or simply pushing through challenges and setbacks with perseverance to succeed. And, as I'll explain shortly, many conveniences of our modern world *exist almost entirely to facilitate* our suck-it-up-and-drive-on addictions. Nonetheless, although the self-determination to power through stressors in this way can be admirable—and during certain immediate life-or-death situations is *absolutely critical* for survival—this way of approaching life can have some dark consequences over the longer term.

In my life, my habitual reliance on suck it up and drive on not only allowed me to meet my dissertation deadline. To name just a few other examples, it also allowed me to achieve a top-5-percent ranking at a physically demanding military qualification course while still recovering from a massive injury to my Achilles tendon; run a marathon in just over four hours (in barely-above-freezing rain, of course!) seven days after accidentally impaling the claw end of a hammer one inch into my right heel; and attain basic proficiency in a new foreign language while working 120-hour weeks before my U.S. Army unit deployed to Bosnia after the 1995 Dayton Peace Accords.

At the same time, I lived for many years an awkward double life: the outward appearance of success (as our society usually defines it) and the inner sense that I was a failure, struggling secretly with symptoms and barely holding it together. As willful as I was, it would eventually take losing my eyesight and leaving a marriage to finally understand that there's an easier way. This book is about how I healed that division in myself—and how you can do the same.

THE GOALS OF THIS BOOK

In the course of my personal quest to understand my self-described mind-body insurgency and the devastating effects it was having on my life, I detoured into a parallel professional quest to understand how life adversity, prolonged stress exposure, and trauma affect us—and influence our decision making and performance. Along the way, I created a resilience training program for people working in high-stress environments, called Mindfulness-based Mind Fitness Training (MMFT)®, about which I'll say much more later in the book. I also collaborated with neuroscientists and stress researchers to test MMFT's efficacy among troops as they prepared to deploy to combat, through four research studies funded by the U.S. Department of Defense and other foundations. In addition to training and certifying others to teach MMFT, I've taught MMFT (pronounced "M-fit") to hundreds of troops before their combat deployments to Iraq and Afghanistan, as well as many other military leaders, service-members, and veterans. I've also taught MMFT concepts and skills to thousands of individuals in other high-stress environments, including healthcare providers, intelligence agents, firefighters, police officers and other law enforcement agents, lawyers, diplomats, social workers, students, teachers and academics, inmates at a maximum-security prison, disaster relief workers, athletes, members of Congress, senior government officials, and corporate executives.

On my journey to wholeness, I engaged in many different tools and therapeutic techniques, including several kinds of therapy, yoga, meditation, and shamanic and mind training. Since late 2002, I've maintained a daily mindfulness practice. I've also completed many long, intensive periods of silent practice, including time as a Buddhist nun at a monastery in Burma. Finally, I sought several years of clinical training and supervision, culminating with certification as a Somatic Experiencing practitioner, perhaps the best known of the body-based trauma therapies.

Despite this wealth of experience, I often found that no one could explain to me, concisely and coherently, *how* or *why* particular techniques worked (or didn't)—or why my responses to them often differed significantly from others'.

Thus, my original intention in creating MMFT—and the first goal of this book—is to share the road map that I discovered with you. I aim to share some of the core scientific and intellectual concepts that undergird MMFT. To be clear, however, this book is *not* the MMFT course—it covers additional topics not addressed directly in MMFT, but also by necessity it can't replicate all of MMFT's experiential practices. I'll draw on recent scientific findings to explain *how* to train yourself to be more resilient before, during, and after stressful and traumatic events. My hope is that after finishing this book, you'll understand your own neurobiology better and thereby make better decisions—without experiencing unnecessary anxiety and without criticizing your imperfections or choices along the way.

Part of why my journey took years is that *there is no quick-fix way to achieve these transformations.* Rewiring the brain and body to improve our performance and build resilience requires an integrated training regimen and consistent practice over time. Just as muscle growth and improved cardiovascular functioning require months of consistent physical exercise, the benefits that can result from mind fitness training require consistent practice over time, too. With consistent practice, we usually see some shifts relatively quickly, while others take longer to manifest. However, you can't just achieve them from reading this book. Thus, I don't want you to take my word for anything in this book—I want you to practice and observe these dynamics in your own life. Rewiring the brain and body is an *embodied, experiential* process. These are basic laws of nature; there are no shortcuts.

This book draws on a lot of evidence from high-stress occupations, such as the military, firefighters, police, medical personnel, and other first responders. That's because much of the peer-reviewed empirical research about stress, resilience, performance, and decision making has been conducted with these groups. Likewise, at other points the book may seem a little heavy with clinical findings about people who've experienced abuse or trauma. Nonetheless, especially if you don't work in a high-stress profession or don't believe you have a history of trauma—and may not feel particularly connected to either category—I want to emphasize: *If you are a human being living in today's world, this book still pertains to you.* Scientific evidence about how our minds and bodies work, and how we make decisions before, during, and after stress and trauma, applies to everyone.

However, I don't just want this book to help you understand and manage your stress better. My second goal is to engage you in a wider reflection about the way that we, individually and collectively, approach stress and trauma. As I've noted, the mind-body insurgency I experienced in 2002 was an outgrowth of my conditioning—and thus, it embodied some deep familial, societal, and cultural beliefs, values, coping strategies, and habits. In this book, I hope to expose such underlying structures, which aggravate our stress and trauma and undermine our performance and well-being. These underlying structures not only affect the strategies we individually rely on to cope with our stress—or not. They also affect the way we interact in our families and relationships; nurture and educate our children; train, incentivize, and reward our employees; and organize our companies and public institutions. They even affect the way our nation interacts with the rest of the world.

Are these strategies aligned with and capable of delivering the desired results? Our culture seems to want it both ways: We want better performance, resilience, and even happiness, yet we don't want to examine the wider blind spots that impede their development. Some of this wanting it both ways manifests in how many of us feel like we don't have choice—that we're powerless in the face of job stress, health problems, rapid technological change, or toxicity in the news. Yet it's possible to change how we interact with these things, to relate to them from a more empowered stance. Ultimately, to feel like we have agency requires clear intentions, consistent practice of the skills that help us develop awareness and self-regulation, and deliberate choices about how we prioritize different aspects of our lives.

I'm not a clinician or neuroscientist. At the core, what I bring to the table is my own lived experience—the stressful and traumatic experiences I've endured during my life, my own journey of recovery, and the observations and insights I've gleaned from teaching thousands of others. Teaching MMFT in many settings, I've worked with people from many different walks of life who've endured a *wide* range of stressful and traumatic experiences. Thus, in addition to my own stories, I'll be weaving some of their stories into this book. I've deliberately changed their names and some details of their stories to protect their privacy.

Although this book is not a memoir, it's necessarily informed by my own experiences with stress, trauma, and recovery. What matters here is the *effects* of those stressful and traumatic events—and the profound shifts that happened once I finally began to recover completely, using techniques in this book. There's nothing in this book that I haven't grappled with and learned from personally in my own mind and body.

I was born in 1970 into a military family, the eldest of three sisters. There's been a Stanley serving in the U.S. Army every generation since the Revolutionary War, including on both sides of the American Civil War. My grandfather saw combat as an infantry noncommissioned officer in Asia during both World War II and the Korean War—in between participating in the postwar occupation force in Germany. My father served thirty years as an armored cavalry officer, including nearly two years in the Vietnam War. I am the first female Stanley to serve in this lineage, followed by one sister. As an Army brat during the Cold War, I moved ten times before college, spending many years overseas and attending German schools. I grew up in an alcoholic, and occasionally violent, household. I'm also the survivor of early childhood sexual abuse, stalking, multiple assaults, and rape, by both strangers and people known to me outside my immediate family, with most of these events occurring before I started college.

After commissioning, I served as a U.S. Army military intelligence officer overseas, in South Korea and Germany and on two deployments to the Balkans. Perhaps a quirk of the post–Cold War military drawdown, I never held a job in the Army commensurate with my actual rank. I always filled positions intended for someone more senior, often by two ranks—an experience gap that was quite challenging. For instance, just after promotion to first lieutenant, I served for several months in a major's billet. In addition to prolonged stress exposure from intensive military training and deployments, I experienced sexual harassment while on active duty, as well as command reprisal after reporting it. That reprisal eventually led me to resign my active-duty commission. During the two-year investigation launched afterward, while I was in graduate school, I was designated as a Department of

Defense whistle-blower statutorily, because of the ranks of the people involved. Eventually, I was exonerated for the false accusations made in reprisal against me, and others were held accountable.

Where did all this stress and trauma go? Predominantly into my body, where I could compartmentalize, ignore, deny, and override the cumulative effects of so many physical and psychic assaults and betrayals. Instead, I threw myself into striving and achieving, such as being student body president and valedictorian of my (fourth) high school; earning degrees from Yale, Harvard, and MIT; and attaining tenure at Georgetown, in one of the nation's most prestigious programs for my field. Thus, as our society usually understands it, I was resilient—capable of tolerating and functioning through an immense amount of stress. In the midst of this compulsive striving, however, I couldn't slow down enough to see what was actually going on: that my choices were inexorably *undermining* my resilience. Suck it up and drive on can facilitate tremendous achievement and success . . . until it doesn't anymore.

For many years while I suppressed the cumulative effects of those assaults and betrayals, my body bore the burden of that denial. (This, too, is a common effect of prolonged stress and trauma without recovery, called *somatization*.) From my early twenties through my midthirties, I experienced chronic respiratory and sinus infections, asthma, coughing blood, insomnia, and migraine headaches. While in Bosnia, I also needed resuscitation from a near-death experience after I completely stopped breathing because of untreated pneumonia and exposure to concrete dust from clearing a bombed-out building with my soldiers. My physical breakdown reached its apex in 2004, when I lost my eyesight during three episodes of optic neuritis, including about three weeks of complete blindness. (It turned out I had untreated Lyme disease, finally diagnosed in 2012, from a tick bite sustained during active duty.) Although some of these issues may initially appear unrelated to stress, they were *all linked to two stress-related root causes*: systemic inflammation and compromised immune functioning. Furthermore, the longer I suppressed my stress and trauma, the greater its toll—eventually leading to both post-traumatic stress disorder (PTSD) and depression, further complicating my physical conditions.

At this point, I want to acknowledge the typically human quality of *comparing mind*—comparing our own experiences to other people's successes and challenges. In fact, a common theme I encounter when teaching mind fitness is how often people measure their own life stressors and events against others'—in the process, inevitably devaluing or writing off their own as "not really that bad." Later chapters will explain how this comparing-mind habit can actually inhibit your recovery from stress. So if you've noticed yourself making such comparisons while reading about my life, please recognize the habit and see if you can set it aside for now.

Indeed, although my life has had its own unique challenges and contours—as every life does—in many ways, my journey through early-life adversity, extreme stress during military service, and PTSD afterward provides *a quintessential example of a human mind and body doing what all dysregulated minds and bodies do*: My window of tolerance to stress arousal was adaptively wired in response to my early social environment. It was narrowed during exposure to prolonged stress and trauma without adequate recovery. It was narrowed further through my habitual overriding of my body's signals to keep on pushing. Finally, I exhibited symptoms in response—with profound consequences for my ability to function in a wise, coherent, healthy, or happy manner.

In fact, although I didn't know it yet in 2002 when I was puking on my keyboard, it would take years of intensive reconditioning to rewire my mind and body so that I could finally access the cooperative mind-body relationship that is actually our human heritage—and yet for many of us has also been conditioned out of reach. It would take years of strong intentions and intensive training to turn with nonjudgmental curiosity toward the parts of myself that I'd been pathologizing—my "weak" and "dysfunctional" body; my intense, uncontrollable, and "irrational" emotions; my secret, shameful coping behaviors—and see clearly, understand, heal, and transform my *whole* self, nothing left out.

Through this process, I came to understand that many of my symptoms resulted from compartmentalized and denied experiences in my past. Because the truth of those experiences had been too overwhelming for my mind when they happened, I'd stored them outside awareness, in my body and in unconscious belief patterns that helped me cope. Only by bringing awareness back into my body could I finally recover and return to a healthy baseline—in the process widening my window of tolerance for better performance during even more stress in the future. I could finally replace my mind-body insurgency with a mind-body alliance—and access the inborn organic intelligence that I'd overridden for so many years. As I learned to trust the cues coming from my whole self, I could finally confront my life completely and honestly—*as it really is*, not as I wanted or expected it to be—and thereby make effective choices in response. I could finally navigate life's flow with more joy, creativity, ease, and connection with others than I ever would've thought possible back in 2002.

SETTING THE STAGE FOR OUR CONVERSATION

Before we move forward together, I need to share several basic definitions and principles. I'll unpack them further later in the book, especially in Part II.

First, I want to introduce an admittedly clunky term—*the mind-body system*—which I'll use as shorthand when I'm talking about our entire human organism: the brain, the nervous system, neurotransmitters (i.e., how the brain and nervous system communicate), the immune system, the endocrine system (i.e., our hormones), and the body, organs, skeleton, muscles, fascia, skin, and fluids.

Our brain was designed to function as a cohesive whole, with each region processing information and protecting us in a particular way. Although these regions actually have overlapping circuitry, I'm going to differentiate them by their respective functions. Evolutionarily the newest is the neocortex; I'll use the term *the thinking brain* for this region. The thinking brain engages in *top-down processing*—our mostly voluntary and conscious cognitive responses to our experiences. The thinking brain is responsible for our conscious decision making; ethical choices; and reasoning, abstraction, and analytical capabilities. It allows us to focus; recall, keep in mind, and update relevant information; and make decisions. To support these functions, the thinking brain has an *explicit* learning and memory system, situating information within space and time, which we can access intentionally. We know the thinking brain is engaged whenever we hear that running commentary of thinking, comparing, judging, and narrating in our head. Its strategy for protecting us is to anticipate, analyze, plan, deliberate, and decide.

In contrast, what I'll call *the survival brain* comprises the evolutionarily older limbic system, brain stem, and cerebellum. These brain regions play key roles with our emotions, relationships, stress arousal, habits, and basic survival functions. The survival brain engages in *bottom-up processing*—our involuntary emotional and physiological responses to our experiences, including emotions, physical sensations, vocalizations, and action tendencies in the body. One of the survival brain's most important functions is *neuroception*, an unconscious process of rapidly scanning the internal and external environment for opportunities/safety/pleasure and threats/danger/pain. In turn, the survival brain's protection plan is quite simple: Approach the former (opportunities) and avoid the latter (threats). To support neuroception, the survival brain has an *implicit* learning and memory system—fast, automatic, and unconscious, bypassing the thinking brain. It constantly acquires implicit memories through *every* experience, without conscious intention or effort. Importantly, the survival brain isn't verbal, so it can't communicate with us via thinking or narrating. Instead, it activates neurotransmitters and hormones, which produce physical sensations and emotional cues, each associated with conditioned impulses to approach opportunities and/or avoid threats. That's why it's called bottom-up processing. Once we become aware of these bodily cues, however, the thinking brain can use that information for conscious decision making. Although we can't know directly what's happening in the survival brain, we can see its effects in our emotional and physiological arousal. Together, the thinking brain and survival brain comprise what many people call "the mind."

One last point about the brain: *Awareness* does not belong to the thinking brain or the survival brain. It functions distinct from the thinking brain's cognitive activity and the survival brain's stress and emotional arousal. Awareness is greater than all of these things—which is why we can pay attention to thoughts, emotions, physical sensations, and the body's posture, temperature, and movements. Mindfulness-based training helps us learn how to direct and sustain our attention—and thereby stabilize awareness—so that we can become aware of, learn from, and modulate these different mind-body experiences.

Linking the survival brain with the rest of the body, the *autonomic nervous system (ANS)* serves as an automatic control for a wide range of bodily functions outside conscious awareness, including the functioning of the organs. I'll explore the ANS more fully in later chapters. For now, it's important to know that the ANS is responsible for stress arousal and recovery—directing the body to focus on immediate survival needs or the longer-term tasks put on hold during stress arousal. The ANS also plays a major role in our patterns of engaging and interacting with others.

Stress is the internal response that our mind-body system creates whenever we have an experience that the survival brain perceives as threatening or challenging. Despite the villainizing and romanticizing of stress in modern society, it's really nothing more than our mind-body system mobilizing energy—which I'll call *stress activation* or *stress arousal*—to respond to a threat or challenge. Indeed, we're wired to produce stress activation and thereby disrupt our internal equilibrium temporarily, so that we can successfully manage that threat or challenge. Then, after it has passed, ideally we'll discharge any remaining activation to recover completely back to our baseline.

This process of perturbing our inner equilibrium and then coming back to our regulated baseline is called *allostasis*. Allostasis allows us to mobilize the appropriate amount of energy and focus for coping well before, during, and after the threat or challenge. However, with *chronic* or *prolonged stress*, our mind-body system doesn't complete a full recovery after a stressful experience—instead, it remains in an activated state.

Trauma is also an internal response, on a continuum with stress. However, not all stress is traumatic. Trauma can occur if, during a stressful experience, we also perceive ourselves to be *powerless*, *helpless*, or *lacking control*. Trauma is especially likely to result if aspects of the current threat or challenge contain cues or triggers related to traumatic events from earlier in our lives.

Without adequate recovery after chronic stress and/or trauma, the mind-body system remains activated and doesn't return to its regulated equilibrium. Instead, over time the internal systems involved in allostasis—the brain, the ANS, the immune system, and the endocrine system—become dysregulated. When this happens, allostasis stops functioning properly and we begin to build *allostatic load*. And, as allostatic load accumulates, we experience *dysregulation*, which manifests as a range of physical, emotional, cognitive, spiritual, or behavioral symptoms. For example, by the time I'd reached my early twenties, I'd experienced decades of chronic stress and trauma without adequate recovery. Thus, I'd built a heavy allostatic load and exhibited many symptoms of dysregulation—including depression, PTSD, insomnia, chronic nausea, racing thoughts, hypervigilance, chronic physical illnesses, and a range of unskillful coping behaviors.

Indeed, if my earlier examples seemed extreme to you—either the achievements or the suffering—that's because in many respects, they are. *Extreme behavior is usually linked to extreme dysregulation*—the *hallmark* of someone masking, suppressing, denying, self-medicating, or coping with extreme dysregulation the best way they know how. As I've suggested, suck it up and drive on was one of *my* core coping strategies, but there are many other strategies that may resonate more deeply with you, such as addictions, tobacco or substance abuse, disordered eating, extramarital affairs, adrenaline-seeking behavior, obsessive-compulsive behavior, self-harming, domestic violence or violent outbursts, isolation, dissociation, and extreme procrastination or paralysis. I'll explore them further throughout this book.

Thus, I hope you can recognize a version of these dynamics in your own life, too. These dynamics affect all of us to varying degrees; they're shared by anyone who fails to recalibrate their mind-body system after a distressing or traumatic event, such as a flood, a car accident, or the loss of a job or a loved one. They're also shared by anyone who habitually overextends their mind-body system during prolonged stress without adequate recovery, such as crashing to meet a deadline or working long hours over an extended period without some days off.

As these definitions suggest, *stress and trauma are a continuum*. In our society, we usually think of *chronic stress* (such as years of workaholism without enough vacation) as very different from *shock trauma* (such as a terrorist attack, sexual assault, or car accident), and both of those categories differ from *developmental and relational trauma* (such as growing up in a violent, abusive, or neglectful home). I don't want to conflate these categories or suggest they're all the same thing. From the perspective of how we individually and collectively understand and make sense of these different kinds of events, they're absolutely *not* the same thing. Nonetheless, from the perspective of our neurobiology—how our brain, ANS, and body experience these different kinds of events—they are indeed quite similar. In fact, the *effects* from being a stressed-out office worker are more closely related to those experienced by a combat veteran with PTSD than the usual societal narrative would have us believe.

Where we find ourselves on the continuum between stress and trauma has everything to do with how our mind-body system (consciously *and* unconsciously) perceives our current situation—and especially whether we feel like we have agency during that situation. *The less agency we perceive we have, the more traumatic the experience will likely be for our mind-body system.* This principle is the foundation on which MMFT is built.

MMFT's goal is to build our capacity to find agency and access choice in *every* situation, no matter how challenging, stressful, or traumatic it might be. Building on millennia of warrior traditions with the same aspiration, MMFT cultivates wisdom and courage—two qualities necessary for finding agency, functioning adaptively during stress, and recovering afterward. In this book, I will teach you how to access agency—both at a micro level, within your mind-body system, and at a macro level, in your interactions with others and the external environment.

LIFE ON THE GERBIL WHEEL

We live in a world where the rate of change is accelerating. Technological innovation, especially in genetics, nanotechnology, neuroscience, robotics, and artificial intelligence, continues to speed up, with profound social, ethical, and philosophical ramifications.

We also see growing political fragmentation, "fake news," and distrust in social institutions, while the Black Lives Matter and #MeToo movements shine light on longstanding social patterns of racism, sexism, homophobia, and sexual violence. Indeed, in 2017, 59 percent of American adults said they consider this the lowest point in our nation's history that they can remember—a feeling shared by Americans of all generations, including those who lived through World War II, the Vietnam War, and the 9/11 terrorist attacks. Beyond these human conflicts, the strains of overpopulation, climate change, and the greatest mass extinction of species since dinosaurs left the earth sixty-five million years ago threaten the biodiversity, health, and capacity of our planet.

None of this is news, but when we take all of it into consideration it's no wonder that our human mind-body systems—initially designed two hundred thousand years ago and left largely unaltered since—often seem too frail and fragile to keep up.

Indeed, by several indicators, the United States today is one of the most violent, stressed, and traumatized countries in the world. The U.S. firearm death rate—3.85 deaths per 100,000, most of them suicides—outranks our socioeconomic peers; it's eight times that of Canada and thirty-two times that of Germany. The United States experienced more than 1,500 mass shootings from December 2012 until October 2017—with at least 1,700 killed and 6,100 wounded. Only Yemen—a borderline failed state embroiled in civil war—has a higher per capita mass shooting rate than we do. More broadly, U.S. residents experienced 5.7 million violent victimizations in 2016—that's 21.1 per 1,000 people—including rape/sexual assault, robbery, and aggravated or simple assault. We also have the world's highest incarceration rate, with more than 2.3 million people in correctional facilities.

At the same time, 89 percent of adults in the United States report having experienced at least one traumatic event in their lives, with most adults reporting exposure to multiple traumatic events.* Of course, not everyone who experiences a traumatic event will develop trauma. Between 4 to 6 percent of men and 10 to 13 percent of women in the United States develop PTSD at least once in their lives; lifetime PTSD rates are more than three times as high among combat-exposed men and sexually victimized women. PTSD rarely occurs alone, most often coexisting with substance abuse, major depression, and anxiety disorders. When someone has PTSD and one of these other conditions, they're also significantly more likely to engage in intimate partner violence and suicidal behavior.

---

* This study included exposure to any Criterion A event for a PTSD diagnosis under DSM-5, such as experiencing a disaster, accident/fire, exposure to hazardous chemicals, combat or war zone exposure, or physical or sexual assault; witnessing physical or sexual assault or dead bodies/body parts unexpectedly; or enduring the threat, injury, or death of a family member or close friend due to violence/accident/disaster.

Other statistics seemingly unrelated to trauma may actually speak to high trauma levels in the United States. About one quarter of American adults currently have a mental illness, and nearly half will develop at least one mental illness during their lifetime—such that mental illnesses account for more disability than any other group of illnesses, including cancer and heart disease.

Moreover, the rate of mental health diagnoses has increased over time, perhaps due to more frequent reporting and the inclusion of more mental illnesses in the diagnostic manual. Longitudinal studies—in which researchers return to the same communities to gather data on multiple occasions—mostly confirm a rising prevalence of depression and anxiety, with a greater lifetime risk in each successive generation. For instance, U.S. young adults in 2007 were six to eight times as likely to be clinically depressed compared to their peers from 1938. Today, the lifetime rate of major depression—between 15 and 20 percent—is *ten times* what it was for Americans born before 1915 (and they lived through the Great Depression and two World Wars). Anxiety rates have also steadily increased over the last seven decades. Today, anxiety disorders are the most common mental illness in the United States, affecting nearly one third of adults—with more Americans seeking medical treatment for anxiety than for back pain or migraine headaches. In fact, compared with people who don't have anxiety disorders, those who do are five times as likely to see a doctor and six times as likely to be hospitalized—in large part because they seek relief for symptoms that mimic physical illnesses, such as heart palpitations, headaches, sleep problems, or gastrointestinal symptoms.

In part to cope with these conditions, substance use and abuse is also growing. One third of Americans have abused or been dependent on alcohol at some point; since 2000, emergency room visits related to binge drinking have increased 50 percent. While Americans constitute only 4 percent of the world's population, we consume 75 percent of the world's prescriptions. More than thirty-five million American adults took antidepressants in 2014, up from thirteen million in 2000—with long-term antidepressant use more than tripling during that period. Furthermore, more than seven million Americans regularly use psychotropic drugs—painkillers, stimulants, sedatives, or tranquilizers—nonmedically. Not surprising, the United States also has the world's highest drug-death rate, which has grown since 1990 by more than 650 percent—leading in 2016 and 2017 to declines in U.S. life expectancy, below many of our socioeconomic peers. Drug overdoses are now the leading cause of death for Americans under fifty—with opioids causing two thirds of them. In 2017, the opioid epidemic killed nearly forty-eight thousand people, more than died annually of AIDS at the height of that crisis.

Along with opioid abuse and alcoholism, suicide has attracted attention as a major driver of the increase in mortality, especially among middle-aged whites with less education and in rural areas. Today, the suicide rate in the most rural areas is almost double that of the most urban areas, largely because of the higher prevalence of guns in rural homes. Overall, suicides in the United States increased by a third between 1999 and 2017, with U.S. deaths from alcohol, drugs, and suicide hitting their highest levels in 2017 since the federal government started collecting this type of mortality data in 1999. Suicide remains one of the ten leading causes of death for Americans.

Of course, these trends don't just affect American adults. Over the last decade, hospital admissions for suicidal teenagers have doubled. In 2016, 62 percent of American undergraduates reported experiencing "overwhelming anxiety" in the previous year, according to an American College Health Association survey, and anxiety is now the most common reason they seek counseling services. More broadly, in 2017, 91 percent of Americans age fifteen to twenty-one—the so-called Generation Z—reported that they routinely cope with stress-related physical or emotional symptoms, such as depression or anxiety.

American adults also report subjectively feeling more stressed and more anxious. For instance, U.S. Google search rates for "anxiety" more than doubled between 2008 and 2016—with higher search rates in places with lower education levels, lower median incomes, and larger rural populations. Likewise, an American Psychological Association survey showed that most Americans believe they live with moderate or high levels of stress—with 44 percent claiming increased stress levels in the past five years, especially at work. Yet despite acknowledging unhealthy stress levels, most Americans also reported feeling thwarted from practicing healthy behaviors, citing being too busy as the primary obstacle.

Consider these U.S. lifestyle indicators: Less than half of American adults engage in the recommended amount of physical activity, with almost 38 percent considered "completely inactive." Roughly one third of American adults are obese, while another third are overweight. One in five Americans report overeating or eating unhealthy foods frequently because of stress. One third of Americans also exhibit at least one insomnia symptom, with 45 percent of American adults reporting in 2017 lying awake at night in the past month.

To explain these rising anxiety and depression rates, researchers point to the growing burden of chronic physical diseases, obesity, carbohydrate- and sugar-heavy diets, and lack of physical activity—all of which arise from the evolutionary mismatch between past human environments and modern-day lifestyles. They also point to inequality, isolation, sense of meaninglessness, and loneliness—paired with declines in social capital and community—that have become hallmarks of modern life. For example, a cross-cultural analysis ranging from rural Nigerian to urban American women found that the more modernized and urbanized the setting, the higher the prevalence of depression. Similarly, adopting an American lifestyle appears to explain higher rates of depression among U.S.-born Mexican Americans, compared with Mexican immigrants. In general, urban dwellers in developed countries have a higher prevalence of anxiety disorders and depression than rural dwellers.

Taken all together, these statistics suggest that the daily life of many Americans *is* subjectively getting worse, and these social shifts relate to the mismatch between the modern world and our Paleolithic neurobiological heritage. We're hardwired like our caveman ancestors, with a neurobiology that was designed to locate and survive immediate, mortal threats (like predators). Importantly, this mind-body system we share with our ancestors was optimized for short bursts for immediate survival, followed by periods of recovery to allow for longer-term tasks like healing, reproduction, and growth.

This wiring hasn't changed, but, of course, the world in which we live has changed radically, as has the kind of decisions we face today. We're rarely in mortal danger, but our mind-body systems are still relying on this wiring to respond to "symbolic threats"—like anxious thoughts about an impending deadline at work, or where and when the next school shooting may occur. Unlike our ancestors, however, we live with this system turned on for days, weeks, months, or even *years* at a time, delaying the longer-term tasks that are naturally deprioritized when stress arousal is turned on.

The paradox is that the conveniences of the modern world exacerbate these problems. We face constant demands, deadlines, time pressure, and technologies that allow us to override nature's constraints. Obviously, 24/7 access to electronic devices feeds the myth of "multitasking"—making it challenging to unplug from our social media feeds or disconnect from our electronic leashes to the office. Yet this constant electromagnetic barrage— usually outside our conscious awareness—is also keeping our mind-body systems turned on. From the velocity of cars, trains, and planes, to the rapidly changing images and sounds of the ubiquitous televisions and radios in public settings, to the computers and handheld devices we turn to in moments of boredom, we are constantly receiving electronic stimulation. Even something as "old-fashioned" as electricity allows us to override our natural biorhythms for sleep, integration, and recovery. *And* we supplement the technological override with a variety of substances—prescription medications, recreational drugs, antacids, laxatives, and sleep aids. Caffeine, nicotine, and sugar help us artificially mobilize energy and focus our attention in the short term, while alcohol helps us wind down at the end of the day. In other words, these conveniences provide ever more creative ways for us to override our wiring, and we end up becoming further and further divorced from our innate capacity for alignment and self-regulation.

Finally, the modern world is much more complex than the world of our cave-dwelling ancestors. Uncertainty, complexity, volatility, and ambiguity are "symbolic threats," meaning that they rarely require decisions involving mortal danger to our physical well-being *right now*. Nonetheless, they still turn on our Paleolithic fight-or-flight wiring, without a straightforward outlet to use all the stress activation we mobilize or an obvious end point when we can finally "stand down." The stressors we're wired to feel *most* threatened by—and therefore mobilize the most energy for—are those we perceive to be novel, unpredictable, and uncontrollable, and these three adjectives provide a pretty good description of modern life!

Is it any wonder, then, that we collectively experience chronic pain, insomnia, constipation, sexual dysfunction, or physical illnesses brought on by compromised immune functioning? That we lose the ability to see the forest for the trees, unconsciously prioritizing the immediate over what's more important for our long-term happiness, success, and well-being? That we feel isolated, disconnected, and dissatisfied in our relationships? That we struggle with apathy, burnout, anxiety, depression, numbness, and mean-inglessness? We're running on fumes, exhausting our mind-body systems by operating them in ways for which they weren't optimized.

Although this misalignment affects us all, it's especially acute for people working in high-stress environments. From tactical, split-second choices about how to react during a crisis, to high-stakes decisions that often have major financial, strategic, or policy consequences—and sometimes even life-or-death consequences—the ability to remain calm, evenhanded, compas-sionate, and resolute in high-stress situations is critical. Yet the more we override our wiring and exhaust ourselves, the further and further effective decision making moves out of reach. No wonder impulsive decision making, poor judgment calls, unethical or violent behavior, and moral injury are growing concerns in many of these environments.

In response to these trends, there's been a surge in transferring informa-tion gathering, decision making, and even task completion as much as pos-sible to technological tools, robots, drones, and computers. This insatiable desire for the technological fix is built in part on the misguided expectation that we can overcome the profound mismatch between our biology and the modern world though *more* technology. It also implies that we're incapable of making skillful decisions given the complexity and barrage of informa-tion in the world today. We secretly yearn to step off the Gerbil Wheel at long last. The semiconscious assumption here is that perhaps we'd just be better off if we could circumvent our emotional, impulsive, and frazzled fallibility.

#WIDENTHEWINDOW

So how can we face challenges our ancestors couldn't have even imagined, and make effective decisions about how to handle those challenges, with this mind-body system optimized for threats of two hundred thousand years ago? How can we navigate these challenges with resilience and resourceful-ness, with wisdom and well-being, without internalizing the limiting belief that our machines can manage it better than we can? These were the ques-tions I faced several years ago, when my body was literally shutting down as a result of my failure to respond to its basic needs.

The answer to these questions, and the basis of MMFT, is learning to use our biology in a new way. By systematically training our attention, we can *widen the window* within which our thinking brain and survival brain work together cooperatively. In fact, our window of tolerance to stress arousal is so critical to the ideas in this book that I even chose it for the book's title. The wider our window, the easier it is to find agency, function effectively during stress, and recover afterward. It's true that this neurobiology we inherited from our cave-dwelling ancestors is misaligned with the modern world. By directing our attention in particular ways, however, we can learn to regulate this wiring *consciously*. When we use awareness to regulate our biology this way, we can access our best, uniquely human qualities: our compassion, courage, curiosity, creativity, and connection with others. We can train ourselves to make wise decisions and access choice—even during times of incredible stress, uncertainty, and change.

Inside our window, we can regulate our stress levels upward or down-ward in order to remain, over time, within the optimal performance zone. We can also consciously integrate input from both the thinking brain and survival brain—neither one overriding the other or hijacking our choices. Inside the window, the thinking brain and survival brain share an *allied, collaborative relationship*, allowing us to access our inner wisdom, what some people call "intuition." For these reasons, the width of our window critically shapes our ability to make agile and adaptive decisions in every situation—from split-second, life-or-death decisions during extreme stress or crisis to the mundane choices about how to spend our time, interact in our relation-ships, and care for our bodies and minds.

The relationships we've habitually conditioned between our thinking brain and survival brain—and between our brain, ANS, and body—have tremendous effect on the width of our window. Our window is initially wired through the interaction between our genes and our early social envi-ronment, starting in the womb and continuing through adolescence. Most of our default strategies for coping with stress and interacting with others were wired in response to this early social environment, with significant consequences on the width of our window. Our window can also become narrowed or widened over time, depending on our repeated experiences. I'll explore these dynamics in Parts II and III.

People with wide windows have greater ability to assess safety and danger accurately, the unconscious survival brain process called *neuroception*. They're more likely to respond flexibly and appropriately during both safe and threatening situations. They're best equipped to meet challenging experiences with their thinking brain functions online and recover completely and efficiently afterward. They're more capable of digesting all information in the current environment rather than finding their attention captured by an immediate, central threat. They're better able to go with the flow and stay connected to other people during challenging events.

However—and this is key—even people with naturally wide windows find that over time their window narrows through chronic stress or trauma without recovery. When we operate with the system always "turned on" without ever turning it off, even the most resilient individuals find that their capacity gets gradually undermined. Importantly, because of the mismatch between our biology and the modern world, this undermining will happen even when we simply leave our Paleolithic wiring to its own unconscious devices. In other words, without some conscious, intentional effort to regulate our biology, each of us eventually narrows our window.

As our window narrows, we increasingly find ourselves outside our window. Outside our window, the thinking brain and survival brain engage in *an adversarial, antagonistic relationship*—with each one trying to override and shout down the other. Here, we might experience *thinking brain override*, when we compartmentalize and suppress our emotions, physical sensations, and the body's needs and limits. Or we might experience *survival brain hijacking*, when emotions and pain drive our decisions and lead us to make impulsive, reactive choices. We're also more likely to self-medicate or mask our distress with food, caffeine, tobacco, other substances, addictions, and violent, self-harming, or adrenaline-seeking behavior. Finally, outside our window, we're more likely to experience trauma.

If you're like many of the people whom I've trained in MMFT over the last decade—or like me, back when I was writing my dissertation—right now your default may be to bottle up stress or try to figure out how to "deal with" difficult situations. Yet, as I'll explore in this book, many strategies you may currently believe (*and our culture promotes*) to be effective ways for dealing with life events—compartmentalizing, suppressing, ignoring, distracting from, reframing as positive or even just as "it's not really that bad," or powering through—could actually be making your stress worse. In this book, you'll learn how these common habits can dysregulate you, undermine your health, distort your perceptions of events, lead you to miss important cues that you're on the wrong track and need to change course, and disconnect you from yourself and supportive relationships.

The next chapter explores how and why we tend to cultivate an adversarial relationship between the thinking brain and survival brain—in the process disregarding the continuum between stress and trauma. Part II then explores the science behind the window—how the window gets wired initially and narrows over time. Part II will help you better understand your thinking brain, survival brain, ANS, and body—and their different cues and functions. Finally, in Part III, I'll teach you *where*, *when*, and *how* to direct your attention to widen your window. I'll help you learn to trust your body's cues as informational resources but then use your thinking brain to make the wisest choice based on that information. Please don't jump right to Part III; you need the information from Part II to make sense of Part III's tools and techniques.

Our neurobiology was designed to function as a cohesive whole, with each part of our mind-body system bringing its unique skills, capacities, and insights to the table. We can unlock this synergistic potential only when the thinking brain and survival brain are working as allies. When we choose to cultivate an allied relationship between the thinking brain and survival brain, we not only operate within our window but can heal and recover from our prior exposure to chronic stress and trauma, and thereby widen that window as well.

CHAPTER TWO

How We Disregard the Continuum between Stress and Trauma

"Greg" was a successful businessman, with a long career acquiring, restructuring, and selling companies—making himself quite wealthy along the way. He had a rich network of professional connections and two gorgeous homes in expensive locations. He was happily married to his fourth wife.

Greg's initial interest in MMFT was not for himself. Rather, he was debating whether to donate significant funding to the nonprofit Mind Fitness Training Institute that I'd founded to share MMFT with others. He saw extensive combat during the Vietnam War. Now he wanted to use his wealth and success "to give back"—to help a new generation of combat veterans coming home from Iraq and Afghanistan. I suggested to him that the best way to know whether to support our work was to experience MMFT himself.

After completing the MMFT course and engaging consistently in the exercises for several months, Greg finally shared with me that since he came back from Vietnam, things had never been quite right. With his new understanding of his neurobiology, he acknowledged that he'd been living for decades with a mind-body system "stuck on high"—with hyperarousal, hypervigilance, excessive startle reactions, insomnia, and high blood pressure. In fact, he admitted, a major reason why he'd accomplished so many things during his storied career was that he rarely slept. He loved his work, especially the "rush" he felt when he closed a major deal, a feeling he'd first experienced in Vietnam. Greg also confided that the reason he was now with his fourth wife was that he'd compulsively cheated on the previous three. At the time, he said, he'd felt like he "couldn't help it." Those affairs— and the lying that went along with them—also brought that "rush" that he loved so much.

"Tanya" was a talented analyst in her midthirties at one of the agencies in the U.S. intelligence community. She was deeply committed to her organization's mission, trying to prevent another terrorist attack from happening on U.S. soil. When I met her, she'd recently been passed over for a promotion, which was awarded instead to a younger male colleague. Since then, she was frequently awake at night with racing thoughts. She was also working even longer hours—leaving virtually no time for a life outside work—so that she could demonstrate her dedication to her superiors and thereby win the next available promotion.

Tanya had gone to college on an athletic scholarship, where she'd deftly juggled intense coursework—a double major in economics and Arabic—and the significant demands of training and competing with a nationally ranked team. Of course, winning that athletic scholarship meant that she had been a masterful juggler for many years before college, too.

While participating in MMFT, Tanya recognized how being passed over for that promotion left her feeling "out of control," a feeling she'd experienced before. During middle and high school, the way she'd coped with this feeling was by severely restricting her diet to four hundred calories per day. Tanya told me that she'd been "successfully treated" for anorexia in high school, returning to her height's normal weight range before college. Since then, whenever things felt hard to handle, she'd relied on demanding workouts to feel "in control." Now, however, she'd recently noticed herself counting calories and restricting her diet again. This surprised her, because she was sure she'd "grown out of that" after high school.

"Todd" was a nineteen-year-old infantryman. He'd already completed one combat tour in Iraq, and his unit was now preparing for another to Afghanistan. Todd didn't like his squad leader *at all*, because the sergeant was always "riding his ass" for forgetting things. Sometimes, for instance, he would show up to formation in the wrong uniform. Other times, he'd forget what he was supposed to do and then "mess things up" for his entire squad. Todd was deeply bothered by his "subpar behavior."

The month before I started training Todd in MMFT, he'd finally sought help for his forgetfulness—not for his own sake, actually, but because it was having such negative repercussions on his squad. Since then, he had been seeing a clinician each week at the health clinic where he was stationed. This clinician diagnosed him with PTSD and gave him three prescriptions: one for insomnia, one for PTSD, and one for the chronic knee and back pain Todd endured from carrying heavy equipment, his ruck, and his weapon. Now that he was working with the clinician, Todd was also experiencing frequent flashbacks, both from Iraq and his childhood.

In fact, Todd had a difficult past. His biological father was in prison, and they hadn't spoken in many years. Growing up, Todd had lived with his alcoholic mom and an extremely strict stepfather. When Todd "messed up" at home, his stepfather would frequently "use a belt." When he was eleven, Todd also had a near-death experience.

Todd had one prize possession: his Harley-Davidson. He *loved* his bike, which he'd saved up for during his first deployment. He loved to ride along the coast, where he could watch the waves and clear his head. After a particularly bad day of "messing up" and getting yelled at by his sergeant, however, he sometimes felt like he *had* to burn off steam. On those days, he'd get on his bike and ride *fast*—which, he admitted, meant at least 150 miles per hour—dangerously bobbing and weaving through highway traffic. In fact, he told me, "the closer the calls, the better I'd feel." Afterward, he'd head to his favorite bar where, after having "one too many," he'd usually pick fights.

On the surface, these three people may seem unalike: They hail from diverse socioeconomic and educational backgrounds. They've experienced radically different kinds of stressful and traumatic events. They cope with their stress in dissimilar ways.

Indeed, if I'd sketched these portraits of Greg, Tanya, and Todd directly to them before they were trained in MMFT—and then asked them whether they believed what they'd endured was in any way similar to the two others' experiences—I doubt they'd have seen much resemblance, besides perhaps "being stressed."

Despite their apparent differences, however, Greg, Tanya, and Todd were having the same experience.

Just like me, all three had developed an adversarial relationship between their thinking brain and survival brain and had accumulated allostatic load. In response, all three were making choices to cope with their dysregulation and internal division the best way they knew how.

Suffering comes in many flavors, but it's still suffering.

GRADING OUR STRESS

As you read these sketches, perhaps you consciously or unconsciously evaluated whether Greg, Tanya, and Todd were somehow *deserving* of our considering them as "being stressed."

Before you continue reading, take a moment to notice whether you judged, compared, evaluated, or ranked their stories (and mine from the last chapter) in some way—such as in terms of their "worthiness" for suffering, the magnitude of the stressful and traumatic experiences they've endured, or the consequences of their coping behaviors on the world around them.

Please know it's *not* a problem if you noticed such comparisons: This is just your thinking brain doing what all thinking brains do. As Chapter 1 explained, our thinking brains have developed some *very deep* conditioning for comparison, evaluation, and judgment—comparing other people's experiences, as well as comparing theirs to our own. It's just how our thinking brains have been wired.

In fact, our deeply conditioned thinking brain habits for comparison and judgment critically influence our collective disregard of the continuum between stress and trauma. In this chapter, I want to explore some of the dynamics that lead us to disregard this continuum—and also powerfully influence our individual and collective behavior. I'm not interested here in judging these dynamics; my intention is merely to unearth and illuminate them.

In our culture, we have collective thinking brain assumptions and expectations about what "stressed" or "traumatized" looks like. Collectively, for instance, we're probably most likely to label Todd's story as "traumatized"—*even though Todd doesn't see himself that way.* Conversely, we're collectively least likely to grant Tanya's story the "traumatized" label—even though, after losing out on that promotion, *Tanya was actually experiencing traumatic stress.*

My keyboard story from Chapter 1 was an expression of trauma, too— although I didn't know it at the time. Instead, I understood my story as an example of "stress hardiness" or "grittiness," something our culture prizes greatly.

Psychologist Angela Duckworth and her colleagues introduced the concept of "grit" to capture the qualities of determination, hard work, and perseverance. Her research with students, West Point cadets, corporate salespeople, and married couples shows that being "gritty"—working hard, being driven, persevering despite adversity, powering through—is a good predictor of eventual *external* success, better than family income and IQ levels. Duckworth's research has resonated deeply in our culture; her book was a bestseller and she even won a MacArthur "genius" award.

It's certainly true that being gritty means we've developed some capacity to tolerate and push through discomfort to reach an important goal. It's also true that gritty people are better able to see failure as a learning opportunity, rather than get discouraged by that failure. They're better at picking themselves back up after adversity and doubling down on their efforts to attain eventual success. These can be admirable qualities.

Interestingly, however, the empirical research about grittiness is silent about its *costs*—and that's part of our cultural conditioning, too.

A universal feature of trauma is *dissociation*—cutting ourselves off from our pain, as well as our shame about that pain. This dissociation may manifest in many different ways, including physical diseases, domestic violence, bullying and harassment of others, addictions, extramarital affairs, and self-harming, violent, or adrenaline-seeking behavior. We can see many of trauma's different faces in Greg's, Tanya's, and Todd's stories.

I chose a form of dissociation—extreme mind over matter—that led to tremendous accomplishments. Thus, although I may have had a relatively heavy trauma load compared with those of my socioeconomic and educational peers, my behavior didn't fit with the conventional understanding of "traumatized." My coping behaviors were socially accepted and rewarded.

To make this point slightly differently: There's no doubt that Greg,

Tanya, Todd, and I were *all gritty* in our own ways. At the same time, I was a compulsive overachiever who lost her eyesight, Tanya an anorexic workaholic, Greg an adrenaline-seeking philanderer, and Todd an aggressive driver and violent binge drinker.

THE DISREGARDED CONTINUUM BETWEEN STRESS AND TRAUMA

How we (individually and collectively) understand, compare, judge, and evaluate "stressful" or "traumatic" events is a task performed by our *thinking brains*. Thus, the meaning that our thinking brains make about "stressful" or "traumatic" experiences is profoundly affected by our familial, organizational, and societal norms, beliefs, and values.

Collectively, we tend to think of "stress" and "trauma" as distinct concepts. Indeed, our thinking brains classify "chronic stress" (such as chronic workaholism) as *different* from "shock trauma" (such as a tsunami, car accident, or terrorist attack), and both of these as *different* from "developmental trauma" (such as growing up in an abusive or neglectful home) and "relational trauma" (such as harassment, discrimination, or relationships built on patterns of abuse or addiction).

Furthermore, researchers and clinicians—the professionals with expertise about these matters—tend to study and treat "stress" and "trauma" separately, too.

For instance, researchers who study *stress* tend to conduct animal research, focusing on the specific biological mechanisms that undergird the stress response and stress-related diseases. Alternatively, they may specialize in "elite performance"—such as conducting research with elite athletes and special operations forces, including SEALs, Rangers, and Green Berets. This second group wants to understand how to improve the mind-body's capacity to function well in high-demand or extreme-stress settings.

At the other end of the spectrum, *developmental and relational trauma* is the bailiwick of family therapists, social workers, child psychologists, and trauma researchers. These professionals try to help people cope with traumatic events in their past (or their present), feel better, and function in their daily lives. Alternatively, these professionals may study the mechanisms by which trauma continues to express itself over many years—and possibly even repeat itself, a phenomenon called *trauma reenactment*. Some professionals at this end of the spectrum may also specialize in domestic violence, criminal recidivism, addictions, eating disorders, or suicidal behavior.

Because these different researchers and clinicians are trained in different disciplines, publish in different peer-reviewed journals, attend different conferences, and focus on *different aspects* of the stress-trauma continuum, it shouldn't be surprising that collectively we'd think of stress and trauma as separate—thus requiring completely different strategies and/or therapeutic techniques. When we divide our collective understanding of the root causes, *of course* we're going to think that we're dealing with different things.

Nevertheless, conceptualizing stress and trauma as separate hides the fact that they share a neurobiological basis. Stress and trauma are not inherent in the event—*they are internal mind-body responses on a continuum*. Where someone falls on this continuum during a threatening or challenging event depends on how their *survival brain* unconsciously appraises that event through neuroception—not on how their *thinking brain* consciously judges, evaluates, or classifies that event.

Thus, whenever we encounter a threatening or challenging event, *whether we'll experience stress or trauma is mostly determined by the current width of our window.*

That's why, for instance, when a thirteen-person infantry squad encounters an ambush, we can be sure that there'll be thirteen different mind-body responses—because there will be thirteen different windows of varying widths meeting that ambush. There will also be thirteen different conditioned responses for coping with the stress or trauma of that ambush afterward.

Furthermore, regardless of whether we experience stress *or* trauma, unless we also experience a complete recovery afterward, we'll be building allostatic load. Over time, without adequate recovery, we'll also eventually experience dysregulation—those physical, emotional, cognitive, spiritual, and behavioral symptoms that occur when the mind-body system is no longer functioning within its regulated equilibrium.

For this reason, although our thinking brains tend to consider chronic stress, shock trauma, developmental trauma, and relational trauma to be different things, *they all create the same effects in the mind-body system.*

If they're so similar in their effects, then why does our culture usually treat them so differently? The short answer is that many powerful and ambitious people have a hard time admitting their mind-body system's vulnerability.

Powerful, high-achieving, and successful people—and the high-status institutions where they work—have no problem acknowledging "stress." Indeed, we tend to consider "being stressed" to be a badge of honor—the evidence that we're successful and accomplished. In our collective understanding, "being stressed" means being overworked, overscheduled, extremely busy, and definitely *important*. It's just a necessary by-product of being a Master of the Universe.

Why else would so many of us boast about how few hours of sleep we got last night? Or how many days have passed since we've seen our kids awake by the time we got home from work? Or how many different activities or demands we're juggling at the same time? Or how many years it's been since we took a proper vacation—or even a full weekend off? In our culture, we *romanticize* our stress, even as we whine about it with humble-brags like these.

Likewise, we collectively engage in society-wide mixed messaging: Although we *profess* that health, relationships, families, communities, and "work-life balance" are important, we simultaneously *reward and admire* people for engaging in imbalanced behavior.

We reinforce this imbalance in our workplaces by setting unrealistic deadlines for ourselves and our subordinates. Giving the bonus to the dysregulated workaholic or the anxious, micromanaging supervisor. Promoting the harassing executive who can't keep his zipper up or the overbearing leader who creates a toxic workplace.

We also teach this mixed messaging to our kids, when we permit their teachers to assign them more homework than is physically possible to complete in the hours remaining in the day—especially if they're to engage in extracurricular activities *and* still have time to move their bodies, play in an unstructured way, and get enough sleep.

On the Gerbil Wheel, parents are encouraged to involve their children in structured programs and lessons *before even starting kindergarten*. In turn, many activities advertise their before-school and after-school programming, which parents appreciate as a cost-effective alternative to daycare. Child-centered, labor-intensive, and financially expensive parenting is now the dominant cultural model for raising children; indeed, mothers who work outside the home spend just as much time tending their children today as stay-at-home moms did in the 1970s.

Most parents I know say they dislike the pace of their children's over-scheduled lives. They dislike shuttling carpools most afternoons, evenings, and all weekend long. They can't *wait* for their kids to get their driver's licenses, so they themselves can finally cease this hectic hassle. Parents and kids alike are both rushed, exhausted, and frazzled.

We *say* we want it to be different—yet we also worry that if we don't fully engage in Life on the Gerbil Wheel, we'll suffer some terrible consequences: Undermining our kids' chances for getting into the most prestigious schools, from preschool through graduate school. Impeding our own opportunities for whichever job, promotion, or professional reward we desire. Degrading our personal and professional network if we don't say yes to every invitation, conference, speaking engagement, cocktail party, neighborhood barbecue, and birthday celebration for our children's classmates. We've even created an acronym for this—FOMO, the fear of missing out.

In equating "being stressed" with being powerful, successful, busy, and important, however, *we also inadvertently disconnect the stress from its eventual consequences.*

Collectively, we tend to ignore how our choices—and our society-wide mixed messaging—lead us to build allostatic load, develop stress-related physical and psychological illnesses, and manifest other symptoms of dysregulation. By divorcing the *lifestyle choices* that correspond to "being stressed" from their *consequences*—on our mind-body systems, our relationships, our communities, and our planet—we're less likely to take responsibility for how we contribute to the eventual outcomes we don't want.

Nowhere is this disconnect more potent than with trauma.

As Chapter 1 explained, traumatic stress occurs when the *survival brain* unconsciously perceives us to be powerless, helpless, or lacking control during a stressful experience. Even though being traumatized is, *itself,* beyond our control, our *thinking brains* don't want to acknowledge it.

In our culture, we're perfectly fine "being stressed" because in our collective understanding that means we're successful, powerful, persevering, gritty, strong, and important. "Being traumatized," however? Well, that equals powerless, malingering, broken, damaged, passive, cowardly, and vulnerable—and nobody wants to be *that*.

As a result, collectively we tend to give credence *only to the most extreme forms of shock trauma*—such as hurricanes, earthquakes, political captivity, torture, terrorist attacks, mass shootings, combat, rape, or kidnapping. If trauma must exist in the world, then collectively we're willing to grade these events as "traumatic." At the same time, however, we're collectively reluctant to accept poverty, abuse, discrimination, harassment, and neglect in their subtler forms as "trauma."

Most thinking brains in our society would probably understand a PTSD diagnosis after combat or rape—but not after sexual harassment or persistent discrimination in the workplace. And certainly not after a childhood of invisibility, emotional neglect, or the withholding of love. Our thinking brains make these evaluations and judgments about other people's experiences, as well as about our own.

In other words, our conventional understanding of "trauma" usually includes only the "Capital T" traumas from shock trauma—not all those chronic, accumulating "little t" traumas in daily life where our survival brains feel helpless, powerless, trapped, or lacking control: A loved one's terminal medical diagnosis. The subtle yet persistent social exclusion or discrimination in our workplace. The dangers of "driving while black." The bullying behavior our children face. These chronic, accumulating "little t" traumas almost always manifest in personal, professional, or community relationships—meaning, they're usually *developmental trauma* or *relational trauma*.

They're also *extremely taxing* on our mind-body systems, because our thinking brains usually devalue and discount their actual effects—while other people's thinking brains do, too.

By way of example, I'd like to briefly highlight the effects of poverty, sexism, heterosexism, and racism—the stigma, prejudice, and discrimination that come with being part of a marginalized group—often leading to both chronic stress and relational trauma. (I'll explore developmental trauma's effects in later chapters.) It's important to recognize that someone may belong to multiple marginalized groups simultaneously. Although overt violence is possible—and we've actually seen an increase in such violence recently—more common manifestations of stigma and discrimination include bigoted attitudes, derogatory remarks, and other microaggressions.

Poor Americans—with family incomes at or below the poverty line—are roughly five times as likely to report "fair" or "poor" health as adults with family incomes above $140,000. They are three times as likely to have physical limitations due to chronic illness or chronic pain. They also have higher rates of obesity, heart disease, diabetes, stroke, and other chronic illnesses than wealthier Americans. A quarter of them are smokers—three times the rate among adults with family incomes above $100,000. They're four times as likely to report being nervous and five times as likely to report feeling sad, hopeless, and/or worthless "most or all of the time." Similarly, mortality rates among whites without a college degree have been rising since the turn of the century—"deaths of despair" driven mostly by increases in suicide, drug overdoses, and alcohol-related liver problems.

Sexism includes acts of disrespect, discrimination, and unfairness due to gender. Many women themselves devalue everyday sexism's effects as less, or not, harmful when compared to overt sexual harassment or rape. Nevertheless, perceived sexism has been linked empirically with depression, psychological distress, high blood pressure, greater premenstrual symptoms, and more physical symptoms, such as nausea and headaches. It's been implicated in women's binge drinking, smoking, and self-silencing in their romantic relationships.

In experimental research, for instance, researchers examined women's stress hormones during experiments with varying conditions of sexism. In one experiment, women were told by a man that they'd been rejected for a job; in a second experiment, women were asked to complete a task with a man they believed was evaluating them for possible employment. These experiments created four different conditions—two with obvious cues for possible sexism, one with ambiguous cues, and one where sexism was not possible. The *only* condition where women did not experience increased stress hormones was when sexism was not possible, after losing the job to a better-qualified woman.

Heterosexism—the victimization, homophobia, discrimination, self-stigma, and sexual identity concealment that gay, lesbian, bisexual, queer, and transgender people experience—has been shown to contribute to psychological distress, anxiety disorders, depression, PTSD, social isolation, and disordered eating. Sexual minorities also experience a higher prevalence of mental health problems than heterosexuals do. In addition, sexual minorities who experience workplace heterosexism exhibit more psychological distress, more health-related problems, decreased job satisfaction, and more work absenteeism.

Racism is significantly associated with poorer health, according to a recent meta-analysis of 333 peer-reviewed empirical studies published between 1983 and 2013. Racism is especially linked with worse mental health, including depression, anxiety, suicidal behavior, PTSD, and psychological distress. It's also linked with weight-related physical diseases—diabetes, obesity, and being overweight. Racism is also linked with economic injustice, with income gaps—at every income level—between black and white Americans, and between Latino and white Americans, persisting consistently across five decades of U.S. household income data.

Importantly, discrimination, prejudice, and harassment—from any -ism—*don't have to actually be experienced to create toxic effects in our mind-body systems.* We can experience a surge of stress arousal while reading or watching the news about events where other members of our identity group are being marginalized. We can also experience this surge while *remembering* or *anticipating* events when we are marginalized ourselves.

One experiment demonstrates this dynamic very clearly: Researchers examined stress arousal among Latinas while they anticipated—and then participated in—an evaluation with a white woman. Latinas who were told that the white woman was prejudiced against ethnic minorities showed greater blood pressure increases and more stress arousal *before* the evaluation—and more threat-related emotions and thoughts during the evaluation—than Latinas who were led to believe that the white woman was not prejudiced. This study highlights how chronic vigilance about and anticipation of discrimination can be just as stressful as actually experiencing it.

Taken together, this growing body of empirical research suggests that everyday relational trauma from poverty, sexism, heterosexism, and racism—which most of our thinking brains write off as "not that bad" or "no big deal"—*actually is*. We may not be as conscious of the stress activation, but these things can still turn our systems on without ever turning it off.

DISOWNING OUR TRAUMA

Before you continue reading, take a moment to write down five adjectives that describe how you see yourself, with one adjective on each line. For instance, you might see yourself as "strong, successful, optimistic, self-reliant, and diligent." Or perhaps you see yourself as "honest, witty, friendly, caring, and compassionate."

Now, across from these five adjectives, write their opposites. Across from "strong" you'd write "weak," for example. Across from "honest" you'd write "dishonest." This second list of adjectives provides a succinct summary of your personality's shadow side.

Identifying with qualities that fit with our self-image means that we'll also naturally struggle with, disown, and fear the qualities that comprise our shadow. Indeed, the more strongly we (individually and collectively) identify with one half of *any* pair, the more we'll disown, deny, suppress, and override its opposite.

What this means is that whenever disowned qualities arise—and as *whole* human beings, inevitably we *will* experience both poles of any pair—the more internal tension we'll experience. The more we'll struggle with reality. If we see ourselves as "honest," we'll likely deny when we act "dishonestly." If we see ourselves as "caring," we'll ignore when we're actually "selfish." If we see ourselves as optimistic and happy, we'll naturally push away sadness, depression, or other negative moods, while also assuming that "something must be wrong with me."

And, of course, if we see ourselves as powerful, self-reliant, gritty, resilient, and strong, we'll struggle with accepting those experiences when our survival brain perceives us to be powerless, helpless, and lacking control.

Nevertheless, when we strongly identify with one half of any pair, the disowned other half is more likely to control us, often unconsciously. As the saying goes, what we resist will persist.

Whether we experience stress or trauma during challenging or threatening events is not actually up to our thinking brain. That's because *neuroception is a survival brain job.* Thus, stress and trauma are never directly related to our thinking brain's classifications and judgments of certain events as "stressful" or "traumatic"—or not.

However, the stress and trauma we experience are *almost always indirectly related* to our thinking brain classifications and judgments. Why? Because when the thinking brain's judgments are *not* aligned with the survival brain's neuroception, we set the stage for an adversarial relationship between them. This internal division *always* makes stress and trauma in our mind-body systems worse.

Whenever our thinking brain only allows and accepts *some* threatening or challenging events, or *some* emotions—usually the ones that align with our self-image—while denying *the rest* of the threatening and challenging events we've actually experienced and *the rest* of the emotions in the whole human repertoire, we create internal division.

Just because our thinking brain doesn't *want* to consciously acknowledge "weak" or "soft" or "irrational" emotions inside us *doesn't mean we won't have them.* Our survival brain will certainly still generate these emotions. In fact, *all* emotions will arise inside our human mind-body system at some point, because the full range of human emotions is an inborn part of our wiring. Whenever our thinking brain refuses to acknowledge emotions it doesn't want, however, we perpetuate the adversarial relationship. In the process, we move ourselves outside our window and build allostatic load. And the longer this continues, the more symptoms of dysregulation we'll experience.

So how does the adversarial relationship between thinking brain and survival brain—and between identifying with one quality while disowning its opposite—actually manifest?

Most commonly, we find some way to distract from, deny, suppress, compartmentalize, ignore, avoid, self-medicate, or mask the disowned quality—and the pain that comes along with it. Our thinking brains don't want to focus on pain. Instead, they tend to focus on the *future*—the life goals we'll achieve, the wealth and fame we'll earn, the possessions we'll accumulate, the beautiful body we'll sculpt, the next high we'll score, the relationships we'll create, or the bucket lists we'll finish.

The problem is, our survival brain, nervous system, and body don't tend to cooperate with this thinking brain strategy of disowning our earlier pain, abuse, stress, or trauma.

You might have noticed a trend in Greg's, Tanya's, Todd's, and my stories. Tanya and I tended to cope with our stress and trauma by *internalizing* it: I threw myself into compulsive overachieving, while Tanya overrode her pain through extreme calorie restriction and excessive exercise. We both also developed internalizing psychological disorders—in my case, PTSD and depression; in Tanya's, anorexia.

In contrast, Greg and Todd tended to cope with their stress and trauma by *externalizing* it: Greg got his adrenaline rush through high-stakes deal making and adulterous affairs, collecting several failed marriages along the way, while Todd got his through aggressive driving, alcohol abuse, and violence.

Nonetheless, *all four of us were doing the same thing*—disowning, suppressing, compartmentalizing, self-medicating, and overriding our stress and trauma with coping strategies *where our thinking brains could feel more in control*. In our own ways, we each relied on socially acceptable outlets and coping strategies that wouldn't threaten our self-identities.

This pattern—*women internalizing* their pain and *men externalizing* their pain—is actually quite common in our culture.

In our culture (and many others), girls and women are taught to "not make waves"—and that anger is not a suitable emotion. Although women experience anger as frequently as men, research shows that women experience more shame and embarrassment afterward. Highlighting how socially unacceptable it is, female anger is more likely to be described as "bitchy," "hostile," "aggressive," and "argumentative." Women are also more likely to cry and experience anxiety when they're actually angry, because sadness and fear are more appropriate emotional territory. It's no wonder we've been conditioned to stuff our pain inside—overriding it for "peace at any price," engaging in self-harming behavior, and manifesting internalizing disorders disproportionately experienced in our society by women, including depression, anxiety, impostor syndrome, eating disorders, and autoimmune diseases.

Conversely, boys and men are taught to be competitive and aggressive—and that fear and sadness are not appropriate emotions. We're all taught that testosterone hard-wires male bodies for aggression—and that playground bullying, sexual harassment and assault, and domestic violence can all be neatly written off with the rationale that "boys will be boys." "Strong" men aren't vulnerable, depressed, or afraid, especially when they can externalize that pain onto others by maneuvering themselves into the "one-up" stance of grandiosity, arrogance, or aggression. Underneath, however, is disowned pain, shame, fear, sadness, or inadequacy. It's no wonder that men have been conditioned to externalize and inflict their disowned pain onto others—disproportionately perpetrating our society's violence, engaging in adrenaline-seeking and risk-taking behavior, and manifesting externalizing disorders disproportionately experienced by men, including intermittent-explosive disorder (IED; i.e., rage attacks), attention-deficit hyperactivity disorder (ADHD), alcoholism, and substance abuse.

Regardless of gender, people who see themselves as successful, capable, gritty, strong, and resilient tend to disown their trauma. Why? Because trauma is the shadow of these self-identified qualities. Their opposites— being helpless, vulnerable, weak, powerless, and lacking control—is what *defines* trauma. This dynamic is most pronounced for men, because in our culture, for a man to admit he's lost his agency, it usually means he's lost his masculinity, too.

Nevertheless, *all of us*, both men and women, are culturally socialized to prefer, value, and identify with "masculine" traits at the expense of "feminine" ones. This polarity appears along several dimensions: strong versus weak, "rational" versus "emotional," powerful versus powerless, perpetrator versus victim, "self-reliant" versus "needy," controlling versus controlled. We also observe this polarity in our collective juxtaposition of mind versus body and "being stressed" versus "being traumatized."

Psychotherapist Terry Real calls this "the great divide," where the first half of each of these pairs get characterized as "masculine" and the second half as "feminine." Culturally, however, the relationship between these two halves is not equal. Rather, the "masculine" holds the "feminine" in contempt, as inferior.

Because all of us struggle with this polarity, *both* genders tend to deny their pain and express it instead via amplified physical symptoms, a process called *somatization*. A large body of empirical research, especially among

combat veterans and survivors of childhood and/or sexual trauma, links somatization with the suppression of emotions, stress, and trauma. This suppressed
pain gets expressed obliquely through physical diseases, chronic pain, gastrointestinal difficulties, back problems, sleep issues, and other physical symptoms. Somatization has also been linked with suicidal behavior. We see
evidence of somatization in Greg's, Tanya's, Todd's, and my stories.

People working in high-stress or high-status environments may be particularly prone to expressing their distress via somatization, because they
may perceive less stigma for seeking help for physical, rather than psychological or emotional, concerns. Perhaps the most revealing research about
this cultural tendency in high-stress environments was a 2003 study with
82nd Airborne Division soldiers, two weeks before they invaded Iraq. Despite similar rates of PTSD, soldiers with and without combat experience
expressed their distress differently. Compared with combat-naïve soldiers,
combat veterans coped with the stress of their upcoming deployment by
generally *denying emotional symptoms*—such as anxiety, irritability, depression, and suicidal ideation—while simultaneously *reporting more physical
symptoms*, including chronic pain, dizziness, fainting, headaches, chest pain,
digestive issues, insomnia, and sexual difficulties.

MARKETS, SUPERHEROES, AND MODERN SCIENCE

So far, we've seen how the common "appropriate" response to stress and
trauma is usually to suck it up and drive on—denying, disowning, masking,
distracting from, suppressing, self-medicating, and compartmentalizing
both stressful or traumatic events *and* the resulting dysregulation. We
tend to equate denial with tenacity, perseverance, and grit. Especially in
male-dominated environments, stoicism and stiff upper lips predominate.
We respect the "walking wounded" and the athletes who "play hurt." The
people who "soldier on" instead of letting their pain, injury, or distress "get
to them."

These unconscious norms and habits are *very powerful* in our society. In
fact, groups tend to replicate individual dynamics. Just as individuals tend
to disown their own trauma, groups disown trauma, too. When the group

disowns trauma, however, it unfortunately sets the conditions for more individuals to become traumatized *and* perpetuates cultural pressures for individuals to continue disowning their own trauma.

In our culture, we collectively tend to assess ourselves—and other people—as the sum of our achievements, tangible creations, inventions, and "value added" to our workplace and community. The United States has thrived as a free market economy, and because the market places great significance on productivity, efficiency, speed, perseverance, grit, wealth, and profits—we do, too. Capitalism also feeds our society-wide mixed messaging: It tends to value and incentivize productivity and profits, while disregarding, denying, and ignoring many costs and consequences of these profits.

We can see this dynamic at work in how most shareholders and Wall Street incentivize market behavior—such as rewarding corporate leaders solely for "the bottom line" while disregarding the "market externalities" of those profits, including low morale, unethical behavior, stress-related diseases among employees, unsafe working conditions, cultures of harassment and discrimination, and detrimental effects on the environment. (In economics, an *externality* is a cost that affects people—or the planet—who did not choose to incur that cost.) For instance, recent research finds that toxic, unethical workers enjoy longer tenures and are more productive, at least in terms of quantity of output—which likely explains why they're selected and able to remain at organizations for as long as they do.

In fact, conceptualizing sexual harassment, environmental pollution, or toxic working conditions as "externalities" is, in itself, a symptom of this dynamic. It's drawing the line in a way that includes, values, and takes credit for one preferred aspect of reality, *while simultaneously denying, disowning, and writing off reality's other aspects that naturally co-arise with it.*

Harvey Weinstein both headed a successful film studio that brought hundreds of blockbusters to the screen *and* was brought down by more than eighty charges of rape and sexual assault. Eric Schneiderman is both the former prosecutor lionized for fighting corporate fraud and corruption *and* a man who resigned after allegations of alcoholism and sexual abuse. David Petraeus—who once told reporters he "rarely feels stress at all"—is both a former CIA director and retired general widely admired for crafting the

U.S. "surge" strategy in Iraq and Afghanistan *and* a man who knowingly leaked classified information to his adulterous mistress.

Since the #MeToo movement, many observers have argued for preserving a separation between "the artist" and "the art"—trying to focus our attention narrowly on people's productive output to ensure that their transgressions remain outside our collective evaluation of their work. The issue with such separation is that *we are actually whole humans*—encompassing both creative genius or courageous leadership *and* the abusive, violent, transgressive, addictive, or unethical choices we make, too.

Our myths of individualism also reinforce norms of separation and denial. Certainly, individualism is deeply embedded in the foundational structures of the U.S. system of governance and its corresponding respect for individual rights. Especially since the Vietnam War and civil rights movement, we've also seen the rise of meritocratic principles in our education system and workplaces.

In the process, we've come to conceptualize life as a journey where we cultivate individual skills and talents, collect accomplishments, and round out our resumes. This myth of individualistic meritocracy creates an exaggerated emphasis on IQ, divorced from other qualities. We emphasize our autonomy, self-reliance, and independence, blinding ourselves to the fact that we're actually embedded in communities. With our imbalanced attention to the *individual rights* of citizenship, we forget that it also comes with *responsibilities*.

This myth of individualism also manifests in our cultural artifacts. Since Homer's *Odyssey*, Western civilization is replete with stories of the heroic individual—the self-reliant hero, usually a man, whose acts of service, sacrifice, and success derive from his isolation (or alienation) from mainstream society. Think Clark Kent, Bruce Wayne, and Peter Parker, superheroes who saved their communities by sequestering themselves apart. Or architect Howard Roark of *The Fountainhead*, Ayn Rand's libertarian primer. Or James Bond. The Lone Ranger. The classic American cowboy. These are rough, tough, rational, hard-charging, *manly* men.

Finally, these norms of compartmentalization also appear in the modern scientific method. René Descartes, Isaac Newton, and other influential scientists and philosophers showed that mathematics was not just the epitome of pure reason but also the most trustworthy knowledge available. As Descartes famously exclaimed: "I think, therefore I am." In other words, Descartes assumed that *thinking* is the central feature of existence—and that man can know everything by way of reason and logic.

These men equated "objective" reality to the conceptual space where measurable results can be attained. This preference for quantified information implied that phenomena that could not be measured—instinct, intuition, emotions, dreams, context—lack reliable information. Underneath this worldview is the belief that the world *can be known*—that there is an objective *certainty* we can attain through enough data, measurement, and/or analysis.

These cultural beliefs created a duality between mind and body, subject and object, and consciousness "in here" confronting things "out there." It appears in our cultural preference for knowledge derived from "rational" thought and "objective" information—rather than from instinct, emotions, imagination, empathy, and other "subjective" sources. In fact, although I'm going to be relying on and sharing empirical evidence and scientific studies in this book, I'm explicitly pairing this "objective" information with my "subjective" lived experience, because science and empirical studies alone always provide an incomplete understanding.

We still see the Cartesian paradigm at work today. This is the cultural basis for our relative disregard for the emotional or physiological aspects of knowing—as well as for our bodies, sensations, and emotions. Even today, we see it in our collective preference for what our *thinking brains* "know" rather than what our *survival brains* "know."

All of these cultural influences, together with our habits to disown our shadow, reinforce the adversarial relationship between our thinking brains and survival brains.

ALL BY OURSELVES

We've been taught to see ourselves as individuals, entirely separate and autonomous from our upbringing, our communities, and each other. We've been taught to identify with our thinking brains' understanding—in the process devaluing input from our survival brain. We've been taught to disown the effects of our choices, both on our own mind-body system and on the people around us. We've been taught that stress, trauma, anxiety, depression, unhappiness, and pain are just a "cost of doing business." We've been taught that our success or our failure is the result *entirely of our own effort.*

This unequal dialogue between thinking brain and survival brain is widespread in our society. We're conditioned to listen to, believe, and make decisions based only on thinking brain input—our own and other people's. We absorb thinking brain narratives from our family, friends, teachers, coaches, bosses, celebrities, the media, advertising, and the news.

We also rely heavily on thinking brain–dominant therapies and techniques. These methods aim to help the thinking brain feel more in control. However, they also tend to disregard, ignore, or attempt to "manage" or "fix" survival brain input, including our emotions, physical pain, and stress arousal. In our society, we often pair thinking brain–dominant tools with prescription drugs—or self-medication via food and many substances—that mask and minimize the symptoms of dysregulation.

Some popular thinking brain–dominant therapies and tools include traditional "talk" therapy and group therapy; cognitive-behavioral therapy (CBT) and cognitive processing therapy, CBT's cousin that dominates U.S. Veterans Affairs clinics; and sports psychology and positive psychology techniques, including goal-setting, "broadening and building" positive emotions, and "strengths-based" positive psychology. All of these techniques rely on thinking–brain dominant strategies: Using thoughts to reappraise and reframe the situation. Finding the positive in the situation. Focusing on gratitude, internal strengths, and other positive qualities. Consciously cultivating positive emotions. Mentally preparing for future situations with visualizations and rehearsals. Focusing on and developing plans for future goals.

Resilience programs widely implemented with the U.S. military and other high-stress organizations are heavily based on these thinking brain–dominant techniques. Not surprisingly, however, none to date has shown empirical efficacy for reducing dysregulation or negative emotions. Later chapters will explain why. Nevertheless, if individuals are unable to regulate their stress and negative emotions effectively when using such techniques, *they may internalize that ineffectiveness as a "problem" with themselves.* In the process, they're likely to double down on socially acceptable coping strategies that build allostatic load and exacerbate the adversarial relationship between thinking brain and survival brain.

In the meantime, our survival brains are still there in the background— feeding our addictions, wrecking our relationships with infidelity and work-aholism, disordering our eating, fueling our overreliance on many substances, damaging our bodies through too much or not enough exercise, driving us to adrenaline-seeking or self-harming behavior, and externalizing our unconscious pain onto the people around us through our violent, abusive, unethical, or transgressive behavior.

And then, when our survival brain hijacks our behavior in these ways, we usually feel shame, self-criticism, judgment, and guilt.

But isn't this just the other side of the same coin? If we insist on taking *all* the credit *individually* for any achievement, success, creative work, or triumph that we have, doesn't that also require that we take *all* the blame for our difficulties, imperfections, addictions, physical and psychological diseases, obesity, relationship problems, and poor choices, too? Seeing ourselves as separate—believing that our thoughts, emotions, pain, behavior, and choices are *entirely up to us*—*of course* our thinking brains will feel bad whenever we inevitably exhibit our human imperfections.

A major reason why we devalue, deny, distract from, compartmentalize, rationalize, mask, or externalize the parts of us that don't fit with our self-image—including "being traumatized" or experiencing distress—is that we believe these things somehow reflect badly on us. We don't *want* to have trauma, negative emotions, addictions, physical or psychological diseases, and unskillful coping behaviors. We assume that "I shouldn't be feeling this way" or "something must be wrong with me." I sure did.

Luckily, these thinking brain beliefs are simply not true.

What I find most fascinating about this unequal dialogue between the thinking brain and survival brain in the realm of stress and trauma is that although we tend to listen more to our thinking brain, *our survival brain actually has much better information about this particular topic.* Remember that neuroception is a survival brain job. The survival brain always has the best information about whether we're stressed, traumatized, or dysregulated. Even more important, as later chapters will explore, *the survival brain controls whether we recover from stress and trauma.* Thus, thinking brain–dominant techniques are always incomplete.

This is where understanding how our mind-body systems are wired can be extremely liberating. Our mind-body systems are wired to connect with others. We are embedded in relationships in our families, neighborhoods, schools, and workplaces—just as the nation is embedded in relationships with other nations, the global economy, and international institutions. These interdependent webs are supported by the planet and its resources. In turn, our choices affect this whole interconnected system.

Importantly, we didn't get to control this wiring—*and it certainly wasn't up to our own effort.* Our mind-body systems were wired through repeated interactions with our environment, especially with the people closest to us during childhood. Through these interconnections, we developed our habitual strategies for interacting with others, enduring adversity, and coping with stress, trauma, and negative emotions. For instance, the different coping strategies that Greg, Tanya, Todd, and I each adopted were the result of our respective life conditioning—strategies that were, by definition, adaptive, because they allowed us to survive.

Therefore, today when any of us experience stress, trauma, negative emotions, cravings, "irrational" impulses, or the urge to make violent or harmful choices, it's really nothing more than our past conditioning playing out. *It doesn't actually say anything about who we really are.*

Part II explores the science behind the window—how our window initially gets wired and how it gets narrowed over time. I want to help your thinking brain better understand and empathize with your survival brain— with what it's doing and *why*—so you can better interpret its cues. By understanding how stress and trauma are a continuum, we can see how we might devalue things that are extremely stressful for the survival brain but "not that bad" to the thinking brain. How we might not prioritize recovery, since we devalued the source of the stress or trauma in the first place. And how we might miss leverage points where we actually have agency for recovery and healing.

Rather than *self-improvement*, the most direct path to feeling better, thriving during stress and trauma, and making effective choices is actually *self-understanding*.

Part II

THE SCIENCE
BEHIND
THE WINDOW

CHAPTER THREE

Grand Canyons in the Mind-Body System— Neuroplasticity and Epigenetics

About 115 years ago, the U.S. military became the first organization in the country to recognize the need for organized physical fitness training (PT). Long before scientific understanding documented such benefits, the military intuited that physically fit troops would have certain capacities—such as strength, stamina, flexibility, and speed—that would enhance individual and group performance during battle.

The first institutionalized PT programs in the United States—at the U.S. Military and Naval Academies—were developed in the 1880s. The West Point PT program rapidly gained traction across the Army by the early 1900s; in 1906, the Army mandated servicewide requirements for garrison and nongarrison PT programs—as well as an annual three-day stamina test to assess compliance. Although Army commanders widely opposed this change, the Army chief of staff turned to President Theodore Roosevelt for support. A strong proponent of physical fitness, Roosevelt issued an Executive Order requiring that all Army officers pass the annual stamina test— and then set the example himself by exceeding the test's standards.

After the U.S. military institutionalized PT programs, it realized, during preparations for both world wars, that most conscripts were incapable of meeting PT standards. Thus, military training during both wars focused heavily on developing physical fitness among conscripts. Concurrently, the federal government passed legislation to improve physical fitness in public schools and piloted a comprehensive PT curriculum in junior and senior high schools. This society-wide campaign to promote fitness among all citizens culminated during the Kennedy administration with the President's Council on Physical Fitness. In addition, several organizations, such as the American Medical Association and the American College of Sports Medicine, made it their mission to disseminate scientific research and educate the public about the health consequences of low fitness levels.

In other words, the military's early institutionalization of PT helped spur physical education in public schools *and* scientific research about physical fitness.

As a result, there's now a society-wide understanding about the benefits of physical fitness and how to train it. We know that with repetition, particular exercises can produce training-specific muscular, respiratory, and cardiovascular changes in the body. We also know that if we want to create specific changes in the body, we can't just talk to a trainer or read a book about it. We must actually engage in the exercises, repeatedly and consistently, for weeks or months. No one can do it for us—we have to be committed enough to do it for ourselves. And should we fall off the exercise wagon, we understand that we're likely to see muscular and cardiovascular atrophy.

Yet even when we engage in particular exercises consistently and repeatedly, the ultimate goal is not simply to get good at those exercises. The goal is to have generalizable capacities we can employ throughout our lives, like strength, stamina, flexibility, and speed. A body strengthened through weight training, for instance, has more capacity to carry a heavy rucksack for long distances, lift a fallen tree limb to release someone pinned underneath, or push a vehicle stuck in the mud back onto the pavement. Likewise, physical fitness is protective. It helps us recover more quickly after physical exertion or injury.

Each of these dynamics also exists with mind fitness training. Just as PT relies on repeated exercises to generate specific changes in the body, mind fitness training relies on specific repeated exercises to create changes in the brain and the nervous system. Thus, when we practice consistently, mind fitness exercises lead to beneficial changes, while also rewiring the detrimental effects from previous prolonged stress and trauma without adequate recovery. This rewiring process is undergirded by the well-documented theories of *neuroplasticity* and *epigenetics*—the major topics of this chapter— which show how repeated experiences change the brain, body, and gene expression.

But just as with physical fitness, we can't simply read or think about mind fitness; we have to actually practice the exercises. As with PT, the goal of mind fitness training is not to get good at the exercises but to develop generalizable capacities we use every day—attention, mental agility, situational awareness, self-regulation, mind-body optimization, and emotional intelligence. Mind fitness is also protective: A fit mind is more likely to bring a wide window to challenging situations and effect a complete recovery afterward—in the process, decreasing the chance of psychological injury while also widening the window for future stress.

NEUROPLASTICITY: CULTIVATING GRAND CANYONS IN OUR MIND-BODY SYSTEM

Until the late 1990s, the field of neuroscience was relatively pessimistic about the adult human brain. The prevailing dogma was that by the time we reached adulthood, we were pretty much stuck with whichever brain we'd developed by that point. The adult brain was thought to be incapable of growing new brain cells (called *neurons*) or reestablishing the networks among these neurons (called *neural networks*) after a serious injury or degenerative disease. In other words, the adult brain was considered to be fixed in structure and function.

It turns out this understanding was wrong.

Neuroscientists now know that the brain changes throughout our lifetime. The brain constantly rewires itself in response to our repeated experiences, with every sensory input, body movement, reward signal, thought, emotion, stress arousal, and association between stimulus and response. This concept is called *neuroplasticity*. Just like young brains, the adult brain can repair damaged regions, rezone regions that performed one task and assign them to perform a new task, grow new neurons (called *neurogenesis*), and grow new neural networks.

For instance, one commonly cited study about neuroplasticity examined the memory and brain structures of London taxi drivers. To become a London cabbie, someone must learn "the Knowledge," the location of every street within a six-mile radius of the city center. After a few years of study, prospective cabbies earn their medallion by demonstrating their ability to drive their fares without referring to a map. That's one heavy spatial memory workout! You can imagine that the longer someone drives a taxi in London, the more consolidated the Knowledge becomes. In fact, brain imaging research seems to validate this: London cabdrivers have larger hippocampi—the brain region that allows us to consolidate explicit memory—than other people matched in terms of age and gender. Moreover, the longer cabbies drive a taxi in London, the more their hippocampi change.

As this example shows, the brain, like the rest of the body, develops the "muscles" it uses most, sometimes at the expense of other abilities. By engaging and repeating certain mental processes—consciously and unconsciously— the brain becomes more efficient at those processes. Over time, the brain regions supporting a certain mental skill rearrange the structural connections between neurons to create more efficient patterns of neural activity. As the Canadian psychologist Donald Hebb put it presciently in 1949, "When neurons fire together, they wire together."

As a result, areas of the brain may shrink or expand—become more or less functional—based on any repeated experience. The brain keeps this power throughout our lives. As science writer Sharon Begley beautifully expresses it, "The very structure of our brain—the relative size of different regions, the strength of connections between one area and another—reflects the lives we have led. Like sand on a beach, the brain bears the footprints of the decisions we have made, the skills we have learned, the actions we have taken."

Interestingly, the brain can also be changed and rewired *without any input from the outside world*. In fact, the brain changes simply from repetitive thought patterns and/or chronic stress arousal. Every time we worry about a negative event in the future, for instance, we activate the amygdala, the survival brain region responsible for neuroception, that unconscious threat-scanning process. Over time, as worrying becomes a habit, the amygdalae— like many brain regions, we have two of them—can actually thicken. In turn, thicker amygdalae then become hypersensitive to worry, prompting even more anxiety and fueling a vicious cycle.

When I teach about neuroplasticity, I often use the metaphor of the Grand Canyon. Admittedly, this metaphor is not geologically correct, but it's clear and compelling, nonetheless. Imagine that back before the Grand Canyon had formed, there was a flat mesa in the northern Arizona desert. When it rained, because the surface was level there was no predictable place where the water would flow. However, let's imagine that at some point, an indentation appears somewhere on the flat surface, creating an irregularity where the ground is slightly lower than the rest of the mesa. The next time it rains, where does the water flow? Into this indentation, of course. As the water flows that way, the dent becomes deeper. As the rains continue, the dent becomes a rivulet—and then an arroyo. As erosion accelerates, the arroyo becomes a ravine. Eventually, many rainstorms later, we have a Grand Canyon. Hence, when it rains today, it is *virtually impossible* for the water to flow anywhere except into the Grand Canyon. The groove is simply too deep. In fact, it would take an engineering feat and monumental effort to direct the water to flow somewhere else.

The human brain is made up of many such canyons—our habitual ways of perceiving, thinking, feeling, responding, and acting. Our canyons can be quite subtle—every time we walk into a room, we first notice everything that's wrong with the situation. Or they can be more obvious—every time we feel lonely or sad, we eat a cookie. Regardless of its particular script, each time we carry out our brain's canyon programming, we reinforce it—making it harder and harder in the future not to carry it out again.

Neuroplasticity boils down to this: The repetition of any experience makes it easier to do—and harder *not* to do—again in the future. That's why it can be so difficult to start a new habit, and why it takes several weeks of deliberate practice before we begin to see the new habit stick. And that's why it's so blasted difficult to stop an *old* habit, especially if that habit gets triggered when we're stressed or dysregulated.

Since the brain is malleable to *any* repeated experiences, neuroplasticity has powerful implications—in detrimental or beneficial directions. In the detrimental direction, prolonged stress, trauma, depression, anxiety, and PTSD are all associated with declines in our cognitive performance, particularly in terms of learning new information or skills, remembering things, and paying attention.

Indeed, a major cognitive symptom of dysregulation is degraded memory, such as forgetting why you walked into a room or frequently misplacing your keys. These aren't just "senior moments"; they can also be signs that we're dysregulated. We saw this effect in Chapter 2 with Todd, who frequently showed up in the wrong uniform and forgot to complete tasks in ways that undermined his squad's performance.

As Todd's experience demonstrates, dysregulation can lead to structural changes in the hippocampus, the brain region associated with explicit learning and memory. In fact, brain imaging studies show that people diagnosed with PTSD—including Vietnam War and Gulf War veterans, civilian police, and survivors of sexual or physical abuse—had significantly smaller hippocampi than people matched for age and gender without such traumatic experiences.

The repeated experience of chronic stress and trauma can also lead to neuroplastic adaptation to stressful circumstances. As I've noted, the brain will develop the "muscles" it uses most—sometimes at the expense of other abilities. For example, one large study of U.S. Army soldiers found that those who served in Iraq outperformed those who had not in terms of faster reaction time on computerized cognitive tests. However, the deployed soldiers also showed significant performance declines in spatial memory, verbal ability, and attention skills—cognitive declines that persisted more than two months after they returned home. In other words, through deployment, their brains built the capacity for quick reaction time—a function more necessary for survival in Iraq—at the expense of other mental skills.

Yet neuroplasticity's effects can be in a beneficial direction, as well. For instance, troops who participated in MMFT before their combat deployments saw improvements in sustained attention and working memory capacity, even during the stressful predeployment training period. They also showed functional changes in brain regions associated with emotion regulation, impulse control, and interoception (awareness of body sensations), which plays a major role in regulating and recovering from stress arousal. Likewise, forty years of empirical scientific research have documented a range of neuroplastic benefits from mindfulness training—from eight-week mindfulness-based interventions to years of extensive practice experience—including improvements in attention; declines in mind-wandering; more left-sided firing of the prefrontal cortex, a part of the thinking brain associated with more positive emotions; and smaller amygdalae, the survival brain region involved with neuroception and worry. All this research shows how the repeated experience of directing the attention in particular ways can lead to beneficial brain changes.

Interestingly, recent research has explored the link between beneficial neuroplasticity and cardiovascular exercise. Research with rodents shows that voluntary exercise grows new neurons in the hippocampus and helps them wire into existing neural networks.

Humans experience these neuroplastic benefits from voluntary physical exercise, too. Research among older adults shows that greater physical activity and higher cardiorespiratory fitness levels are linked with better brain oxygenation, healthier brain activity patterns, and greater gray matter volume in the prefrontal cortex and hippocampus, brain regions involved in executive functioning and explicit memory. For instance, when elderly adults were randomly assigned to walking or stretching programs for a year, walkers grew their hippocampi while stretchers did not. Walkers also performed better on cognitive tests and had higher blood levels of brain-derived neurotropic factor (BDNF), which increases neurogenesis and strengthens the connections between neurons. Likewise, research with children shows significant links between physical exercise and improvements in cognitive performance—including perceptual skills, IQ, verbal tests, mathematics tests, memory tasks, and academic achievement.

NEUROPLASTIC CONSEQUENCES OF THE GRAND CANYON

Neuroplasticity's consequences will result from *any* repeated experience—this is a law of nature. To a large degree, the choice about whether to access its detrimental or beneficial consequences is up to us. Since many of us spend a lot of time on autopilot, however, we're actually *choosing to allow our unconscious habits and patterns to drive most of our repeated experiences.*

Importantly, all repetitive habit patterns—including thinking brain habits like worrying, to-do list planning, fantasizing, and comparing our-selves to others—have material effects on the function and structure of our brains. These habits can also drive our mind-body system's stress arousal levels. Thus, when we let such common thinking brain habits occur uncon-sciously and repetitively, they can actually work against our self-regulation, our situational awareness, and even our happiness.

There are several common "Grand Canyons" that we might judge as "not all that bad" but which actually have significant detrimental effects for an allied relationship between our thinking brain and survival brain. The first is *mind-wandering*. Not surprisingly, a significant body of empirical research demonstrates that mind-wandering is linked with attentional lapses and declines in performance. Especially in high-stress professions—where indi-viduals might need to patrol through a crowd to locate a possible terrorist or notice subtle shifts in the wind that might signal that a wildfire pattern is about to change—attentional lapses can even be deadly. Attentional lapses can also undermine our ability to access and learn from real-time feedback, compromising our ability to adjust our behavior or take corrective actions in a timely manner.

Troops sometimes say to me, "It's okay, I can check out now while I'm driving to work or standing in formation, but when the shit hits the fan while I'm out on patrol overseas, I'll be paying attention." Yet neuroplasti-city suggests that's not what's going to happen. True, it's likely that they'll be extremely aware during their first patrol, because adrenaline will help them focus. Yet after consistently reinforcing an autopilot default mode dur-ing daily life, it's highly likely that after a few days patrolling the same vil-lage, their minds will get complacent and default back into autopilot. The more we reinforce autopilot, the more likely mind-wandering becomes—even during situations when paying attention really matters.

Furthermore, even when our attentional focus may not have immediate life-or-death consequences, habitual mind-wandering has some pretty un-fortunate survival brain costs. Neuroscientists Matthew Killingsworth and Daniel Gilbert created an experience-sampling app that contacts research participants through their smartphones at random moments with a "pop quiz" to ask questions about their current activity, mood, and mind-wandering. Their study of 2,250 American adults, who each received fifty pop quizzes, showed that mind-wandering is common, regardless of what people are doing. During pop quizzes, 47 percent reported that they weren't thinking about their current activity. Surprisingly, the nature of what people were currently doing had little impact on whether their minds wandered— in every single activity (except making love) at least 30 percent would report they weren't thinking about their current task. Now, here's the really interesting part: *People reported feeling less happy when their minds were wandering than when they were on-task, and this was true during all activities—even if they were currently doing something unpleasant.*

What's amazing is that this finding holds regardless of where the mind wandered to! Of course, it's easy to understand how someone whose mind wandered to something *unpleasant*—like worry—might feel unhappier than someone whose mind stayed on-task. Nonetheless, compared to on-task minds, people *still* reported being unhappier if their mind wandered to something *pleasant*, like a fantasy or a happy memory.

Overall, what people were thinking about at the moment of their pop quiz was a better predictor of their happiness than was what they were doing. Previous research demonstrated that negative moods are known to cause mind-wandering, but this study suggests that mind-wandering also causes negative moods. How's that for a vicious cycle?

A second common thinking brain habit that works against an allied relationship between the thinking brain and the survival brain is "multitasking." I've put that in quotation marks because our brains can only truly multitask with highly automatic behaviors, like—you guessed it—walking and chewing gum at the same time. For activities that require thinking brain attention, there's really no such thing as multitasking—instead, we're actually task-switching and dividing our attention. So while it may *subjectively feel* like we're doing two things at once, we're actually only toggling back and forth rapidly—with significantly less skill and accuracy than if we'd simply focused on one thing at a time.

For example, college students using Instant Messenger while reading a textbook took 25 percent longer to read the passage—not including IM time—compared with students who simply read. Similarly, an experiment tracked the work patterns of twenty-seven Microsoft employees over two weeks. Researchers found that when a worker interrupted his primary task to reply to an email, he'd get diverted, on average, for ten minutes—not just replying to that email but cycling through a range of other apps before returning to his primary task. Sometimes the diversion lasted hours. Other research shows that people who interrupt their work flow to answer email tend to work faster but also experience more stress, higher frustration, and more time pressure.

Likewise, drivers using cell phones while driving have slower reactions, with a greater chance of running red lights and more difficulty staying in their lane and maintaining appropriate following distance. One observational study of fifty-six thousand drivers found that drivers talking on their cell phones were more than twice as likely to fail to stop appropriately at intersections. These researchers concluded that "a person who drives while talking on a cell phone . . . is a worse driver than an individual at the legal limit of alcohol intoxication."

Especially in the digital era, media multitasking is a major work habit for many people. The stimulation that comes from frequently checking our social media feeds, texts, or email queues gives us a small burst of dopamine—one of the neurotransmitters involved with our brain's reward system—which can make it addictive. (Most people check their smartphones on average 150 times each day, which averages to once every six minutes.) Not only does the dopamine rush make us feel good, but it also increases our confidence that we're effective multitaskers, reinforcing the habit further. The dopamine rush may also make us overly optimistic, so we're less careful completing our current task and more likely to make mistakes.

Yet once we've trained our brains to the multitasking canyon—as well as to our phone's alert noises signaling a new text or email—we find that we're bored or restless when we try to focus solely on one primary task. As a result, frequent multitaskers tend to be more impulsive and sensation-seeking. In other words, multitasking contributes to that mismatch of our Paleolithic neurobiology with the modern world from Chapter 1.

Although many people say that multitasking makes them more productive, empirical research shows otherwise. One study assessed how much people multitask and then tested their performance on three cognitive tests.

The researchers measured their ability to focus attention and ignore irrelevant information, switch rapidly and accurately between different categorization tasks, and remember a sequence of letters that they'd seen before, which tested working memory. On all three cognitive control tasks, high media multitaskers performed *less well* than low multitaskers. *The high multitaskers actually had a slower task-switching speed than the low multitaskers!*

A related study asked three hundred participants to rate their multitasking frequency, as well as their perceived ability to do so, and then complete a multitasking test. High self-reported multitaskers had lower working memory capacity and were more impulsive and sensation-seeking. They also tended to rate their own multitasking ability as higher than average—meaning their *perceived* ability and *actual* ability to multitask were inversely related.

Heavy multitaskers also tend to search for *new* information rather than draw on information they already have that may be more relevant, and thus more valuable, to their current activity. In other words, their survival brains—especially their amygdalae—seem to have hypersensitive neuroception: They survey the external environment more actively, which can contribute to greater stress arousal. Thus, multitasking can work against an allied relationship between the thinking brain and survival brain.

This hypothesis was corroborated by a recent brain imaging study, which found that heavy multitaskers had less dense gray matter in the anterior cingulate cortex (ACC), the brain region involved in impulse control and emotion regulation. According to the researchers, this smaller ACC density may explain why heavy media multitaskers have lower cognitive control performance, more difficulties with emotion regulation, and greater impulsivity.

The takeaway: Every repeated experience matters. Therefore, whichever experiences we choose to repeat—either consciously or unconsciously—are changing our mind-body system. Armed with this understanding, we can choose to engage in consistent physical fitness and mind fitness training regimens, in order to intentionally rewire our brains and entire mind-body systems in a beneficial way.

EPIGENETICS: HOW THE GRAND CANYON AFFECTS OUR GENES

Through our repeated experiences, neuroplasticity changes the structure and functioning of our brain and nervous system. There's a parallel process called *epigenetics*—how our repeated experiences affect whether our genes get "turned on" or "turned off."

When I teach, I often encounter people who tell me that they "can't help it" that they have an anxiety disorder—or depression, diabetes, heart disease, an addiction, or some other physical or psychological illness—because "those genes run in the family." At the most extreme, these individuals tell me that mind fitness training "could never work" for them, because they "can't fight their genes."

Of course, when someone's thinking brain has such beliefs, they're setting themselves up for a self-fulfilling prophecy. Believing that their genes are fixed and the disease or addiction is inevitable, they're much less likely to engage consistently in the beneficial habits that could counteract their genetic tendencies.

Nonetheless, recent research has dismantled the old idea that "having" a particular gene will definitely produce a particular behavior or disease. Indeed, many genes work together to produce any outcome, so the effect of any one gene is typically very tiny. But even more importantly, a wide range of experimental and empirical research convincingly demonstrates that whether genes express or not—whether they are "turned on" or "turned off"—depends on our repeated experiences.

In other words, while we may have a genetic tendency toward a particular trait, *whether that tendency actually manifests through gene expression is strongly influenced by our environment and our habits.* To put it simply, environmental inputs and habits can lead to changes in our DNA or surrounding proteins—creating what's called an *epigenetic change*. These epigenetic changes can turn gene expression on or off, leading to persistent effects in our mind-body system. Throughout our life, we may accumulate epigenetic changes in both directions—turning genes on that were previously off, and vice versa. For instance, research shows that chronic sleep deprivation and shift work can adversely affect the genes regulating the circadian rhythm as well as immunity.

Just as the brain is malleable to any repeated experiences—both detrimentally and beneficially—epigenetic changes can also occur in both detrimental and beneficial directions. It all comes down to which repeated experiences our mind-body system has.

Not surprisingly, detrimental epigenetic changes have been linked with stressful or traumatic experiences without adequate recovery, especially during childhood.

During autopsies, for instance, people who died by suicide and had been abused during childhood showed distinct epigenetic changes in their brains, compared with people who died by suicide and did not have child abuse histories and with people who didn't die by suicide. Another study investigated people who'd experienced trauma in childhood versus adulthood. Trauma survivors with active PTSD showed significant, detrimental epigenetic changes, while trauma survivors without PTSD did not. More importantly, the active-PTSD group with childhood trauma showed *up to twelve times as many* epigenetic changes compared with the active-PTSD group without childhood trauma. Both studies—and several others—highlight how early life adversity can leave lasting neuroplastic and epigenetic changes on our mind-body systems.

One of the most common epigenetic changes from chronic stress or trauma, without adequate recovery, shows up in immune system functioning. Chronic stress arousal affects the programming of important cells in the immune system, called *macrophages*. (Depending on where they function in the body, macrophages can also have specialized names. For instance, macrophages in the brain and spinal cord are called *microglia*.)

Macrophages, including microglia, are responsible for recognizing and destroying the "bad guys" in our mind-body system, including infections, accumulated damage, and dead cells. For this reason, they also play an important role in aging. Macrophages do their work by producing *cytokines*, proteins that play a role in cell signaling. For optimal immune functioning, we need a balance of inflammatory and anti-inflammatory cytokines.

However, chronic stress arousal, especially during childhood, programs the macrophages in a dysregulated way. They become extremely effective at producing inflammatory cytokines—turning inflammation on—and less effective at producing anti-inflammatory cytokines—turning inflammation off. Moreover, these hyperreactive macrophages continue to release inflammatory cytokines long after the infection, toxin exposure, injury, or physical trauma that triggered them is gone. Why does this matter? As Gary Kaplan, a physician who clinically treats and writes about these processes, explains: "Once [macrophages and microglia] have sustained a hyperactivated state, they remember it. They are quicker to flare up and harder to quiet down."

In other words, chronic stress leads to epigenetic changes in the immune system that result in chronic inflammation in the mind-body system.

In turn, chronic inflammation may manifest in many different ways—including chronic pain, fibromyalgia, chronic fatigue syndrome, chronic headaches and migraines, arthritis, back pain, eczema, psoriasis, cardiovascular disease, asthma, allergies, irritable bowel syndrome, and insulin resistance, a precursor to type 2 diabetes. Chronic inflammation of the microglia is also responsible for neurodegenerative diseases, such as depression, anxiety disorders, PTSD, multiple sclerosis and other autoimmune diseases, Alzheimer's disease, and schizophrenia.

For instance, Todd and I both experienced this epigenetic effect after our respective experiences with childhood adversity. My chronic inflammation eventually manifested as asthma, allergies, migraines, and inflammation of my optic nerves, while Todd's manifested as chronic knee and back pain.

Detrimental epigenetic changes can also be passed to offspring. Much of this research has been conducted with rodents, whose shorter life spans allows researchers to observe transgenerational effects more easily.

Several experiments show that rat pups with attentive mothers—rat moms who nursed, licked, and groomed them—grew up to be more resilient. As adults, offspring of attentive rat moms showed less fear and lower stress hormone levels during stress. They were also better learners and showed delayed aging in the hippocampus. These changes were epigenetic, meaning that the early life experience of repeatedly being licked by an attentive rat mom altered the on/off switch for genes regulating the pups' stress response. Even more, the female rat pups grew up to become attentive moms themselves—thereby passing their stress-hardiness and attentive mothering skills epigenetically to *their* offspring.

Conversely, female rat pups separated from their moms during infancy grew up less resilient. As adults, these rats showed attention problems, higher stress reactivity, and lower-than-normal gene expression in brain regions associated with maternal behaviors. Not surprisingly, when they had babies, the female rats were less attentive mothers—licking, grooming, and crouching over their pups less than other rat moms.

Detrimental epigenetic changes aren't just passed down via mothering styles. For instance, one study exposed male mice to traumatic conditions in early life and then compared them with nontraumatized males. Afterward, the traumatized mice showed depressed behaviors and lost their natural aversion to open spaces; their faulty neuroception left them less self-protective. Their metabolism also became dysregulated, with lower levels of insulin and blood glucose than nontraumatized mice. These epigenetic changes were passed on to their offspring—via sperm, of course. Incredibly, the dysregulating effects of that first generation's early life trauma on both metabolism and behavior *persisted in the third generation.*

Just as with neuroplasticity, however, epigenetic changes can also occur in a beneficial direction. Indeed, the window-widening habits I'll teach you leverage this scientific principle. I'll address this more fully in Part III; here I want to preview a few examples.

I've already mentioned the beneficial neuroplastic consequences from cardiovascular exercise. However, consistent physical activity also leads to epigenetic changes in how the brain responds to stress. For instance, experiments with mice help us understand why regular exercise reduces anxiety. In these experiments, one group of mice received unlimited access to a running wheel, while the other group did not. After six weeks, both groups were exposed to cold water, a stressful experience. The brains of the sedentary mice leaped immediately into a reactive, excited state as soon they came into contact with the cold water, turning on genes that make neurons fire rapidly. In contrast, the brains of the running mice did not show these genes, which helped them control their reaction to the cold water. At the same time, running mice released more GABA (gamma-aminobutyric acid), a neurotransmitter that tamps down neural excitement. Researchers posit that these two epigenetic changes contribute to the anxiety-reducing effects of exercise. In other words, consistent physical exercise reorganizes the brain, via epigenetic changes, to become more stress-resilient.

A second example of beneficial epigenetic changes comes from mindfulness meditation. For instance, telomeres—the protective caps at the ends of chromosomes—are essential for cell division and, as we age, shorten over time. For this reason, telomere length is used as a proxy measure for biological aging. In empirical research, people who reported low mind-wandering and more present-moment awareness had longer telomeres on their immune cells than people who reported high mind-wandering, even after controlling for stress. Conversely, shorter telomeres have been significantly linked with depression and chronic stress. These studies show how chronic stress and mind-wandering both hasten the cellular aging process, while mindfulness can slow it down.

Other studies have shown how mindfulness meditation has a significant buffering effect against inflammation. For instance, one study used laboratory-induced inflammation to create skin blisters. In this research, individuals who'd completed an eight-week mindfulness-based stress reduction (MBSR) course had significantly smaller blisters than individuals who'd completed a similarly matched health enhancement program without mindfulness practices. Another study showed that compared to a control group, elderly adults who participated in MBSR showed reduced expression of genes related to inflammation, which the researchers measured by drawing blood to sample immune cells. After MBSR, elderly participants also reported fewer feelings of loneliness, which has been linked with chronic inflammation in other research.

A third pathway to detrimental or beneficial epigenetic changes comes from our sleep habits and diet. Chapter 9 will explore the detrimental epigenetic effects of chronic sleep deprivation. In addition, our diet radically affects the health of our *microbiome*, the microorganisms that live in our gut and intestinal tract. I'll explore the epigenetic powers of these gut flora in Chapter 17.

STRUCTURE VERSUS AGENCY AT THE MICRO LEVEL

Why have I dedicated a whole chapter to neuroplasticity and epigenetics? Two reasons. First, we need to understand just how much these dynamics affect our neurobiological structures. Especially when we aren't aware of their effects, neuroplasticity and epigenetics can have tremendous consequences on our brains, nervous systems, and bodies—down to the cellular level. Even those habits we collectively think of as "not that bad"—checking out on autopilot, worrying, multitasking—can over time have detrimental effects on our mind-body systems.

Many contributors to the width of our window—such as our genetic heritage, our early social environment, and the stressful or traumatic experiences we've endured throughout our lives—weren't up to us. Indeed, as this chapter's research shows, some of these structuring dynamics are passed down through many generations. How our mind-body systems were wired in response to these experiences was uniquely adaptive, because it allowed us to survive. At the same time, however, the neuroplastic and epigenetic changes from chronic stress or trauma set the stage for narrow windows over the longer term.

Second, while a legacy of chronic stress and trauma—especially during childhood—has profound enduring effects on the width of our window, *it's not fate.* Narrow windows are not destined to stay narrow, any more than wide windows are destined to stay wide. We can change the width of our window through *any* repeated experiences, for good or for ill, throughout our lives. In fact, the only thing that's truly under our control is where, when, and how we repeatedly direct our attention—and whether we're directing it consciously.

Whatever the current structure of our mind-body system—and any symptoms of dysregulation we may be experiencing today as a result—*we always have agency in shifting that structure through our repeated choices.* These neurobiological structures are not fixed; they're simply stabilized for now. In fact, the less awareness we have about these structures—and the habits that reinforce them—the more power they have over our lives.

Nevertheless, the scientific research about neuroplasticity and epigenetics shows us that these mind-body structures are malleable. In every choice we consciously or unconsciously make, we either reinforce or change these structures that then shape our future choices. We can choose to let our earlier neurobiological structures play out their programming. Or we can choose to interrupt this programming when it's no longer serving us, and choose instead to intentionally rewire these structures and widen our windows. No matter how difficult our past may have been, the choice today is entirely up to us.

Learning to use our biology in a new way requires taking responsibility for our choices. Whatever we're doing repeatedly has big effects on our mind-body systems. A fit mind and body not only have more capacity for thriving during stress, trauma, uncertainty, and change *today*—they also set the structural conditions for thriving during stress, trauma, uncertainty, and change *tomorrow*. In other words, neuroplasticity and epigenetics help us understand how we can influence the mind-body structures that will shape our lives in the future. Armed with this understanding, we open the possibility for profound change—and for wider windows.

CHAPTER FOUR

The Body during Stress and Trauma

The next two chapters explore how our mind-body systems experience stress, chronic stress, and trauma. This chapter describes the body's responses, while the next chapter will describe the brain's responses.

Before you continue reading, you'll want a notebook or journal for the reflective writing exercises throughout the rest of this book. Part II exercises will guide you in evaluating the current width of your window—and appreciating how it got this way. Your reflections from the Part II exercises will be crucial for tailoring exactly how you will go about widening your window in Part III. What you put into these exercises will critically influence what you get out of this book. No one needs to see your reflections but you.

After you have procured your notebook, I encourage you to explore stress arousal concretely, in your own mind-body system. To do this, make a list of some of the sources of stress in your life right now.

After writing down the sources of stress, rate each item in terms of its stress intensity—1 for things that feel mildly stressful to you, 10 for things that feel almost unbearable. Now, pick an item you've assessed as falling around a 5. If your list has nothing but 10s, pick whichever item feels the least stressful. It might be financial worries, a challenging relationship, an overbearing boss—it doesn't matter what it is, as long as it's a source of *moderate* stress. (For all those Type A overachievers in the house, I really mean it—pick a 5, not a 10!)

Once you have one in mind, complete this short exercise. Read all the instructions first. Place your feet flat on the floor and, if it feels comfortable for you, close your eyes. Visualize a scenario involving your chosen source of stress while you bring nonjudgmental curiosity to investigate what happens inside your mind-body system. This is an experiment in your mind-body laboratory, and it's time to gather some data.

First, scan your attention throughout your body. In particular, notice what's happening in your chest and stomach, the muscles in your legs and arms, your hands, your neck, your jaw, your eyes, and inside your mouth. Notice if there are any changes in your heart rate or breathing rate. Also notice your body's temperature and whether your posture shifts.

Next, bring your attention to what's happening in your mind. Is the mind calm and quiet, or alert and focused? Or do you notice a lot of cognitive activity, like racing thoughts? If you notice the mind is thinking, see if you can discern any pattern to the thoughts, like planning or worrying. Or is the mind fuzzy, foggy, and distracted? Finally, notice if there are any emotions present, such as anxiety or sadness. When you visualize your stressful scenario, you may not notice anything in your body and mind, and that's perfectly fine, too.

After you've noticed and cataloged all of the physical sensations, thoughts, and emotions, open your eyes and feel the support of the chair underneath you. You might direct your attention to the places of contact between the back of your body and the chair. As you do this, notice if anything shifts in your body and mind.

Afterward, jot down everything that you noticed—physical sensations, thoughts, emotions. I promise you'll get much more out of these chapters if you do.

(NOT) SEEING IS BELIEVING

The first time I completed this visualization exercise *with awareness*—as you just did—was accidental. I'd been visualizing this way *unconsciously* my entire life, through that deeply conditioned habit known as "worrying." However, my first conscious experience with this visualization exercise occurred in late 2004, during a three-month silent meditation retreat.

At the time, I was on a medical leave of absence from Georgetown because of problems with my eyesight. Earlier that year, I'd experienced two episodes of optic neuritis—inflammation and atrophy of my optic nerves—which manifested as intermittent migraine headaches and months of blurry, double vision. In addition, several weeks before my silent retreat began, my husband and I separated.

About a month into the retreat, another neuritis episode began. Possibly because I chose to forgo treatment at a hospital and work with a local acupuncturist, this episode was much worse. Soon I could only see shadowy patches of light and dark, with few shapes. Eventually, I experienced three weeks of almost total darkness. As my eyesight deteriorated, I stayed in my room. I was alone in the dark, literally and figuratively.

One day, I was lying on the bed noticing the sensations of throbbing and burning—deep in my eye sockets and at the base of my skull—interspersed with hot ice-pick daggers of pain. Whenever I became exhausted from observing these sensations, I'd take a break and direct my attention to sounds outside, to birds and the wind.

Then, out of nowhere, came the following thought: *What if, this time, my eyesight doesn't come back? What if I'm going to be blind for the rest of my life?* (Notice how these questions started with *what if . . . ?*)

With this thought, my mind-body system was off to the races. Pounding heart. Chest pain so intense I was surely having a heart attack. Rapid, shallow breathing. Butterflies in my stomach. Clammy hands. Dry mouth. At the same time, my thinking brain was flooded with racing thoughts: *How would I make a living while divorced, alone, and blind? Would I need to get a roommate? Give up being a professor? Would I need to learn braille?* These racing thoughts proliferated as my body became increasingly anxious. Only when I was on the verge of puking did I finally snap out of it and notice: *Oh, I'm feeling anxious right now. Anxiety is like this.*

With that realization, I decided to direct my attention into my mind-body laboratory for a spontaneous experiment. *So what happens when I feed this anxiety by visualizing the rest of my life without ever being able to see again?* As a Type A overachiever myself at the time, I didn't exactly pick a 5 for my first time with this exercise. Immediately, everything amplified—the ice-pick daggers and pain in my eyes, head, neck, and shoulders; the racing thoughts; and the panicked planning.

And now what happens when I disengage from the thoughts and simply focus on the ice-pick daggers? Immediately, the pain intensified. However, the racing thoughts decelerated, and my heart rate and breathing rate slowed. Interesting. Although the pain got more intense, things were now calmer overall. *Thoughts clearly make things worse.*

And now what happens if I focus instead on the back of my body touching the bed? Over the next few minutes, my heart rate and breathing rate returned to normal. My hands and face got warmer. My nausea passed. I started yawning. The pain lessened. My eye sockets still throbbed but became itchy and hot. My mind calmed down. Soon I could easily redirect my attention to hearing the birds and the wind once again.

As my story suggests, there are strong interrelationships between how our survival brain unconsciously appraises our situation and how our body experiences stress arousal or recovery in response. That's where we're headed in this chapter.

WHAT IS STRESS?

Stress has become the all-encompassing term for anything we don't want to experience—being stuck in traffic, relationship problems, health concerns, depression, or anxiety. For many of us, stress means something harmful—to be avoided, reduced, or managed.

One downside to this conventional understanding, however, is that it can perpetuate an aversive relationship with stress—as well as the unconscious belief that we're relatively powerless in influencing stress and its effects. In turn, this belief leads many of us to try to manage stress and trauma through denial, avoidance, compartmentalization, self-medication, or distraction.

I want to encourage you to develop a more empowered relationship with stress—to trust it's something that we *can* influence, through the ways we choose to direct our attention. Understanding the Stress Equation (see Figure 4.1) creates the possibility for us to proactively influence our stress activation levels and performance during stress and trauma.

The first component of this equation is the *stressor,* an internal or external event that our survival brain perceives to be challenging or threatening. *External stressors* can be anything that generates changes in our lives and our

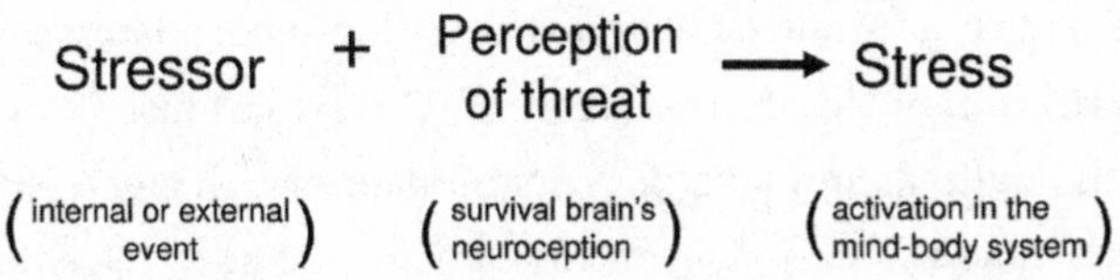

$$\text{Stressor} \quad + \quad \begin{array}{c}\text{Perception}\\\text{of threat}\end{array} \longrightarrow \text{Stress}$$

$$\left(\begin{array}{c}\text{internal or external}\\\text{event}\end{array}\right) \quad \left(\begin{array}{c}\text{survival brain's}\\\text{neuroception}\end{array}\right) \quad \left(\begin{array}{c}\text{activation in the}\\\text{mind-body system}\end{array}\right)$$

Figure 4.1: The Stress Equation

The Stress Equation explains how the mind-body system produces stress activation/stress arousal. Whenever we experience a (1) *stressor,* an internal or external event that (2) the survival brain perceives as threatening or challenging, then our mind-body system turns on (3) stress arousal, which is physiological activation in the body and mind.

social status: Traffic. Bills. A fight with a loved one. An upcoming exam or surgery. Harassment or discrimination. External stressors can also include challenges we usually consider "positive," such as buying a new house, getting a promotion, or having a baby. *Internal stressors* include illnesses, physical injury, hunger, thirst, sleep deprivation, chronic pain, intense emotions, flashbacks, nightmares, and intrusive thoughts.

Stressors can also be classified as acute or chronic, and physical or psychological. *Acute stressors* are major stressors that happen during short time periods, such as surgery, a natural disaster, a mass shooting, or the death of a loved one. In contrast, *chronic stressors* affect us over an extended time period, such as financial concerns, job demands, relationship problems, or a chronic medical condition.

Physical stressors, such as an infection, physical assault by another person, or exertion during an athletic competition, affect the whole mind-body system. In contrast, *psychological* or *symbolic stressors* arise in our thinking brain without creating immediate physical danger for the body. The visualization exercise was a symbolic stressor.

One of the most common symbolic stressors is *anticipation,* feeling stressed or anxious about events that may occur in the future. Compared with our animal cousins, being able to imagine and prepare for future contingencies is one of humanity's unique talents—or curses. As my story suggests, we know an anticipatory stressor has been triggered when we find our thinking brain using those two little words: *what if . . . ?*

The second component of the Stress Equation is *appraisal of the stressor* via the survival brain's neuroception process. The survival brain is constantly scanning the internal and external environment—each smell, sight, sound, touch, taste, physical sensation, mental image, thought, or emotion—and checking to see if the stimuli are *threatening or challenging.*

This internal appraisal process is unique to each of us, based on our survival brain's unconscious learning and conditioning from earlier experiences. That's why two people confronting the same stressor may differ dramatically in their respective neuroception—and then experience different stress arousal levels in response. Research shows that when confronted with identical stressors, individuals with Type A personalities—individuals who are driven, oriented toward achievement, and internally pressured by time urgency—will exhibit greater stress arousal than individuals with Type B personalities.

Our survival brain will perceive greater threat—and thus generate more stress arousal—if it perceives stressors as *novel, unpredictable, uncontrollable, and threatening to our ego, our sense of identity, or the survival of our mind-body system.* Likewise, if aspects of the current stressor contain cues or triggers related to past traumatic events, the survival brain is also likely to perceive greater threat.

When we've experienced a stressor many times before, we know how it will feel and what will be required for successfully navigating through it. Similarly, when a stressor is predictable, we generally know when it will happen, how long it will last, and—crucially—that it won't be sprung on us without warning. Together, familiarity and predictability help us know which internal coping strategy will likely work best for this stressor. As a result, our survival brain is more likely to feel a sense of agency and self-efficacy, which reduces stress arousal.

During the Nazi bombings of Britain during World War II, for instance, London was hit every night while suburban areas were bombed sporadically. In both areas, initially hospitals saw an increase in ulcers, a stress-related disease. Yet the suburban area experienced a greater increase in ulcers. The unpredictability of attacks made them more stressful. Within three months, however, ulcer rates in both areas had returned back to normal—suggesting that even suburban residents adapted to the now more familiar, if still unpredictable, bombing schedule.

Believing that we have some control—over the situation, our environment, or even ourselves, through a sense of agency—can also lead the survival brain to appraise stressors as less threatening. However, the effects of feeling a sense of control are context dependent. With mild or moderate stressors, feeling a sense of control can decrease our stress arousal levels.

With catastrophic or traumatic stressors, however—such as getting a fatal medical diagnosis or losing a loved one—feeling a sense of control can be counterproductive. The thinking brain might assume that "it's all my fault, I should have prevented that," even with situations that are actually outside our control. Individuals who believe that events and outcomes are the result of their own effort are most susceptible to such perceptions. Known as having a strong *internal locus of control*, this dynamic is quite common in American culture. Yet when confronting something that's actually uncontrollable, individuals with a strong internal locus of control experience more stress arousal than individuals with an *external locus of control*, who believe that events and outcomes are controlled by fate or chance.

The third component of the Stress Equation is *stress arousal*, the physiological, cognitive, and emotional activation in our mind-body system. This component is a law of nature: *Whenever the first two components of the equation arise together, there is no way around experiencing stress arousal. It's just not possible.* I've italicized that, so that your thinking brain can really take this in!

One of the most common, counterproductive thinking brain habits is to criticize or judge ourselves when we're stressed, or to devalue the source of our stress. Devaluing thoughts can be the most insidious, because our thinking brain usually tries to spin the situation positively, such as thinking "It's just a flat tire, after all. At least it wasn't a car accident!" Remember, however, that neuroception is a *survival brain job*! Despite its best intentions, the thinking brain can't offer thoughts like these and expect the stress arousal to magically disappear. The thinking brain is trying to help, but it may actually make the stress arousal worse.

Return to your notes from the visualization exercise. While some stress activation symptoms are common to everyone, each of us has conditioned our own unique constellation of stress symptoms in our mind-body system.

In the body, for example, you may have noticed a faster heart rate. Faster or shallow breathing, or holding your breath. Constriction in your chest.

Muscle tension in your arms, legs, buttocks, shoulders, neck, or back. A hunched or collapsed body posture, or the body leaning forward. Shoulders raised up by your ears. Butterflies in your stomach. Dry mouth or clenched jaw. Glaring, squinty eyes. Clammy hands or sweating.

In the mind, you may have noticed racing thoughts. Thoughts related to planning, worrying, or trying to figure out how to fix the situation. Comparing thoughts, such as "I need to get over this already; my situation isn't nearly as bad as Person X has it." Devaluing thoughts, such as "Even though this stressful thing is happening, I still have my family [car, job, health]." Critical thoughts, such as "It's just a silly email! I shouldn't feel stressed about that!" or "If only I had done X or Y, I could have prevented this." Escapist thoughts, such as fantasizing about what you'll do this weekend.

Emotionally, you may have noticed feeling distracted, frazzled, pressured, anxious, terrified, impatient, restless, disrespected, angry, tired, overwhelmed, ashamed, guilty, or burned out. Alternatively, you might have noticed sleepiness, fuzziness, or numbness. Or you might have felt disconnected, unable to notice anything in your mind-body system.

HOW STRESS AROUSAL WORKS

Stress activation, which comprises these many different symptoms in the mind-body system, is how we mobilize energy to respond to a threat or a challenge. *Stress activation is all about shifting energy from long-term to immediate needs.*

Stress arousal was optimized for coping with the environmental threats that our cave-dwelling ancestors faced. I'll call this the Saber-Tooth Tiger Threat Template. In a contest with a saber-tooth tiger, the first ten minutes are critical—thus, the stress response was optimized for surviving these ten minutes. In the caveman's world, either he made it through these ten minutes alive or he didn't. If he did, he had plenty of time afterward to hide out in his cave, rest, and recover before venturing out for his next challenge.

Stress arousal was intended as an immediate response to handle change or crisis, followed by recovery—and a return to baseline equilibrium with no ill effect—afterward. As Chapter 1 explained, this process is called *allostasis*. Allostasis allows us to vary internal conditions, including heart rate, breathing rate, temperature, and blood sugar levels, so we can galvanize the appropriate amount of energy and focus for coping well before, during, and after change or crisis. To accomplish this, allostasis relies on interactions between (1) the brain; (2) the hormone (endocrine) system, especially the system that controls stress hormones, which is called the hypothalamic-pituitary-adrenal (HPA) axis; (3) the immune system; and (4) the autonomic nervous system.

Therefore, even today, at the moment of perceiving threat, our mind-body system is designed to focus on danger and mobilize energy to react *fast*. Of course, receiving emails that leave us upset or getting stuck in traffic is quite different from being chased by a saber-tooth tiger! Nevertheless, we still mobilize the same response that cave people used to survive the Saber-Tooth Tiger Threat Template.

When the survival brain neurocepts threat, it sends messages to the endocrine system to release hormones needed for immediate survival—and to inhibit hormones used for long-term needs. These hormonal changes controlled by the HPA axis occur in two waves.

First, once the survival brain perceives threat, it directs the endocrine system to release adrenaline. Within seconds, adrenaline increases our heart rate, to quickly pump blood to the organs and big muscles in our limbs, so that we can move fast. In the lungs, adrenaline dilates our bronchial tubes to increase our breathing rate so we can take in more oxygen. Adrenaline also mobilizes the body to release glucose, so we'll have a ready source of energy.

At the same time, blood flow shifts away from the digestive system, which we may experience as nausea or butterflies in the stomach. After all, if we don't survive the next ten minutes, digesting our last meal is irrelevant. Adrenaline also constricts the blood vessels supplying our skin, decreasing the chance that we'd bleed out should a tiger scratch us. That's why, when we're stressed, our skin may feel cold and clammy, our palms may get sweaty, and our hair may stand on end. Adrenaline also triggers fibrinogen, which speeds up blood clotting, as a further defense against blood loss.

Everything about stress activation is initially aimed at transporting oxygen and glucose, since we need energy and brain focus right now.

After this first wave, the survival brain and HPA axis controlling stress hormones work together to adjust our stress level so that it corresponds with the particular stressor that we're facing. Now the HPA axis provides a fine-tuning second wave of stress activation.

At this stage, the survival brain conducts a secondary appraisal focused on the question "Do I have the *resources* to address this stressor?" If the survival brain recognizes that we have internal or external resources, the HPA axis can dial back stress activation levels. However, if the survival brain feels powerless, helpless, or lacking in control—that is, if the stress is traumatic—the HPA axis is likely to amplify stress activation. Thus, *whether we perceive ourselves as having agency in the situation plays a critical role in our second wave of stress activation.*

For instance, in my story, when I focused on the *content* of my thinking brain's anxious anticipation, my survival brain felt powerless, which amplified my second wave. In contrast, when I directed my attention solely to physical sensations of pain, my survival brain felt less out of control, which dampened the second wave. And, when I directed my attention to the contact of my body against the bed, my survival brain registered the bed's support as a resource, reducing the second wave even further—eventually allowing my mind-body system to experience some recovery.

During the second wave, the HPA axis activates hormones that mobilize energy and focus for the current emergency, while inhibiting hormones for longer-term tasks. Later, after the stressor has passed, it activates hormones to facilitate recovery. During both the second wave and recovery, the HPA axis works with the immune system and the autonomic nervous system.

As part of the second wave, the HPA axis activates hormones to mobilize energy by increasing the circulating glucose in our mind-body system. Glucose provides fuel for the muscles and enhances the thinking brain's focus and short-term memory. The most important of these energy-mobilizing hormones is *cortisol*.

Cortisol has two jobs during the second wave. First, it replenishes energy stores that got depleted during the first wave's adrenaline rush. Second, cortisol boosts immune functioning. It sends white blood cells to "battle stations" at vulnerable places in the body, such as the skin and the lymph nodes, where they're most likely to be needed in case of injury or infection—think early defense against a tiger scratch or bite! For this reason, stress arousal enhances immune functioning in the short term. After an hour, however, sustained stress arousal suppresses immunity, down to 40 to 70 percent below our normal baseline. This is why we're more susceptible to catching colds when we've been chronically stressed.

The HPA axis also activates other hormones that prioritize immediate needs. For instance, it releases *endorphins*, internal opioids that blunt our perception of pain. It also releases *vasopressin*, a hormone that regulates the cardiovascular system during stress and sets the autonomic nervous system into its defensive mode.

Conversely, the HPA axis inhibits hormones related to long-term needs, including *growth hormone* and the sex hormones *estrogen, progesterone*, and *testosterone*. It also inhibits *insulin*, which directs the body to store energy for later use. After all, who needs long-term "projects" like digestion, reproduction, tissue repair, energy storage, or growth if we won't live to see the end of the day?

THE HIERARCHY OF HUMAN DEFENSES: HOW THE NERVOUS SYSTEM CONTRIBUTES TO OUR SAFETY

We have three hierarchical levels of "defense in depth," with each defensive strategy supported by a distinct neural circuit between the brain and the ANS.

Say you're walking down an empty street, alone after dark. Someone jumps out of the shadows, demanding your wallet. First, you're likely to negotiate with and placate the assailant, perhaps even offering him your wallet, while you evaluate the threat. You're likely to check to see if he's armed, mentally ill, drunk, high, or physically larger and stronger—any of which could make your escape more difficult. You're also likely to look around to locate escape routes or help.

However, imagine it rapidly becomes clear that he's armed with a knife, physically stronger, and interested in assaulting you. As your stress arousal increases, fear or anger takes over. At this point, you're likely to scream, hoping someone will come to your rescue or at least call the police. Most likely, you'll feel an instinct to run and flee from this dangerous situation. However, depending on your own conditioning, as well as whether you're armed or have any martial arts training, you may feel an instinct to stand your ground and fight instead.

As you run to get away, he grabs your arm and your stress arousal increases further. Soon, the man has you pinned from behind, pressing your face and the front of your body against a nearby wall. You've realized he actually wants to sexually assault you. As you continue screaming and writhing to break free, he pushes the knife into your neck. He tells you that unless you stop screaming and behave, he's going to kill you. While he rips at your clothes, you begin to feel dazed and confused. Your body stops fighting. You feel strangely disconnected from what happens next. By having "given up," however, you've actually increased the likelihood that you'll survive the event.

How does our mind-body system automatically know how to take these sequential actions to keep us safe? It comes from interactions between the survival brain and the ANS. Although the ANS generally receives less attention than the brain and stress hormones, it plays a central role in MMFT. In fact, MMFT draws from two lineages—mindfulness training and body-based trauma therapies with techniques for regulating the ANS after chronic stress and trauma. Thus, understanding how the ANS works is a critical part of widening the window.

The ANS serves as the bridge between the brain stem—the most primitive part of our survival brain that controls stress arousal and recovery—and the rest of the body. Thus, it provides an *automatic control system* for many bodily functions outside conscious awareness. However, the ANS is not entirely outside conscious control. The thinking brain can consciously alter some automatic functions, such as suppressing a yawn.

The ANS has wide-ranging authority, influencing the eyes; salivary glands; head, neck, and facial muscles; larynx and pharynx; heart; lungs; stomach, intestines, liver, pancreas, and kidneys; rectum and bladder; and genitals. Thus, it shouldn't be surprising that the ANS plays a major role in stress arousal and recovery.

The ANS has two branches that send messages from the survival brain to the organs—to direct the body to focus on immediate survival needs during stress arousal or to focus on recovery after the threat has passed. The *sympathetic* branch turns the stress system on, while the *parasympathetic* branch turns the stress system off and prepares the body for digestion, recovery, growth, reproduction, repair, and rest.

In addition, the ANS also has a feedback loop from the organs *back to* the survival brain—so it can receive and learn from physical sensations. This feedback loop, called the *visceral afferent system*, plays a significant role in the survival brain's detection of internal stressors. It explains why gastrointestinal distress—such as acid reflux, nausea, or constipation—can create a vicious cycle. Through this system, the survival brain "hears" what's going on in the gut, perceives a threat, and directs the mind-body system to mobilize more stress arousal. Of course, with more stress arousal, we deprioritize the "long-term projects" of digestion and elimination, further exacerbating our gastrointestinal distress.

In addition to its role in stress arousal and recovery, the ANS also plays a major role in our patterns of interacting with others. Stephen Porges, the researcher who coined the term *neuroception*, developed the polyvagal theory to explain how the nervous system unconsciously mediates our capacity for trust and intimacy. The mammalian nervous system didn't just develop to ensure survival during life-threatening emergencies. It also developed to promote social interactions and social bonds in safe environments.

The three stages of our hierarchical defense developed evolutionarily, with each new defensive strategy building on evolutionarily older ones. Ideally, this means we'll rely first on the most recently developed defense, our *social engagement system*. In the physical assault example, you relied on social engagement when you tried to negotiate with the assailant, looked around to locate safety and support, and screamed for help.

If that strategy fails to provide safety, the survival brain and nervous system then "fall back" to the two evolutionarily older defenses. The second defense is *fight-or-flight*, which in this story involved trying to get away from the assailant.

Finally, if the survival brain perceives that fight-or-flight won't work, it and the nervous system fall back to the third line of defense: *freeze*. In this story, this happened when you stopped struggling with the assailant and felt dazed and disconnected from what was happening.

I want to note that while fight-or-flight and freeze are fully online before we're born, the social engagement system continues to develop into our teenage years. As a result, it is highly sensitive to our early social environment—with significant implications for the initial wiring of our window.

The survival brain's neuroception process automatically determines which of these three defenses is activated at any particular time.

Whenever the survival brain neurocepts safety, we're inside our window of tolerance. Inside the window, we can access all branches of the nervous system in "well-being mode." However, when the survival brain neurocepts threat or challenge, it moves the nervous system into "defensive mode." Whether our nervous system is in well-being mode or defensive mode largely depends on which hormones we've turned on.

The social-bonding hormone called *oxytocin* keeps the nervous system in well-being mode. Oxytocin is released only when we're inside our window. In well-being mode, we can mobilize energy for enjoyment and play, without turning on defensive behaviors. We can also handle "long-term projects" like digestion, elimination, rest, recovery, sex, growth, and tissue repair. And we can connect with and support others.

In contrast, when our survival brain neurocepts threat or challenge, it turns on stress arousal and directs the HPA axis to release stress hormones—including *vasopressin*, which takes the nervous system out of well-being mode and into defensive mode. I mentioned vasopressin earlier, as the hormone that inhibits digestion, elimination, and reproduction.

Once we are in defensive mode, the wider our window, the more likely we can defend ourselves successfully with the first or second lines of defense—social engagement and/or fight-or-flight—without having to "fall back" to freeze. The first line of defense, social engagement, is available only when we're inside our window. The more stress arousal we experience during the second line of defense, fight-or-flight, the more likely we'll end up outside our window. Finally, freeze occurs only after we've moved outside our window.

Table 4.1 explains the different branches of the ANS and their functions in well-being and defensive modes. You'll notice that the parasympathetic nervous system (PSNS) is divided into two branches, because it uses two forks of the vagus nerve. The *ventral* PSNS branch is located along the front side of the body, while the *dorsal* PSNS is located on the back side of the body.

Table 4.1: Branches of the Autonomic Nervous System and Their Functions during Neuroception of Safety and Threat

ANS Branch	Well-Being Mode: Neuroception of Safety	Defensive Mode: Neuroception of Threat
Ventral Parasympathetic Nervous System (Ventral PSNS)	• Social engagement/ social bonding/ attachment • Cardiovascular regulation (vagal brake) • Recovery functions for heart and lungs	• Initiating social engagement—first line of defense • Orienting to the external environment, to identify/locate threats and recruit allies • Engaging, negotiating, and cooperating with others to end the threat • Yelling for help
Sympathetic Nervous System (SNS)	• Energy mobilization for exercise, dancing, joyful movement, and play	• Energy mobilization for fight-or-flight (mobilized defenses)— second line of defense
Dorsal Parasympathetic Nervous System (Dorsal PSNS)	• Sleep • Digestion and elimination • Reproductive functions • Deep relaxation • Recovery functions for visceral organs	• Freeze (immobilized defenses)—third line of defense • Oxygen conservation

The ventral PSNS controls three functions. The first, the *vagal brake* on the cardiovascular system, allows us to make rapid, nuanced, adaptive adjustments to our heart rate and breathing rate. When it's working properly, the ventral PSNS allows us to regulate our cardiovascular system and manage minor stressors simply by engaging and removing the vagal brake—without having to turn on all the stress hormones.

To test how flexible and efficient our vagal brake is, researchers use a measurement called *heart-rate variability (HRV)*. When we inhale, we stimulate the SNS, which increases our heart rate, and when we exhale, we stimulate PSNS, which decreases our heart rate; for this reason, the interval between any two heartbeats is never exactly the same. HRV is used to test the flexibility of this system. High HRV—also referred to as high vagal tone—means our vagal brake is working efficiently. Low HRV (low vagal tone) means we've removed the vagal brake to increase our heart rate and turn on stress arousal. However, chronically low HRV means the vagal brake is always off and no longer functioning properly. In effect, in this state, we turn stress on without ever turning it off. Without a functioning vagal brake, the cardiovascular system is always experiencing allostatic load. Over time, people in this state are likely to manifest cardiovascular disease, high blood pressure, and increased risk for atherosclerosis and heart attacks.

The second function of the ventral PSNS is recovery after stress arousal. Once the threat or challenge has passed, the ventral PSNS triggers the release of *acetylcholine* to reengage the vagal brake. In turn, this slows our heart and breathing rates, relaxes our muscles, and facilitates digestion. The ventral PSNS dampens stress hormones and modulates our immune functioning, so that we can fully recover, calm down, digest and eliminate our food, heal, grow new tissue, and rest.

The third function of the ventral PSNS is controlling bodily functions for the *social engagement system*. These include our head, neck, and eye muscles, which allow us to look around, orient to our external environment, and make eye contact. Our facial muscles, which allow us to smile, nod, frown, and use other expressions to connect socially with other people. Our larynx and pharynx, which allow us to modulate our voices. The muscles that help us chew, suck, and swallow. The muscles in the middle ear, which play a role in our ability to listen. The ventral PSNS also plays a role in the release of *oxytocin*, the social-bonding hormone. As a result, when the ventral PSNS is turned on, we can feel calm, centered, and connected with other people. We can signal our internal state to others through our facial expressions and body postures, while also attuning to subtle emotional shifts in people around us.

These three ventral PSNS functions are available to us when we're inside our window. Inside the window, we can communicate effectively, nurture relationships, and turn to others for support with challenges. Someone with a wide window may even be able to access the social engagement system in defensive mode during life-threatening situations, such as when she tries first to negotiate with or "talk down" an assailant or calls out for help. Thinking evolutionarily, the social engagement system developed so our cave-dwelling ancestors could cooperate, nurture their offspring, hunt together, and live safely in nomadic tribes. As you might imagine, this first line of defense involves a lot of social communication and thus requires significant thinking brain and survival brain allied interaction—which is available only when we're inside our window.

Because the ventral PSNS is deeply involved with both social engagement and recovery functions, one important implication is that *if we experience difficulty regulating our stress arousal, we're also likely to have trouble creating and maintaining workable, supportive, and satisfying relationships*, in both personal and professional settings.

With this background, you can see how the ventral PSNS plays the first line of defense when our safety and social connections become threatened: We increase our heart and breathing rates. We turn our head and neck, and move our eyes, to orient to our environment. We reach out for support or help from others. We signal our distress through changes in our facial expressions and tone of voice. Often these shifts are enough to manage the challenge successfully, in which case the vagal brake gets reengaged and keeps us inside our window.

Should we find ourselves in immediate danger or discover that no one comes to our assistance, however, then the threat increases. At this point, the survival brain neurocepts danger, kick-starting our second line of defense.

When we "fall back" to our second line of defense, the SNS in defensive mode takes over. It works with the HPA axis to mobilize energy and stress hormones. SNS arousal is called the fight-or-flight response for good reason: It increases metabolic activity and cardiac output to facilitate active, mobilized defenses—either to fight off an attacker (fight) or to run away to safety (flight). Remember that we can also turn on the SNS in well-being mode, such as when we exercise, play, or dance.

SNS arousal mobilizes a lot of energy and stress hormones. However, many defensive SNS behaviors will not expend all this mobilized energy. On the fight side, we may get defensive, demand attention, justify our behavior, or yell at someone. On the flight side, we may withdraw into ourselves with incessant worry or try to please other people. Although these behaviors are associated with stress activation, by themselves they won't expend all the arousal. Over time, this can contribute to turning stress on without ever turning it off.

Although fight and flight are both controlled by the SNS, they are different defensive strategies. Each has its own associated emotions, physical sensations, and motor impulses. The *fight response*—associated with anger or a sense of being energized, powerful, or in control—involves movement *toward* the stressor. It's also associated with increased saliva production and a narrowed attentional focus, such as tunnel vision—think of a cheetah pursuing its mark with a single-minded focus. We tend to access fight when we participate in athletics or competitions related to social dominance, when we perceive ourselves to be stronger than the attacker, or when fighting back is expected, such as during a bar brawl or combat, or as a member of a street gang.

In contrast, the *flight response*—associated with fear, anxiety, dread, or a sense of being powerless or helpless—involves movement *away from* the stressor. This could mean running away from the danger or toward safety, such as enlisting help from someone stronger or wiser. In contrast to the fight's tunnel vision, with the flight response, attention is usually wide, unfocused, and distracted, as we scan to locate a viable escape—think of a rabbit frantically searching for a hole to get away from its predator.

In many situations both fight and flight get activated at the same time, which may facilitate the survival brain and nervous system perceiving that neither strategy can succeed. This dynamic may occur in any situation where we are asked to be both predator and prey at the same time. A trial attorney may feel anxious before making her case in a courtroom. A firefighter must enter a burning house to rescue someone, putting himself in danger. The quintessential example of this dynamic may be combat.

If these active SNS defenses fail—such as when we're trapped and no one's available to help us—then we "fall back" to our final line of defense. At this stage, the SNS continues to remain turned on, but we also turn on the dorsal PSNS in defensive mode, too.

The dorsal PSNS in defensive mode often kicks in during life-threatening situations—such as when we've been physically restrained or pinned down by an assailant, immobilized by gunfire, or in a car accident with no option but to brace for impact. Children may also activate the dorsal PSNS in defensive mode when their survival brain perceives no escape from an abusive care provider. When fight-or-flight is no longer possible or would make things worse, not moving becomes the best strategy. The *dorsal PSNS in defensive mode is associated with trauma*, because the survival brain perceives us to be powerless, helpless, and lacking control.

The first dorsal PSNS function in defensive mode is the *freeze* response, which is what happened in the assault example when you submitted to the assailant and felt dazed and disconnected from what was happening. Unlike the active defenses of SNS, dorsal PSNS defenses are immobilized—freeze, submission, or "playing dead."

The second is a massive conservation of oxygen and energy, by dramatically slowing down the cardiovascular system (known as *bradycardia*) and other internal organs. The dorsal PSNS also radically reduces metabolic activity and stops our digestive system—so we may lose control of our bladder and bowels.

In well-being mode, dorsal PSNS plays a major role in digestion, elimination, sleep, and reproductive functions, because it controls most of the neural pathways into our visceral organs. However, well-being mode is available only when the survival brain neurocepts safety and the body is releasing oxytocin, the social-bonding hormone. One important implication of this: We cannot have social engagement and freeze turned on at the same time.

Thus, *if the dorsal PSNS in defensive mode has been turned on, then by definition the ventral PSNS has been turned off—and we lose access to ventral PSNS functions*, including the ability to relate effectively to other people and to our external environment in the here and now. Once the dorsal PSNS in defensive mode takes over, our body is preparing for shutdown—at its most extreme, to protect us from a painful death. Therefore, conscious awareness and other thinking brain functions get degraded and may go offline completely. We may no longer register fear or physical pain. We may stop taking any self-protective actions. Instead, we "give up," collapse, faint, or feel like we'll faint.

In freeze, we lose the capacity to make eye contact and move our head, neck, and eyes to orient to our surroundings. We might experience freeze as extreme tunnel vision, in which we lose parts of our visual field. We may see sequential freeze-frame images, or lose the ability to see in color. We can no longer modulate our voice, speaking instead in a monotone. We lose control of our facial muscles, creating a facial expression that looks deadened, pale, slack, or confused. Our hearing becomes less responsive to human voices and more sensitive to threatening sounds. We may even experience complete hearing loss.

We may also feel like everything's happening in slow motion or we've entered some altered state of reality. Alternatively, we may lose track of time; afterward, we may not be able to remember whole segments of time having passed. We may feel confused and dissociated, as if we're observing events through fog or cotton. We may also experience ourselves subjectively as being outside our body, as if watching ourselves from the ceiling or from across the room. Each of these descriptions of freeze comes from my own experience and the experiences of people I've trained during the last decade.

Thus, *although freeze may not present obvious outward visual cues of arousal, internally it is a highly activated state*, with both the SNS and the dorsal PSNS turned on in defensive mode. In fact, freeze has the *highest* stress arousal levels, by definition outside our window, which makes sense because it's usually associated with traumatic stress.

WHICH DEFENSE WILL OUR SURVIVAL BRAIN CHOOSE?

Let me draw out some important implications from all this science. When allostasis is working properly, our survival brain and nervous system collaborate to evaluate our level of risk in every moment—working with our hormone system to adjust our stress levels and energy mobilization and with the immune system to proactively protect us.

Thus, if the survival brain neurocepts safety—or a level of risk within our window of tolerance—our nervous system will mostly stay in well-being mode. We'll be able to connect and cooperate with others and stay oriented to our surroundings. Or, if the survival brain neurocepts threat, the nervous system will move into its defensive mode and employ the hierarchy of defenses. The more danger our survival brain neurocepts, the more stress arousal it will mobilize—and the further we will "fall back" through the hierarchy, from social engagement to fight-or-flight and eventually to freeze.

Regardless of how much stress arousal we mobilized to cope with the situation—even if we had to fall "all the way back" to freeze—when allostasis is functioning properly, after the threat has passed, the survival brain will once again turn on the ventral PSNS so we can recover completely. This process goes on instinctively, usually without any thinking brain input—although the thinking brain can influence this for good or ill.

When allostasis isn't working properly, however, we may not have access to this full range of adaptive responses that are our inborn repertoire. In general, the narrower our window, the smaller the range of responses we'll be able to employ.

For instance, if someone's vagal brake isn't functioning properly—meaning the ventral PSNS isn't fully available—they won't be starting from a regulated baseline. Thus, when their survival brain neurocepts *any* threat or challenge, even a minor one, they'll immediately mobilize stress activation—even if stress arousal isn't the most effective choice for the situation. They may show signs of a fight response—overreact to the slightest provocation, storm off, or lash out with a temper. Or they may default into a flight response—withdraw or get flooded by a wave of anxious thoughts. Or they may default right into a freeze response—dissociate, collapse, become paralyzed with extreme procrastination, or feel spacey, numb, and overwhelmed.

In these states, they'll not only have difficulty downregulating their stress activation, they'll also have less capacity to recruit thinking brain functions, such as creative problem solving, perspective taking, situational awareness, and impulse control. They'll be less able to detect positive social cues and connect effectively with other people. When they are operating from this activated state, their relationships may no longer be a source of safety, trust, connection, and support. Instead, they may feel misunderstood or isolated.

Neuroplasticity plays a role in which strategies our survival brain will choose. Since *any repeated experience* changes the brain and nervous system, it makes sense that our mind-body system develops habitual defensive and relational strategies, the ones it relies on routinely. As the same strategy gets employed over and over, it becomes deeply encoded in the survival brain's implicit memory. Each time we unconsciously choose a strategy, it becomes both easier to default there again *and* harder to access any other strategies. Over time, we condition ourselves to default to only a few strategies—in the process, losing our ability to access and employ the full range of options innate to our mind-body system.

Some of this default wiring comes from our early social environment. Some of it comes from repeated experiences with threat and safety earlier in our lives. And some of it comes from socialization in our schools, workplaces, and communities.

Thus, it's not an accident that many men default to strategies on the fight spectrum. They're often taught that "boys don't cry"—they should be tough, stand their ground, and "take it like a man." High-stress professions tend to enculturate and reinforce a fight default, for both genders. Indeed, most training in high-stress environments is *explicitly designed* to socialize and wire this particular default. For someone with this conditioning, any little provocation may trigger a fight response.

To give another example, let's say someone had repeated experiences during childhood of relying on freeze. Now, later in life, they may still default to confusion or numb acquiescence during threatening situations—even when standing their ground or getting the hell out of there would be more effective. This default often occurs among people who experienced physical, emotional, or sexual abuse as children. From the perspective of the survival brain and nervous system, this default makes sense: Given how young and powerless they were when they were abused, freeze was very adaptive—it allowed them to survive. Clearly, however, it's probably not adaptive in every situation today.

In effect, the survival brain and nervous system may become stuck in a few programming loops, neurocepting danger and then inflexibly cueing that programming's particular defensive and relational strategies—whether they're appropriate for the current situation or not. When we've wired default strategies that rely on fight, flight, or freeze, these neural pathways become well-traveled, highly sensitized superhighways that get easily triggered. This wiring is influenced by neuroplasticity and epigenetics—and also by how the survival brain learns and remembers, something I'll explore in the next chapter.

In fact, until the survival brain is allowed to complete its bottom-up processing and fully recover, its default strategies will unconsciously get triggered, again and again. We need conscious intentionality to support the survival brain's recovery process and develop new neural pathways for accessing the entire hierarchy of human defenses. This is partly why thinking brain–dominant techniques and therapies are always incomplete.

CHAPTER FIVE

The Brain during Stress and Trauma

A Marine corporal, "Julio," was preparing for his first combat deployment. At the time, I was training his platoon in MMFT. I was immediately struck by his conscientiousness and curiosity.

Julio's unit had leaders who had shown resistance to MMFT—an attitude that trickled down to some in Julio's platoon. As a result, I was teaching forty Marines who assumed silly postures during movement exercises, or started burping and farting contests with their buddies. Others refused to make eye contact or speak with me. Others interacted with me in an excessively polite manner, then mocked MMFT when they thought they were out of my range of hearing. For the first few weeks, in other words, I routinely encountered the peacocking, sulking, stonewalling, and dominance behaviors that often manifest among emotionally immature groups when they are asked to do something they don't want to do.

In contrast, Julio, a team leader for his squad, was diligent in keeping his Marines respectful and engaged in class. He listened, asked excellent questions, and quickly connected the dots between different aspects of the material. With a long-standing martial arts background, he also took to MMFT's exercises with ease. After each class, he'd approach me privately to talk through the implications of what he was learning.

During the fourth session, MMFT participants learn about the freeze response and how it fits within the hierarchy of human defenses. I show video clips of animals and humans experiencing freeze, so that participants can observe how the freeze response appears. During this discussion, participants also frequently talk about their own previous experiences with freeze—the traumatic event(s) that led up to freeze and what they remember noticing, at the time, in their own mind-body system.

Understandably, this discussion *itself* can trigger stress arousal among participants—and in many groups, some people default into a freeze response during the discussion. For this reason, it's a challenging module to teach. The instructor must facilitate the entire group to downregulate their stress arousal—while also offering additional support to those who may have defaulted into freeze, without drawing attention that might lead them to feel stigmatized and ashamed.

During this platoon's class, the freeze discussion was especially intense. Many platoon members shared vivid personal stories. Marines who'd never spoken before eagerly stepped into the conversation. It was the first time that the entire group actively engaged in the class without any resistance.

However, many Marines experienced high stress arousal levels during the discussion, and three people—including Julio—defaulted into freeze. I watched their eyes become glassy and vacant, staring off without seeing into the distance. Their shoulders hunched forward. Their bodies became still and unmoving. The Marines in freeze did not speak.

With that much stress activation present, I needed to improvise. Through a variety of techniques, I was able to support most Marines in downregulating their arousal levels back inside their windows. During extra class breaks, I surreptitiously worked with Julio and the two other Marines who'd defaulted into freeze; eventually they, too, stabilized back inside their windows.

Nevertheless, I could tell that Julio's freeze had been severe—he'd both been the first one to experience freeze and the longest to stabilize afterward. Intentionally, I'd already scheduled every Marine to meet with me individually during the next few days, to discuss their experience to date with the MMFT exercises. Now, I could easily shift the interview schedule around to meet with Julio that day, without singling him out unnecessarily.

Once alone, Julio told me that during the videos about freeze, he'd unexpectedly found himself flashing back to an event from childhood—something he hadn't thought about in a very long time. For more than forty-five minutes together, Julio and I patiently worked through what happened, allowing him to reexperience the same flashback from inside his window and then discharge the stress activation. Through my external cuing of his attention, together we helped his survival brain complete some necessary bottom-up processing and recovery. I'll explore this recovery process further in Part III. In this chapter, however, I want to explore what was happening in Julio's thinking brain and survival brain during the flashback.

Julio is nine years old. He's playing ball in the street with friends and several older relatives who belong to a gang. Suddenly, a truck turns the corner and speeds down the street. As people in the truck start shooting, Julio's group realizes the shooters belong to a rival gang. He feels an older cousin grab his arm and run in search of cover. He sees his cousin fall forward, still holding on to Julio's arm. Julio feels himself falling, too, getting dragged down by his cousin.

Suddenly, things become completely silent. Everything appears to be in black and white. Julio looks over and sees his arm still caught in his cousin's grip. He wishes he could take his arm back. He's falling forward in slow motion, and it feels like the fall will go on forever.

At some point he feels himself being pulled to his feet. There's a sharp pain in his shoulder. The next thing Julio notices is looking down at his cousin's dead body. His cousin is covered in blood and guts.

At this point, sound rushes back in—and it's deafening. Julio hears sirens wailing, and people yelling and crying. He realizes someone is yelling at him urgently, but he can't figure out what they're saying. He feels woozy, like he's going to faint.

THE SURVIVAL BRAIN DURING STRESS

Just as stress arousal involves our nervous system and body, it also affects our brain, especially learning and memory. Taking an evolutionary view, this makes sense: Being able to remember and learn from stressful events is important for survival. The thinking brain and survival brain each have their own forms of learning and memory, which allow them to perform their respective functions.

Yet stress and trauma have different effects on the thinking brain and survival brain, with rippling consequences on their respective learning and memory processes. This chapter explores these differences.

Stress and trauma originate in the survival brain with neuroception, that unconscious process of appraising internal and external stimuli as either threats/danger or opportunities/safety. The part of the survival brain responsible for neuroception is called the *amygdala*, which you may remember from Chapter 3 as the brain region that thickens with repeated worrying. Neuroception cues a mostly unconscious repertoire of conditioned responses, which follow a tendency toward approaching opportunities and avoiding threats. Because neuroception leads to conditioned responses, the survival brain engages in "bottom-up processing"—our mostly unconscious emotional and physiological responses to experiences. Economist Daniel Kahneman describes this as System 1 thinking—"thinking fast"—because it operates automatically, with no effort or sense of voluntary control.

Because the survival brain isn't verbal, it doesn't communicate with us about this process through thoughts or narratives. Rather, it generates emotions and physical sensations to cue these conditioned defensive and relational strategies. Since the survival brain is outside conscious awareness, we can't see or know what's happening there directly. We can only see its *effects* in the symptoms of stress activation in our mind-body system.

To support neuroception, the survival brain needs a learning and memory system that's fast. Thus, its learning system is reflexive, unconscious, and involuntary—bypassing the thinking brain completely.

The survival brain's learning system is called *implicit learning*, or System 1 learning. Implicit learning occurs predominantly in the amygdala, generalized from all previous experiences of neuroception. Most of this conditioned learning is processed and stored in the survival brain.

Implicit learning is supported by *implicit* (or *nondeclarative*) *memory*. Every experience we have constantly builds implicit memory. When implicit memory relates to gaining motor skills or body-based responses, it's called *procedural memory*. For instance, procedural memory includes learning how to play an instrument, ride a bicycle, walk or run, and shoot a weapon.

While the thinking brain gets degraded by prolonged or extreme stress levels, survival brain learning and memory functions occur unconsciously at *any* level of stress arousal. Moreover, *the greater the stress arousal, the more the survival brain learns and remembers.*

At moderate stress levels, the amygdala works with the hippocampus to create explicit memory, with the amygdala providing the emotional component of the memory. However, at high stress levels, the amygdala still provides the emotional component of memory, with more intensity, but hippocampus function is disrupted. Thus, at high arousal levels, we may never consolidate *conscious* memories, even as the survival brain remembers a great deal. Because the hippocampus can even be, in effect, offline at high stress levels, conscious memories of extremely stressful or traumatic experiences are often incomplete, contradictory, or fragmented. Julio's story provides an example of this dynamic—some details in his narrative are especially vivid while other parts are missing. Yet the amygdala learns and generalizes the most from extremely stressful and traumatic situations.

Most things that make us anxious were conditioned through implicit learning—either because our amygdala unconsciously associates it with something previously neurocepted as threatening, or because the amygdala has *generalized based on its similarity* to something previously neurocepted as threatening. In fact, when implicit and procedural memory get encoded during a life-threatening situation, these unconscious memories become more permanent and resistant to decay—as well as more susceptible for generalization to other situations.

Implicit memory is not just facts or information. It involves nervous system responses, physical sensations, muscle and myofascial tension, body postures, emotions, and patterns of motor movement used in the act of defense. These sensory and motor responses become conditioned as part of the survival brain's repertoire for when we encounter similar threats in the future. These conditioned responses may get evoked without us even being conscious of them, as they were for the Marines who experienced freeze in class.

In Julio's case, as he watched the video of an animal getting pinned down by a predator, his survival brain likely generalized to recall Julio's experience of falling down with his cousin. While Julio wasn't actually "pinned down" during that fall, the *felt sense* of being trapped, helpless, and unable to get away from his cousin's grip was quite similar. This similar felt sense in Julio's body—traveling to his survival brain via the ANS visceral afferent system—is likely what spurred his survival brain to start a freeze response.

To give another example: I once taught "Sam," a Marine who'd previously deployed to Afghanistan. Years later before another deployment, Sam woke up early each morning with a pounding heart, shallow breathing, his body tensed into a fetal position, and a strong impulse to jump out of bed. He couldn't understand why he was having this reaction.

I explained to Sam that even if his thinking brain couldn't understand it, we had to trust that his survival brain had an important reason for behaving this way. The next time it happened, I asked him to set his thinking brain's frustration aside and get out of bed to complete an exercise for down-regulating stress arousal. (I'll teach you this exercise in Part III.)

After a week using this exercise each morning, Sam's thinking brain figured it out: During his previous deployment, he had been awakened most mornings before dawn to his forward operating base getting shelled. Understandably, his survival brain developed strong implicit and procedural memories for taking cover from incoming mortars upon awakening. Today, anticipating his next deployment, Sam's survival brain generalized from having been awakened repeatedly under fire in Afghanistan to create his predawn anxiety attacks at home.

THE THINKING BRAIN DURING STRESS

Just as the survival brain uses neuroception and implicit learning to help keep us safe, the thinking brain also performs functions for our survival. The thinking brain's strategy for protecting us is to analyze, plan, deliberate, and decide. It engages in "top-down processing"—our mostly voluntary and conscious cognitive responses to our experiences. Daniel Kahneman describes this as System 2 thinking—"thinking slow." System 2 is indeed slow and effortful, characterized by concentration, conscious deliberation, and a sense of agency.

Parallel to the survival brain's neuroception, the thinking brain is responsible for *executive functioning*, which occurs predominantly in the prefrontal cortex. Executive functioning allows us to focus, pay attention, and recall task-relevant information, while holding distractions at bay. Executive functioning also supports conscious decision making. We use executive functioning for "top-down" regulation of stress arousal, impulsive behavior, cravings, and emotions. Thus, as you can imagine, it also plays a big role for harnessing "willpower."

Executive functioning may be impaired in several ways. We can deplete executive functioning through multitasking, as Chapter 3 explained. We can also impair it through alcohol and drugs, which is why we have less inhibition when we're under the influence. It's also impaired by stress.

Executive functioning is like a credit bank: We can deplete it through heavy use in two ways. On the one hand, we might deplete it through what are called "cold" cognitive tasks—mental tasks that require detailed attention and focus—such as reading dense text, writing a report, or completing detailed calculations. On the other, we might deplete it through "hot" regulatory tasks—conscious, top-down efforts to curb cravings, reframe or compartmentalize negative emotions, and manage or suppress stress arousal.

Whenever we deplete executive functioning via "cold" or "hot" tasks, there is less credit in the bank to pay the costs of either. That's why, for instance, when we've been reviewing detailed financial documents all day, a "cold" task, and then someone cuts us off in traffic on the way home, we're more likely to indulge our frustration and give them the finger, or when we get home we're more likely to cheat on our diet. Nothing is left for "hot" regulation.

Conversely, when we're experiencing chronic tension in an important personal relationship or persistent discrimination at work—both situations requiring consistent "hot" regulation—we may have to read the same paragraph seven times until it finally begins to sink in because nothing is left for "cold" comprehension.

Regardless of how executive functioning gets impaired or depleted, in this state, stress and emotions are more likely to drive our decisions. The way our thinking brain makes sense of situations will be more biased by stress and emotions. We're also more likely to engage in habitual, impulsive, reactive, violent, or unethical behavior.

To support executive functioning, the thinking brain needs a learning and memory system that places a significant weight on *context*—situating information within space and time. The thinking brain's learning system is *conscious learning*, or System 2 learning. Conscious learning occurs predominantly in the hippocampus.

Conscious learning is supported by *explicit* (or *declarative*) *memory*—such as the memory of events, facts, faces, words, or information. In contrast to implicit memory, we access explicit memory intentionally, such as when we recall our life story or try to incorporate new information in our "knowledge bank." And because nerve fibers in the hippocampus don't develop the fatty sheath that allows them to conduct electricity until we're about two years old—a process called *myelination*—it's rare to have explicit memories from our earliest years.

Explicit memory is influenced by our intelligence and other individual differences, but it's also profoundly affected by our stress arousal levels, as Julio's story shows.

Mild to moderate stress arousal enhances explicit memory and conscious learning in the short term. With mild to moderate stress, higher cortisol and blood sugar levels means the hippocampus has access to ready energy, which focuses our attention and helps with explicit memory formation, storage, and retrieval. In fact, the brain is quite glucose-greedy—it's 3 percent of our body weight but uses 20 percent of circulating glucose—with the hippocampus among the most glucose-greedy regions.

That's why drinking caffeine sharpens our attention—it causes a spike in our cortisol levels. That's also why low blood sugar levels, such as when we're hungry, are associated with declines in executive functioning and explicit memory and increases in irritability. For example, one study examined Israeli parole board judges and their food breaks. It found that the more time that passed after the judges' previous meal, the more likely they ruled against granting parole. As they got "hangry," they had less ability to modulate their irritability—and less ability to clearly and compassionately evaluate details of the cases before their bench.

In contrast to mild to moderate stress arousal, *executive functioning and explicit memory functions may be impaired or damaged with prolonged or high stress levels.* Why? Excessive or chronically elevated cortisol levels cause neurons in the hippocampus to lose their bushy dendrites—the branches that allow them to connect to neighboring neurons—which lead neural networks to atrophy. As prolonged stress arousal continues, existing neurons die, new neurons stop growing, and the hippocampus shrinks in volume. In effect, the brain puts beneficial neuroplasticity and neurogenesis in the hippocampus on hold during extreme or prolonged stress.

We usually experience these brain changes as memory problems. We may also have trouble paying attention, learning new information, and planning and executing tasks. We're likely to have problems screening out distractions. Each of these symptoms suggests depletion or disruption of executive functioning and explicit memory.

Remember that study in Chapter 3 about the soldiers who deployed to Iraq? Compared with soldiers who didn't deploy, these soldiers' reaction times got faster—a sign that their survival brains worked overtime during their high-stress deployment. Two months after coming home, however, the cost of this survival brain shift showed up in thinking brain declines in attention skills, executive functioning, and explicit memory.

Research with high-stress occupations—including internal medicine residency programs, law enforcement work, firefighter drills with fires, and military deployments and field-training exercises—shows how individuals tend to experience greater anxiety and distress in these environments. They also experience more symptoms of cognitive degradation, including dissociation, confusion, problem-solving deficits, attention lapses, difficulties with visual pattern recognition, and declines in working memory.

Since stress arousal may also occur with symbolic threats—such as the anticipation of negative future events or the "kindling" of past traumatic memories—*thinking brain declines can also happen even when we aren't facing direct, physical harm.*

Anyone can experience declines in executive functioning and explicit memory during prolonged or extreme stress, even if they don't work in life-threatening professions. These declines are just as likely after chronic sleep deprivation or during challenging life transitions, such as changing jobs, moving, getting married, or having a baby.

Aging may also exacerbate these effects. Prospective research with healthy elderly people showed that the seniors whose cortisol levels increased most over the years of the research study had the greatest memory declines and greatest loss of hippocampal volume.

Empirical research shows that excess stress and/or elevated cortisol levels are linked with smaller hippocampi and memory problems in many different circumstances: Intercontinental flight attendants with long careers switching time zones and experiencing chronic jet lag without enough recovery between flights. People taking prescription steroids over long periods, including hydrocortisone cream, cortisone injections, or oral/inhaled steroids. People with prolonged major depression. People with PTSD after repeated trauma, such as survivors of repeated childhood abuse or prolonged combat exposure. In each case, the more intense or prolonged the stress, the greater the memory problem and the smaller the hippocampus.

HOW THE WINDOW AFFECTS THE RELATIONSHIP BETWEEN THE THINKING BRAIN AND THE SURVIVAL BRAIN

The relationship between stress arousal and performance—including thinking brain functions—is an inverted U–shaped curve (Figure 5.1). At low levels of stress arousal, we may not experience enough stress activation to be alert and motivated to complete tasks effectively. In fact, *eustress*—which uses the Greek prefix *eu* for "good"—provides attentional focus and energy. *Thinking brain functions are enhanced by eustress*, as the mild to moderate stress arousal provides increased levels of circulating glucose and cortisol to focus our attention and recall task-relevant information.

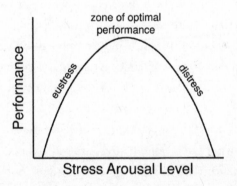

Figure 5.1: The Yerkes–Dodson Curve

In 1908, psychologists Robert M. Yerkes and John D. Dodson were the first to posit the inverted U–shaped relationship between perceived stress levels and performance, known today as the Yerkes–Dodson curve. Performance on a task improves as someone approaches moderate stress arousal, but steadily decreases past this point until it drops off completely with freeze. Eustress ("good" stress) describes low stress arousal levels, while distress describes high stress arousal levels.

Conversely, at high arousal levels, our performance degrades steadily. Not surprisingly, at the extreme end of this spectrum is the freeze response. On the distress side of the curve, thinking brain functions become impaired—with difficulties in paying attention, screening out distractions, remembering task-relevant information, downregulating our stress and negative emotions, and making effective decisions.

Thus, optimal performance, conscious learning, and effective decision-making are most likely to occur at moderate stress levels, where there's enough stress activation to keep us alert and focused, but not enough to enter our distress zone. With this in mind, our neurobiological *window of tolerance* to stress arousal is *the window within which we are capable of adjusting our stress levels upward or downward to remain, over time, within the optimal performance zone of moderate arousal.*

Inside the window, the thinking brain and survival brain work together as allies, with their functions fully online. Inside our window, we're more likely to engage in *accurate* neuroception—that is, if we neurocept danger, the situation is truly, objectively dangerous. We're more likely to perceive relevant internal and external cues—and then absorb, objectively assess, and integrate all these new data with information we already know that affects the current situation. We're more likely to consider and evaluate the costs and benefits of all of our options. Then, we're more likely to make the choice best aligned with our values and goals. Afterward, we're more likely to consciously assess the consequences of this decision and thereby adapt and learn for the future. *Thus, the wider our window, the more likely we can maintain accurate neuroception and effective integration of thinking brain and survival brain processes*—even during levels of high stress arousal and emotional intensity.

Peak performance requires the capacity to adjust our arousal levels to match the task at hand. Different tasks need different arousal levels. For instance, when trying something new, we're more likely to succeed at lower arousal levels. That's because the survival brain appraises novelty as more threatening, as Chapter 4 explained. Conversely, tedious or familiar tasks may require greater stress arousal to create focus and motivation. This is actually one of the reasons why people procrastinate with unpleasant tasks: As the deadline looms, their stress arousal increases, eventually creating enough stress to motivate them to handle it.

Conversely, when we experience stress arousal or emotional intensity beyond our window, we're more likely to engage in *faulty* neuroception. Outside our window, we're more likely to neurocept danger even when the situation is not actually harmful—and then stimulate protective behaviors in response. Alternatively, we may neurocept safety/opportunity when a situation is actually quite dangerous, which can lead us into high-risk situations. As a result, we may mobilize too much or too little stress arousal for the situation at hand.

Outside our window, the survival brain is more likely to be driving information search, assessment, and decision making—with thinking brain processes operating in a degraded or impaired manner. High stress levels tend to narrow our perception, so that we unconsciously focus on and prioritize the immediate over what's really important for our long-term success. We tend to make meaning of things in a biased manner. We focus on information we perceive to be psychologically central, becoming absorbed by our stressors and with stress itself. In the process, we lose perspective. When we're stressed, we also tend to gather less information. We're more likely to draw consequential and sweeping conclusions from small amounts of information, often taken out of context. In addition, we become biased toward the negative—negative information is more likely to capture our attention, receive more processing, and be more salient for recall. This makes sense evolutionarily. Our Paleolithic ancestors were more likely to survive if they could remember and learn rapidly from negative, stressful events.

A similar narrowing and rigidity occur with decision making. We're more likely to accept the first workable option, rather than consider the full range of alternatives. We rely more heavily on stereotypes, oversimplifying assumptions, and past experience—rather than seeing clearly the unique contours of the current situation. Our ability to analyze complex situations, strategic interactions, and the consequences of our decisions declines. We're also more likely to make mistakes.

Outside our window, the thinking brain and survival brain interact in an adversarial relationship, which shows up in three different ways. First, degraded thinking brain functions may manifest as inaccurate situational awareness, memory problems, distractedness, waves of anxious planning thoughts, or defensive reasoning.

Second, we may experience *survival brain hijacking*, where emotions and stress arousal bias our perceptions, absorb most of our attention, and drive our decision making and behavior. Here, the survival brain and nervous system are likely to end up in default programming, even if it's not the most appropriate choice for the situation at hand. While this programming plays out unconsciously, our thinking brain may try to blame others or ourselves for it.

Third, we may experience *thinking brain override*, where we "live in our heads"—disconnected from survival brain signals, including our emotions and physical sensations. Thinking brain override is almost certainly present when we engage in suppression, denial, compartmentalization, and suck it up and drive on.

Of course, we may experience all three facets of this adversarial relationship in different situations. Until a complete and effective recovery occurs in the survival brain, these patterns will continue to play out and narrow the window further. Over time, these dynamics build allostatic load and lead us to develop symptoms of dysregulation.

THE SURVIVAL BRAIN DURING TRAUMA

The pathways by which trauma and chronic stress narrow the window are related, but distinct. Trauma involves additional dynamics that further complicate the relationship between thinking brain and survival brain.

The most important additional dynamic is that during trauma, the survival brain's implicit memory system becomes corrupted. It perceives us to be powerless, helpless, or lacking control. Afterward, this perception of helplessness becomes deeply seared into implicit learning. It doesn't matter that we actually survived. The survival brain learned that we didn't successfully defend ourselves.

For instance, we might have been trapped or physically restrained, unable to fight back or escape, or overpowered. We might have experienced a car accident, natural disaster, terrorist attack, or mass shooting, where events were completely outside our control. More subtly, we might have experienced the double bind of discrimination or harassment at work—where speaking up to defend ourselves or someone else might have cost us our job, so we remained silent or acquiesced. When the survival brain encodes implicit memories from such traumatic events, it unconsciously links the high levels of stress arousal with immobility, helplessness, and lack of control.

In Julio's flashback, he was unable to break free from his cousin's grip and stop himself from falling. He was also unable to prevent his cousin's death. His survival brain likely generalized to learn that he couldn't successfully defend himself whenever he felt "pinned down."

The implicit memory system stores learning from traumatic events differently from learning after "successful" (completed) defenses. Without a successful defense and recovery, the stress activation mobilized during the traumatic event was never discharged. In turn, the traumatic memories get encoded in a way that ensures their persistence.

As a result, *the survival brain believes the traumatic event was never completed.* And—this is key—until the survival brain, nervous system, and body have an opportunity to finish the incomplete defensive strategy and discharge its associated stress activation, the survival brain continues to perceive the event as ongoing. It also perceives itself as powerless in successfully defending against the ongoing threat—which increases the likelihood of defaulting into freeze again, just as Julio did during the animal video.

This corruption in the survival brain's implicit memory after traumatic events has several consequences. First, because the survival brain believes the traumatic event is still ongoing, it's unable to fully process and learn from that event. At the same time, it continues to rely heavily on the incomplete or unsuccessful defensive strategy that it used during the original traumatic event. It's as if the survival brain unconsciously believes, "If I try that same defensive strategy again in this current situation, maybe it will work this time, and then I can finally defend myself successfully."

What this means is that *a traumatized survival brain loses its ability to distinguish between the past and the present, so it cannot learn and adapt.* Without complete recovery, the survival brain remains frozen in time, back when the original traumatic event occurred. Understandably, this dynamic feeds into trauma reenactment, unconsciously re-creating situations or relationships that echo aspects of earlier traumatic events. It's also why trauma survivors, especially those traumatized during childhood, are vulnerable to reacting to dangerous situations today with survival brain–initiated responses that are irrelevant—and often counterproductive or harmful—for what's actually happening right now. The survival brain's only available learned response for danger today is to repeat the default conditioning from that original traumatic event.

The incomplete or unsuccessful defensive strategy encoded during a traumatic event becomes the new generalized default programming for situations the survival brain perceives to be similar to the original traumatic event. In Julio's case, while he watched an animal "pinned down" in the video, it evoked the same sensations in his body that he'd experienced when he was "pinned down" by his cousin's grip during the fall. That similarity was enough for his survival brain to trigger the default programming of freeze once again in the classroom.

The traumatized survival brain remains primed to use its default strategies long after environmental conditions change. These default strategies preempt relevant information about the current situation and take precedence, even when other strategies might be more appropriate. The survival brain will continue to rely on such default programming until it experiences a complete recovery. Only then will the survival brain finally recognize that the traumatic event truly is in the past—and that the mind-body system is safe today.

The second consequence of the corrupted implicit memory system is that the survival brain becomes hypersensitized to any cues associated with the original trauma.

All information that the survival brain captures during a traumatic event—*much of which is outside thinking brain awareness*—gets stored together in the implicit memory system as part of a *memory capsule*. Memory capsules include sensory input (sights, sounds, smells, tastes, touch), body postures, physical movements, emotions, and physical sensations, especially related to SNS and dorsal PSNS stress arousal. Memory capsules may also include perceptual alterations, such as the perception of watching the body from the ceiling or feeling as if time has slowed down.

The strength of each memory capsule depends on the intensity of the traumatic event and whether that event occurred repeatedly over time. And the stress activation and emotional intensity we experienced during the traumatic event, which we have not yet discharged through a complete recovery, serves as the "glue" holding the memory capsule components together. Memory capsules remain active and vulnerable to getting triggered until a complete recovery.

When the survival brain perceives a cue today similar to a component of one of our unresolved memory capsules, that cue may trigger that memory capsule. In turn, the mind-body system reacts as if that trauma is happening again right now—with the same level of terror and helplessness that it felt when the trauma originally occurred. It's like having many different unconscious doorways into a room that holds the unresolved trauma.

That's why, for instance, combat veterans or survivors of gun violence may experience flashbacks when they hear a car backfire. In effect, their amygdala generalizes the backfiring sound and associates it with gunfire, a sound stored in a memory capsule.

Importantly, because memory capsules are stored in the survival brain's implicit memory system, the thinking brain may not understand how or why a memory capsule gets triggered. Indeed, the thinking brain may not even be aware that a memory capsule *is* triggered. Because memory capsules are stored at very high arousal levels—when the amygdala remembers best—the thinking brain's explicit memory may have been disrupted or offline. That's why conscious memory of the traumatic event may be vague, fragmented, and contradictory. Certain details may be seared in conscious memory with crystalline clarity, while others are completely missing.

A third consequence of the corrupted implicit memory system is that whenever a memory capsule gets triggered, the survival brain and body create symptoms that demonstrate the survival brain's belief that the traumatic event is still ongoing. These symptoms include flashbacks, nightmares, intrusive thinking, worry, rumination, and physical symptoms of stress activation out of proportion with the current situation. By creating these symptoms, the survival brain is trying unsuccessfully to complete the original traumatic event and restore allostasis. These symptoms highlight how the survival brain and body don't understand that the event is actually in the past—but instead continue to mobilize stress arousal to cope with this misperception of an ongoing threat.

One example of this third consequence is *kindling*, when previously benign events increasingly tend to trigger stress activation. Kindling occurs when unresolved memory capsules get repeatedly triggered, such as through flashbacks, each time intensifying the symptoms. Eventually, *internal* cues of stress activation—such as shallow breathing, muscle tension, nausea, or certain body postures—may become a source of stress arousal *completely independent of any external cues.* That's why it's called kindling, like the small, flammable twigs used for starting fires.

Each time kindling occurs, it "ups the ante"—increasing the mind-body system's sensitivity to old cues. Over time, the mind-body system can become self-activating—increasingly less attuned to the present surroundings.

In effect, the traumatized survival brain projects the ongoing danger it perceives inside the mind-body system—related to unresolved memory capsules—outward onto the external environment. It gets caught in the past, overlaying unresolved memory capsules of danger onto the present situation, even when it may actually be completely safe. This distortion is survival brain hijacking at its most extreme. In this way, unresolved memory capsules unconsciously bias the traumatized person's perceptions, absorb and capture their attention, drive their default defensive and relational strategies, and color their relationships and decision making.

Over time, the survival brain neurocepts these internal stressors—the physical sensations, pain, distressing thoughts, and emotions associated with the original trauma—as ever more threatening and challenging. This is why symptoms from earlier traumatic events can actually worsen over time, almost as if they "had a life of their own." They do—through kindling.

THE THINKING BRAIN DURING TRAUMA

Traumatized humans have another layer that complicates this picture. That's because after trauma, the integration of information processing between the different parts of the brain—on the cognitive, emotional, and sensorimotor levels—is often compromised. This exaggerates an adversarial relationship between thinking brain and survival brain—and adds additional impediments to complete recovery.

Humans have more and bigger neural circuits running from the survival brain (amygdala) to the thinking brain (prefrontal cortex) than the other way around. This makes evolutionary sense, since rapid threat assessment and reaction increased our chance of survival. Furthermore, during prolonged or traumatic stress, neural circuits originating in the survival brain get extra workouts, even while neural circuits originating in the thinking brain are likely impaired or degraded. Together, these two imbalances help to explain why survival brain hijacking may occur when we're outside our window.

More important, these two imbalances also clarify why it's difficult for the thinking brain to correct the survival brain's corrupted implicit memory system. This process—called *fear extinction* or *trauma extinction*—relies on the thinking brain circuits that get degraded when we experience prolonged or traumatic stress. Extinction actually involves forming a new memory, rather than erasing an existing memory.

After a traumatic event, the thinking brain understands that the event is behind us and that we survived. Since it knows the event is over, the thinking brain usually relies on its protection tools—thinking, analyzing, planning, deliberating, and deciding—to try to keep us safe in the future. It may analyze and learn from the event, assign meaning to it, blame others, engage in self-critical thinking for how we contributed to the event, or plan how to prevent such events from ever happening again.

In other words, for the thinking brain, the trauma is in the past, and its course of action is to use its tools to move on.

The thinking brain's course of action is diametrically opposed to the survival brain's post-traumatic understanding and protection plan. After all, the survival brain believes the traumatic event is still ongoing. Thus, when the survival brain perceives cues related to unresolved memory capsules—including physical sensations of stress arousal, via kindling—it continues to neurocept danger and mobilize stress. Paradoxically, therefore, since the survival brain doesn't support the thinking brain's "reasonable" belief that the trauma is over, *this mismatch between their respective understandings may actually lead the survival brain to feel less safe.*

It's as if the survival brain says, "I notice all this stress activation in the body. Since the body's activated, there must be a threat, right? But I'm not seeing a threat out there. Therefore, it must be that I'm about to get blindsided by something and the danger will be even worse than I expect." This dynamic helps explain why trauma survivors often feel like they're "walking on eggshells" and "waiting for the other shoe to drop." It's how their survival brain unconsciously tries to square a relatively safe external environment with the internal feelings of danger.

No wonder, then, that this process often fuels a vicious cycle of kindling—leading the traumatized survival brain to perceive neutral and even positive stimuli as threatening. In fact, the traumatized survival brain may grow to distrust *all* forms of activation in the body—even pleasant arousal, such as during exercise, dancing, or sex. During pleasurable events, for example, the thinking brain may enjoy what's happening while the traumatized and hypervigilant survival brain is preoccupied with locating a threat to "explain" the arousal.

With such a tremendous post-traumatic mismatch between the thinking brain's and survival brain's understanding of the current situation, it's no wonder that the survival brain continues to sense danger. Yet because the thinking brain knows that the traumatic event is over, *it usually has no idea why the mind-body system is behaving this way.*

The thinking brain may try to analyze our behavior and symptoms, using adversarial thinking brain habits. For instance, it may have critical or comparing thoughts, such as "I should be over this by now, what's wrong with me? Other people have dealt with much worse than this." The thinking brain is also likely to engage in self-judgment, guilt, and shame. It may experience anxious thoughts, worrying that symptoms will never end or get worse. All these adversarial thinking brain habits only fuel even more stress activation.

In response to such "analysis," the thinking brain may then decide that the stress arousal and symptoms of dysregulation are a "problem" to be fixed. The likely result is *thinking brain override*—such as through suppression, compartmentalization, suck it up and drive on, and powering through. Thinking brain override can also manifest when we live "in our heads," disconnected from our emotions, intuition, and bodies.

Some therapeutic techniques—such as cognitive behavioral therapy, positive psychology, exposure therapy, and other talk therapy, cognitive reappraisal, or goal-setting techniques—may also inadvertently encourage thinking brain override. Thinking brain–dominant techniques aim toward effortful, top-down self-regulation by strengthening the ego, inhibiting stress arousal, or desensitizing us to cues in our unresolved memory capsules—since stress activation is "the problem" that needs to be managed.

However, when these techniques aren't integrated with bottom-up processing, from inside our window, the survival brain cannot update the corrupted implicit memory system. Thus, using these techniques alone, it's possible for a traumatized person to rigidly hold their stress arousal levels within their narrowed window and thereby avoid triggering their unresolved memory capsules. At the same time, however, they remain compartmentalized and dissociated from the parts of themselves that believe the trauma is still ongoing. In turn, the survival brain and body continue to overlay unresolved memory capsules onto the present, unconsciously rely on the trauma's default defensive programming, and, through kindling, create increasingly severe symptoms of dysregulation.

Thus, as trauma clinician Pat Ogden and her colleagues explain, although thinking brain–dominant techniques "offer effective *management* of hyperarousal and provide significant relief, they may not fully address the problem."

To be sure, thinking brain override can be exceedingly skillful in certain situations and for short periods of time. Nonetheless, these methods by themselves do not facilitate the complete recovery needed to widen the window. *In fact, while the thinking brain may feel more in control by relying on these methods, the survival brain's perception of helplessness and lack of control will likely continue.* Thus, chronic reliance on these methods alone is likely to perpetuate unresolved memory capsules—and add to allostatic load.

Finally, thinking brain override may also actively inhibit the survival brain's recovery attempts. Most of us were never taught what the discharge of stress activation looks like in our mind-body system. (You'll know in Part III!) Thus, when these sensations and behaviors spontaneously occur, we may override and inhibit them—especially when they play into thinking brain narratives and cultural norms, such as "boys don't cry."

The more extreme the thinking brain override becomes, the more stress arousal the survival brain may create to inform the thinking brain: "I am not safe right now!" Trying to get its message through, it turns to survival brain hijacking, such as flashbacks, nightmares, keyboard vomiting, and other symptoms of dysregulation. In response, the thinking brain doubles down on its suppression, compartmentalization, suck it up and drive on, and override of the survival brain's signals. In effect, it thinks, "What's wrong with me? Those events are in the past. I should be over this by now."

Every time the thinking brain thwarts the survival brain's attempts at recovery, it intensifies our symptoms of dysregulation. More importantly, it also reaffirms and reconditions the link between stress arousal and the survival brain's perception of helplessness—that core neurobiological coupling that defines traumatic stress. In effect, the survival brain generalizes from its traumatic belief that "I cannot defend myself successfully" to "I cannot recover successfully, either." This helplessness for completing a full recovery further locks the traumatic patterning in place. It's a vicious cycle.

For a traumatized person, it's one hell of a Gordian knot: stress arousal in the body, unresolved memory capsules, default programming of the incomplete/unsuccessful defensive and relational strategies used during the traumatic events, and the learned helplessness associated with being unable to recover completely now. Until this Gordian knot is severed, the survival brain and body will continue to believe the traumatic event is still ongoing and thus rely on traumatic default programming.

And yet, because the thinking brain and the survival brain have profoundly different understandings of prior trauma, *the thinking brain often unwittingly serves as one of the primary obstacles to a complete recovery ever happening.* Instead, to manage increasing symptoms of dysregulation, most traumatized humans cope with a range of behaviors that are socially acceptable—while tragically only narrowing the window further.

Before starting the next chapter, I encourage you to reflect, with nonjudgmental curiosity, on whether you engage in either thinking brain override, such as compartmentalization, or survival brain hijacking, such as flashbacks and kindling. In your journal, you might ask whether there are particular situations, relationships, or memories that trigger these mental habits. When these habits have been triggered, how do you typically work with them?

The first step to healing an adversarial relationship between the thinking brain and survival brain is to become aware of its presence. With an understanding of these neurobiological dynamics, over time we can support the survival brain in effecting a complete recovery—and extinguishing the adversarial relationship between them, as well.

CHAPTER SIX

Parents and Attachment Styles

The last three chapters covered the basics of how our mind-body system functions during stress and trauma—and highlighted how the width of our window determines how much capacity for skillful response we'll have during stressful or traumatic situations. People with wide windows are more likely to neurocept danger accurately, respond flexibly with thinking brain functions online during both safe and threatening situations, and recover completely afterward. In contrast, people with narrow windows are more likely to neurocept danger inaccurately and then inflexibly cue the defensive and relational strategies of their default programming—whether or not these strategies are appropriate for the current situation. Outside their narrow window, they're also more likely to react with either thinking brain override or survival brain hijacking—both of which impede complete recovery afterward.

With this groundwork in place, now it's time to ask: What determines the width of our window?

To answer this question, this chapter explores how our window initially gets wired; the rest of Part II will explain how our window becomes narrowed over time. In these chapters, you'll notice that I emphasize childhood a great deal, since our early-life experiences have a profound, enduring effect on our mind-body system throughout our lives. Yet for many of us, these childhood influences on our window remain outside our awareness or disowned as unimportant.

Nevertheless, if you take anything from this chapter, let it be this: Early-life experiences—especially with our parents and other important care providers—powerfully reverberate *throughout our lives, especially in how we interact with other people and cope with and recover from stress.* Even people with "happy childhoods" characterized by close familial relationships may have started life with a narrow window. In turn, a narrow window from childhood compromises our ability to create satisfying personal and professional relationships and to recover completely after stress over the longer term.

Repeated childhood experiences set in place the neurobiological structures that influence the width of our window, even in adulthood. Chapter 3 explained that we can choose to interrupt their programming when it's no longer serving us and choose instead to intentionally rewire them and widen our windows. No matter how difficult our past may have been, the choice today is entirely up to us.

Yet before we can rewire any neurobiological structures and habits, *we must become aware of their existence.* Thus, through this and the next chapter, I want you to become aware of how your early life continues to shape your life today—especially how it influences your relationships and your capacity to recover from stress and trauma.

WIRING THE WINDOW INITIALLY

The parts of the brain that most distinguish humans from other primates continue to mature into our thirties, leaving them subject to environmental shaping the longest. These brain regions include the *frontal lobes*, which control the thinking brain functions of language, executive functioning, reasoning, and thinking. They also include the *parietal lobes*, which integrate sensory stimuli, spatial awareness, and input from the visceral afferent system, which is the communication channel from the organs back to the survival brain.

Relatedly, we "bring online" our human hierarchy of defenses in reverse order. As Chapter 4 explained, we are born with the second line (SNS, or fight-or-flight) and third line (dorsal PSNS, or freeze) of defense already fully developed. However, our first line of defense, the ventral PSNS circuit, is still underdeveloped when we are born. It begins its development during the last trimester of pregnancy and continues through adolescence. Thus, ventral PSNS functions—our ability to interact with other people, regulate our heart rate and breathing rate without turning on stress arousal, and recover completely after stress arousal—continue to evolve into our late teenage years. By extension, *this means the development of these ventral PSNS functions is exquisitely sensitive to our childhood social environment.*

A newborn infant gets her first ventral PSNS workouts from nursing and feeding. How? To nurse, the infant unconsciously learns how to disengage the vagal brake to increase metabolic activity. The infant also flexes the ventral PSNS by sucking, swallowing, and bonding with her mother during nursing. Afterward, the infant unconsciously learns how to reengage the vagal brake to support digestion and sleep. Interestingly, colic—when babies cry unconsolably for hours, which affects roughly one in five babies—may be a sign that the infant's ventral PSNS circuit is having difficulty learning how to regulate these processes.

By about age six months, the infant's ventral PSNS expands its capacity to use social cues for regulating the vagal brake and recovering from stress, such as when the mother's smiling face and cooing voice help calm the baby down. As the ventral PSNS continues to evolve, the child becomes ever more capable of engaging with and being soothed by other people, as well as downregulating his own stress arousal and negative emotions. Thus, positive bonding during nursing and early life begin the infant's learning trajectory toward self-regulation, recovery, and social engagement, the three facets of the first line of defense.

Premature birth, infant illness, and neglect or abuse can all interrupt the development of an infant's ventral PSNS circuit. (Preterm infants born before thirty weeks are especially compromised, because they are born without a functioning vagal brake.) In these situations, infants are more likely to be hypersensitive and hyperreactive to negative or ambiguous environmental cues. As a result, their survival brains are more likely to neurocept danger, even when the situation is safe, and then turn on stress arousal and fall back to the second line of defense (fight-or-flight). They're more likely to have difficulty self-soothing and downregulating stress and negative emotions. They're also more likely to experience difficulty learning social skills and creating social bonds.

In one study, nine-month-old infants completed attention and social interaction tasks while researchers measured how well their vagal brakes functioned. Researchers tested these infants again when they were three years old. Compared with infants with properly functioning vagal brakes, infants who had difficulty regulating their vagal brakes at age nine months were significantly more likely to exhibit social and behavioral problems years later. As three-year-olds, these toddlers experienced significantly more withdrawal, depressed behavior, and aggressive behavior than the children whose vagal brake had functioned appropriately when they were infants.

In other words, infants with ventral PSNS developmental difficulties begin a trajectory toward a compromised first line of defense—with significant lifelong implications for the width of their windows.

SECURE ATTACHMENT WITH "GOOD ENOUGH" PARENTS

Mastering all of the capacities that make up the first line of defense—social engagement, attachment, the vagal brake, and recovery after stress arousal—depends to a large degree on how attuned and harmonious early interactions with our parents or other important care providers are. Since children implicitly acclimatize themselves to the emotional communication they receive from their parents, the *attachment style* they develop is well matched—and extremely adaptive—for interacting with and getting needs met by these early attachment figures.

John Bowlby, the British psychoanalyst who developed attachment theory, emphasized that infants and children are programmed by evolution to bond with one or, at most, a few adults. Given how helpless we are as infants, it makes sense we'd create such bonds to ensure our survival. An infant's attachment style develops through their social and emotional communication with their primary attachment figure, usually the mother. As trauma researcher and clinician Bessel van der Kolk puts it, "Early attachment patterns create the inner maps that chart our relationships throughout life, not only in terms of what we expect from others, but also in terms of how much comfort and pleasure we can experience in their presence."

Attachment theory developed in two parallel communities. Early developmental psychologists and trauma clinicians have focused on early childhood development and attachment bonds with our parents; much of this research is conducted through clinician observation and evaluation of children during a protocol of standardized situations. In contrast, social psychologists have focused predominantly on attachment styles in romantic adult relationships; much of this research is conducted through adults completing self-report questionnaires about their relationships. Both communities refer to Bowlby as their theoretical forebear, but their respective research remains distinct and not well integrated.

Since I'm focusing here on the early neurobiological wiring of the window and the defensive and relational strategies that get wired in our early social environment, I draw mostly from and use the names for attachment styles from the first community. However, to give you some context about typical adult relational strategies and attachment styles, I've also included some empirical research from the second. Please keep in mind that the adult self-report research is probably less accurate than the clinician research with children. For instance, adults may answer self-report questions in ways that make their attachment style seem more secure or more self-reliant than it actually is. Indeed, the evidence about adult attachment styles varies widely between studies.

Our primary attachment style usually remains stable throughout our life, becoming generalized to other relationships. It's possible that we developed different attachment patterns with each attachment figure we frequently interacted with during childhood, such as our father or another relative. When this happens, those alternate attachment patterns may be triggered by similar relationships or situations in the future.

Empirical studies suggest that about three quarters of adults keep the same attachment style through their lives. Nonetheless, there isn't a direct linear relationship between our childhood and adult attachment styles. Many life experiences—including parental divorce, traumatic events, romantic relationships, and therapeutic work—may intervene and shift our attachment style. For example, someone may have wired a secure attachment style in childhood but then as an adult experience relational trauma, abuse, infidelity, or a traumatic end to a romantic relationship; such experiences might then shift their relational strategies in subsequent relationships toward an insecure attachment style. Conversely, someone may have initially wired an insecure attachment style but then with intentionality, over time, develop a secure attachment style as an adult, such as through intensive therapeutic work, supportive relationships, and consistent practice of securely attached relational strategies.

Attachment styles include emotional communication patterns and relational strategies, which get encoded in the survival brain's implicit and procedural memory. Our attachment style manifests in our patterns for proximity seeking, such as how much physical closeness and contact with others feels comfortable to us. Our attachment style also affects our social engagement behaviors (such as when and how much we smile, nod, lean in, or make eye contact) and defensive expressions (such as when and how much we frown, withdraw, cross our arms, or tense our bodies).

Bowlby emphasized that the basic task of an infant's first year is forming attachment—creating a "secure home base" from which the child can eventually move out into the world. So where does this secure home base come from?

Ideally, if the infant's primary caregivers are regulated themselves—inside their own windows of tolerance—they attune to their baby's physiological and psychological states and have the capacity to deal with these states effectively. Regulated parents can sense their baby's needs and meet them, such as changing her diaper when it's wet, feeding her when she's hungry, and holding her when she needs to be soothed.

In effect, the parents provide a "holding environment" for the baby's needs and growth. A critical part of providing this holding environment is tolerating and staying present with the infant through his dysregulated states, such as fear, anger, frustration, hunger, or exhaustion. Well-attuned and regulated parents also modulate their baby's arousal—calming him when his stress arousal is too high or stimulating him when his arousal is too low. In the process, they help the infant learn how to stay inside his own just-developing window of tolerance.

With well-attuned and regulated parents, the child's survival brain and nervous system become conditioned to associate distress with the subsequent experience of quickly being soothed after distress. Through such repeated experiences—where the parent *consistently* and *accurately* perceives and responds to the infant's body-based needs—the infant's ventral PSNS circuit develops. In effect, the baby internalizes her parents' external soothing process, so that her ventral PSNS circuit learns how to downregulate after distress by itself. With repeated experiences of stress arousal followed by soothing and recovery, the baby also develops her *orbital prefrontal cortex*, the brain region involved in self-regulation of stress arousal. The child's brain and nervous system experiences beneficial neuroplasticity through repeated attuned interactions with her parents.

As the child grows, the parents' "holding environment" extends from the child's body-based needs to his emotional and psychological states. Regulated parents help the child learn to tolerate negative and positive emotions, both in himself and in others. Ideally, the child learns to rely on his parents as a "safe home base" from which he can venture out to explore new things. His survival brain and nervous system learn to trust that should the child experience any distress, he can always come back to his parents for soothing and support. This process—stretching outside his comfort zone and recovering afterward—will also become internalized in his mind-body system, further developing the ventral PSNS circuit.

Of course, even the most attentive parents are not always perfectly attuned to their children's needs, nor are they always regulated themselves. So what happens when the parents occasionally "lose it"—because, let's be clear, all parents inevitably will?

Attachment researchers stress that having what they call "good enough" parents ensures that the child's developing ventral PSNS circuit will not be harmed. When distress occasionally occurs in their relationship, such as when there's been an attunement breach between the parent and child, the "good enough" parent can provide *interactive repair*—taking conscious steps to restore their attuned relationship.

Say, for instance, the parent must interrupt play because it's time for the toddler to go to bed. The child doesn't like this and expresses that dislike, perhaps even starting to tantrum. With interactive repair, the parent could calmly help the child work through and resolve her frustration, while still holding the bedtime boundary. Or, say the parent comes home frustrated about a setback at work, unrelated to the child, and unintentionally snaps at the child. With interactive repair in this situation, the parent would first calm himself back down and then apologize to the child, helping her work through and resolve her anger, anxiety, or shame after being snapped at.

With interactive repair, "good enough" parents embody smooth transitioning between negative and positive emotional states—and assist their child's survival brain and nervous system in doing the same. In this way, they help their child to develop self-regulation skills, as well as flexibility and resiliency in relationships.

Thus, with generally attuned and regulated parents, the child learns how to regulate stress activation, bodily sensations, emotions, and impulses. Essentially, by imitating and internalizing his parents' cognitive, physical, emotional, and relational patterns, the child's brain and nervous system get shaped by his parents' more mature brains and nervous systems. This ideal scenario is what allows someone to initially wire a wide window.

Children with parents who provide "good enough" attunement develop what's called a *secure attachment style*, characterized by a relatively wide window. Their capacity for employing social engagement, regulating their vagal brake, and fully recovering after stress arousal gets fully developed. As a result, securely attached children develop the ability to access and adaptively use all three lines of defense. Most important, perhaps, they also learn *to access agency*—the sense that what they do can change both how they feel inside and how other people respond to them.

From such an upbringing, securely attached adults have learned how to display congruence between what they're feeling and how they express themselves and behave in the world. Securely attached adults can unambiguously share their intentions, moods, and desires with others. Their internal physical, cognitive, and emotional states are usually aligned with their speech, facial expressions, body language, and other social engagement behaviors. Securely attached adults are comfortable being alone and autonomous. They instinctively know how to self-regulate their arousal levels. They also know how to be "in sync" with others around them, and they are comfortable giving and receiving support. Given their wide windows, they can tolerate high-intensity arousal states in themselves and others, as well as enjoy emotional intimacy and physical and sexual contact.

Between 50 and 63 percent of adults have a secure attachment style, according to large empirical studies. However, secure attachment is less common among adults from disadvantaged and lower socioeconomic backgrounds. Older adults, from all socioeconomic backgrounds, may also have lower rates of secure attachment: Only 22 to 33 percent of late-middle-aged and elderly adults were shown to have secure attachment styles in at least five empirical studies.

WIRING AN INSECURE ATTACHMENT STYLE

So if roughly half of adults are securely attached, what about everyone else?

I want to preface my answer with four general points.

First, when we're born, we are neurobiologically programmed to attach to *someone*. So it doesn't matter whether our parents and care providers were attentive, attuned, and loving, or distant, neglectful, abusive, or inconsistent. Either way, it was most adaptive for us to develop defensive and relational strategies *that specifically corresponded to our attachment figures.* This was literally the only way to survive and get at least some of our needs met.

Second, at one time our parents were infants and children themselves. Their eventual parenting styles were profoundly influenced by their own initial wiring, as empirical and experimental studies suggest. Our parents unconsciously nurtured us with the window they had wired. To say this differently, *our parents cannot help us develop brain and nervous system capacities— and defensive and relational strategies—that they themselves have not developed.*

For instance, parents with insecure attachment styles are often less skilled with interactive repair, likely because they didn't experience interactive repair in their *own* families of origin. These parents may have never observed what it's like to work through attunement breaches effectively. They may have experienced a different pattern for dealing with attunement breaches, such as one that was common in my house growing up—an emotional or violent outburst followed by a period of withdrawal, silent treatment, and "walking on eggshells," followed by everyone coming back together as if nothing had happened, without ever acknowledging or discussing the attunement breach. In turn, that relational strategy gets passed to their children.

Thus, *our parents' own attachment styles will strongly affect the initial wiring of our window*: Children whose parents have secure attachment styles are more likely to develop a wide window, while children whose parents have insecure attachment styles are more likely to develop a narrow window.

Third, attachment style is more related to parental sensitivity and attunement than to the child's temperament, gender, or birth order. A child's gender and temperament have little effect on their attachment style, according to large empirical studies; thus, children with "difficult" temperaments don't necessarily have insecure attachment styles. Moreover, in roughly one third of families, siblings develop different attachment styles—but birth order does *not* explain this variation. Instead, attachment style is more related to parental (especially maternal) sensitivity and attunement, which are context dependent. Mothers may have varying attunement capacity across their different children, depending on the mother's own stress arousal and depression levels during each child's time in the womb and early years. Family attunement can also vary depending on different stressors during each child's first several years of life.

Finally, *stress arousal is contagious. Parents' stress arousal levels get conveyed to their infants and children.* Several recent studies found evidence of similar stress hormone levels between a mother and her child—from infancy through adolescence—especially when the mother is anxious, depressed, or coping with intimate partner violence.

A recent experiment with sixty-nine mothers and their babies illustrates this stress contagion effect elegantly. Both mom and baby wore sensors to track their stress activation. The mothers were separated from their twelve- to fourteen-month-old infants for a ten-minute interview with two "expert evaluators," during which the mothers gave five-minute speeches about their strengths and weaknesses, followed by five minutes of Q&A. The mothers were randomized into three conditions. Some moms experienced a "positive evaluation," with the "evaluators" offering progressively more positive feedback, smiling, nodding, and leaning forward. Conversely, some moms experienced a "negative evaluation," with the "evaluators" getting progressively more negative, frowning, shaking their heads, crossing their arms, and leaning back. Finally, some moms experienced a "control evaluation," delivering their speech and answering written questions aloud while sitting alone in a room.

After this "evaluation," all mothers and infants were reunited and the mothers were instructed to help their babies relax. The researchers found that each mother's stress arousal level after her "evaluation" was quickly transmitted to—and embodied by—her infant upon their reunion. Even though the baby was never directly exposed to their mom's stressors, the baby's stress levels quickly mirrored their mom's. Maternal stress transmission had the greatest impact when the mom experienced a negative evaluation. The researchers concluded that "infants may be predisposed to attend to their mothers' heightened-arousal states, such as reactions to negative, threatening, or angering events."

Interestingly, when a mom experienced either a positive or negative evaluation—meaning she met with live "evaluators"—her baby showed significantly less social engagement afterward. Instead, during subsequent interviews with mom and baby together, the baby was more likely to avert their gaze or actively twist their body away from the stranger.

This study has some powerful implications. Once reunited, the baby soon mirrors their mom's stress arousal. Since the baby's ventral PSNS circuit is still underdeveloped, however, the baby has less capacity for social engagement or stress recovery, two facets of the first line of defense. Thus, faced with an interviewer—a stranger and a new stressor—the baby immediately "falls back" to flight response, trying to get away from the stranger by averting their gaze and actively twisting their body away.

Now, imagine a child experiencing much higher stress levels than those experienced by the infants in this study. And imagine that this child's high arousal levels persist for weeks, months, or years at a time—either because of stress contagion from dysregulated parents, or even worse, because the parents themselves are a source of fear.

Realistically, in such conditions, how many repeated experiences is this child going to have for practicing social engagement skills, regulating the vagal brake, recovering completely from stress arousal, and fully developing the ventral PSNS circuit?

Probably, not many—but certainly, not enough.

In these situations, the child will likely have *many more* repeated experiences of practicing and employing the second and third lines of defense—the defensive and relational strategies associated with SNS (fight-or-flight) and dorsal PSNS (freeze). Of course, such repeated experiences will have neuroplastic consequences: many opportunities for the survival brain to neurocept danger and turn on stress arousal, and many opportunities to bypass the ventral PSNS strategies completely while falling back to fight, flight, and/or freeze instead.

THE AMAZING ADAPTATIONS OF INSECURE ATTACHMENT

Our human neurobiology can find its way through *any* conditioning. Thus, as I discuss the three insecure attachment styles, I want you to keep in mind how *resilient and adaptive* they are, for their unique social environments.

Insecure-Avoidant

Infants are likely to develop an *insecure-avoidant style* if they have a mother (or other primary care provider) who actively thwarts physical and emotional connection. These mothers tend to dislike holding, talking with, and making eye contact with their babies. More benignly, these mothers may be distant and preoccupied because of caregiving obligations for another chronically ill sibling or elderly relative, work outside the home, or their own health problems. In essence, this mom transmits to her baby the message of "I'm not available to you, physically or emotionally."

In response, the child seamlessly adapts by expressing little need for proximity and—at least from outside observation—little interest in physical, emotional, or eye contact with the parent. These adaptations are called *deactivating strategies*, which protect the child from unwanted feelings of pain and parental rejection. These infants and children often don't cry when Mom leaves the room, don't seem to care when she comes back, and tend to focus their attention more on toys and other inanimate objects instead of the people around them.

Underneath this surface appearance of little separation distress, however, insecure-avoidant children experience a lot of stress arousal. During experimental separation from their parents, these children show chronically elevated heart rates suggestive of constant SNS activation. *This paradox—seemingly unaffected by others on the surface, highly aroused underneath—creates a core patterning* in the avoidant mind-body system that, over time, leads to major disconnections between their inward states and outward behaviors.

In other words, avoidant children and adults often experience a large incongruence between what they're feeling emotionally and physiologically inside and how they express themselves outwardly in terms of eye contact, facial expression, posture, and speech. In fact, they're often unaware of what's actually going on inside—which fuels an adversarial relationship between their thinking brain and survival brain, too. For this reason, some trauma researchers call this pattern "dealing but not feeling."

The predominant relational strategy for the avoidant attachment style is "I can take care of myself." In school, avoidant children are likely to be loners or bully other kids. Avoidant adults tend to be compulsively self-reliant and try to deal with stress on their own. They tend to keep other people at arm's length—often believing that personal and professional relationships are entangling, unpleasant, or "too much work." They tend to withdraw when they're stressed and undervalue relationships as a source of support.

Often unaware of their own emotional landscape, they also tend to be less sensitive to social and emotional cues in others. When someone tries to engage socially or emotionally with an avoidant adult, they may withdraw, "armor up" with muscle tension and closed body postures, send mixed messages, or use other deactivating or distancing strategies.

Avoidant adults usually operate within a narrowed window, often coping with thinking brain override of the body's sensations and emotions. As a result, avoidant adults don't usually rely on the first line of defense (social engagement). Instead, their default programming generally tends toward strategies associated with the third line of defense, the dorsal PSNS—withdrawal, lack of eye contact and emotional expression, and responses on the freeze spectrum. When provoked, however, they may also rely on fight responses (SNS).

If they have a partner, avoidant adults tend to be reluctant to seek support from them during stressful situations. They also tend to view their partner as needy and overly dependent. They may be completely unaware of their own needs and fears in the relationship. Not surprisingly, avoidant adults tend to be less satisfied than other attachment styles in their romantic relationships. While insecurely attached adults are significantly more likely to engage in infidelity than securely attached adults, some research shows avoidant adults to be the most unfaithful attachment style. They're also significantly more likely to get divorced and have multiple marriages during their life span. For instance, Greg's story from Chapter 2 suggests he has an insecure-avoidant attachment style.

About one quarter of all adults have the insecure-avoidant attachment style. Some studies suggest that this attachment style is more common among older age groups—with some research finding 37 to 78 percent of older people with avoidant attachment. Another study found avoidant attachment as the most common style among older urban African Americans (83 percent). (It is also possible that these groups may be more likely to portray themselves in self-report questionnaires as more self-reliant than they actually are.)

In my experience teaching in high-stress environments, insecure-avoidant attachment is also likely overrepresented in high-stress professions—because occupations like firefighter, special operative, surgeon, or Wall Street trader usually require, and also reward, a strong capacity for self-reliance and emotion suppression.

Insecure-Anxious

The second insecure attachment style is called either *insecure-anxious* or *insecure-ambivalent*. Infants will develop this attachment style if their mother interacts with them unpredictably and inconsistently—responding to *her own* emotional needs and moods, not her baby's. Rather than meeting the baby's arousal levels and needs, Mom may overarouse her infant through stress contagion (if Mom's anxious or angry) or fail to engage with her infant for nursing or play (if Mom's depressed or down). Since Mom's emotional needs and arousal levels usually drive the relationship, she may inconsistently attune to and often override her baby's needs and arousal levels. In essence, this mom transmits to her baby a message of "I may be available to you—but I may not—depending on what's going on with me right now."

Once again, the infant seamlessly adapts to this inconsistency. Because he is unsure about his parent's reliability, this infant is likely cautious, ambivalent, anxious, angry, or inconsolable. As Chapter 4 explained, when the survival brain perceives something to be unpredictable and uncontrollable, it's more likely to neurocept greater threat.

Without consistent attunement and parental assistance in downregulating arousal, an anxious child's nervous system will likely experience chronic stress arousal and difficulty with wiring the recovery facets of ventral PSNS.

Moreover, since Mom's needs and moods often override the baby's internal states, this infant's survival brain and nervous system will internalize this conditioning—unconsciously learning how to override her own physical and emotional needs. Compared with avoidant children, anxious children may have more awareness of their internal states and more congruence between internal states and their outward behavior. They may outwardly show sustained or exaggerated negative emotions. For this reason, trauma researchers often call this style "feeling but not dealing."

The predominant relational strategy for the anxious attachment style is "I want more from my relationship with you," which can lead to a *preoccupation* with relationships and unfulfilled relational needs. This preoccupation can manifest in many different ways, such as people pleasing, with the hope of having their efforts reciprocated. Acting out, screaming, withdrawing, pouting, sleeping around, or engaging in other "protest behaviors" to try to get someone's attention and draw them close. Obsessing about relationships. Underneath all of these strategies, the anxious person may be unable to recognize safety within a relationship—even when the relationship is "objectively" safe.

People with an anxious attachment style are more vigilant to changes in others' emotions—and more sensitive to and accurate in reading others' cues. This makes sense, because their survival required learning to read the needs and moods of their inconsistent caregivers very carefully. Yet because they're usually chronically activated themselves, their thinking brain functions may be degraded, which may lead them to misinterpret these cues, jump to conclusions, and then overreact.

Insecure-anxious people usually operate with a narrowed window, often coping with survival brain hijacking. They generally have a low threshold for stress arousal, easily becoming overstimulated with wide mood swings. Their default programming tends toward strategies associated with the second line of defense, fight-or-flight (SNS). They may focus excessively on their internal distress—which only amps it up further—and then lean heavily on others around them to help them manage their hyperarousal. They often experience difficulty "coming down" after stress or negative emotions. They may also believe that they're incompetent at regulating their emotions or that they can't take care of themselves.

In relationships, anxiously attached adults tend to prefer intense and enmeshed relationships. They tend to find isolation stressful, and they fear abandonment. They also tend to seek excessive reassurance from their partner. Perhaps unconsciously, they may use chronic stress and negative emotions to pull their partner closer—and worry that showing signs of recovery or calm will lead their partner to pull away. While not as unfaithful as avoidant adults, they are significantly more likely than securely attached adults to engage in infidelity—especially for fulfilling unmet emotional intimacy needs, getting their partner's attention, or "getting back at" their partner. At the same time, they are the most likely of all attachment styles to stay in unhappy relationships or marriages.

Research shows that 6 to 22 percent of all adults have an insecure-anxious attachment style.

Insecure-Disorganized

The third insecure attachment style—*insecure-disorganized*—is sometimes called *insecure-anxious-avoidant*, highlighting how it may manifest as exaggerated aspects from both of the other insecure styles.

Most adult attachment research doesn't measure and include this attachment style—grouping adults into the two other insecure styles, depending on which traits predominate—so it's generally assumed to be relatively rare. Yet disorganized attachment is not uncommon among youth. For example, a meta-analysis found that 15 percent of infants and children from "normal" middle-class families had disorganized attachment styles. Among children in clinical treatment or from lower socioeconomic backgrounds, the rate was two to three times as high.

Infants will develop an insecure-disorganized style when their parents are some combination of neglectful, abusive, depressed, or traumatized themselves. As a result, the child experiences severe misattunement from her parents to her arousal levels and needs. Sometimes the parent is homeless or raising the child alone. Sometimes the parent is depressed and preoccupied with his own unresolved loss or trauma—a recent death of a loved one, trauma from his own childhood, or more recent or ongoing trauma from sexual assault, domestic violence, or war. Often these parents are alcoholics or substance abusers. Some parents may be enraged, intrusive, or abusive with their child. Others may be confused, afraid, and fragile, as if they don't know how to be the adult and want their child to comfort them in role reversal. In other words, the parents themselves become a source of fear, distress, terror, and confusion—not of safety and security. Alternatively, these children may have been abandoned to an orphanage or the state's care.

Parents of disorganized children are not necessarily abusive or neglectful all the time. Indeed, researchers who study this attachment style argue that these parents may exhibit "bouts of frightening behavior, having themselves endured major loss or maltreatment . . . [but] are otherwise sensitive and responsive." Thus, disorganized attachment "can be viewed as a second-generation effect of the parent's earlier trauma."

Once again, the infant adapts. Because their parents are necessary for their survival but also radically inconsistent in caregiving—and a chronic source of fear or terror—these children develop contradictory "approach-avoid" patterns, both in their bodies and in their defensive and relational strategies. It's not really safe to get close, it's not really safe to run away, and yet, as powerless children, they have to rely on someone. Thus, their behavior may appear contradictory. For instance, a disorganized infant might show excessive distress while his mother is away but then act indifferent upon her return. Or a disorganized child might seek proximity, but then immediately freeze or withdraw. In effect, when a disorganized child tries to activate her underdeveloped first line of defense for attachment, the survival brain immediately neurocepts danger—conditioning the second and third lines of defense to arise concurrently or in rapid succession.

Disorganized infants and children often experience chronic hyperarousal and hyperactivity (SNS fight-or-flight), followed by periods of hypoarousal, collapse, or numbness (dorsal PSNS freeze). Compared with other attachment styles, they also show more dysregulation of the vagal brake, immune system, and stress hormone system (HPA axis).

These children rarely experience interactive repair after finding themselves on the receiving end of their parents' rage, terror, abuse, or neglect. Since their parents may switch states rapidly, they may experience a pervasive sense of having to "walk on eggshells," never knowing what might set their parents off. For this reason, disorganized infants and children may also engage in unusual behaviors to self-soothe, such as repeated rocking, head-banging, or trancelike motions.

Disorganized children often have to "grow up fast"—assuming responsibilities for themselves and other family members way beyond their current stage of neurobiological development. Compared with other attachment styles, disorganized children are more likely to exhibit depressive symptoms, shyness, social phobias, attention and learning problems, and aggressive behavior. As teens, they may become sullen, angry, disconnected, and resistant to their misattuned parents—especially when the parents expect their child to look after them. Disorganized teens often manifest poor impulse control, violence, and high-risk and/or self-harming behavior—such as binge eating, cutting, reckless driving, stealing, promiscuity, substance abuse, gang membership, and suicidal behavior.

The predominant disorganized relational strategy is "I need contact, but I can't let down my guard." Compared with other attachment styles, disorganized children and adults are often the least capable of accessing and regulating their internal states. They tend to show the greatest incongruence between their internal states and their outward behaviors. They're the most likely to develop dissociative disorders and other chronic freeze responses. They're also likely to develop the faultiest neuroception—such as naïvely trusting strangers and high-risk situations and distrusting loyal allies and safe situations. After all, without any *inner* sense of security, it becomes almost impossible for the disorganized survival brain to distinguish between true safety and danger. This faulty neuroception sets them up for either being traumatized or traumatizing others in the future—sowing the seeds for further trauma for themselves and others.

About 3 to 5 percent of all adults have a disorganized attachment style. Compared to the general population, this attachment style is likely overrepresented in traumatized, mentally ill, homeless, incarcerated, and lower socioeconomic populations.

WIRED TO CONNECT

You may have noticed yourself resonating with more than one attachment style. If that's the case, remember that although we develop our primary attachment style with one person (usually our mother), we may also develop different relational strategies for interacting with other care providers (such as our father or another relative).

For instance, several attachment researchers and clinicians highlight how "opposites attract"—with avoidant and anxious adults often ending up together. Children who grow up in *this* social environment may develop strong skills in self-reliance (in response to the avoidant parent) while also being chronically hyperaroused, anxious, or depressed; acting out to get attention; or developing strong people-pleasing skills (in response to the anxious parent). They'll also likely be hypervigilant and extremely adept at reading both parents' emotions and the relational dynamics between them. Later in life, although they'll have a primary attachment style, they may manifest defensive and relational strategies from both parents' attachment styles, depending on the particular context or relationship. They may also re-create similar relational dynamics with their own partners as adults.

Moreover, even securely attached individuals are likely to exhibit insecure defensive and relational strategies occasionally—especially when we're stressed, dysregulated, outside our window, or getting triggered by someone else. If you can bring awareness to this indicator, you can choose actions to bring yourself back inside your window—and bring ventral PSNS fully back online. I'll talk more about how to do this in Part III.

Since ventral PSNS is deeply involved in our ability to interact with other people, regulate our cardiovascular system, and recover completely after stress arousal, *our attachment style affects both our relationships and our ability to recover from stress and trauma.*

Recent empirical research highlights how insecure attachment is linked with lifelong difficulties with *both* aspects of the ventral PSNS—social engagement/attachment/supportive relationships *and* recovery/self-regulation.

Compared with securely attached adults, insecurely attached adults are significantly more likely to use destructive and coercive behaviors—and less likely to use constructive and cooperative behaviors—during conflict in personal and professional relationships. They're more likely to worry about negative relationship outcomes. They're more likely to experience relationship discord, engage in infidelity, and experience violence or abuse within romantic relationships.

Insecurely attached adults are also significantly more likely to experience sleep disturbances, severe pain and disability from arthritis, and medically unexplained chronic pain—even after controlling for life histories of psychological illness and substance abuse. They're at greater risk of developing diseases and chronic illnesses—especially conditions involving the cardiovascular system, because of the lack of a properly functioning vagal brake.

Finally, insecurely attached adults are at greater risk of developing mental illnesses associated with emotion dysregulation—and experiencing more severe symptoms. Insecure-anxious and insecure-disorganized adults who tend toward anxious relational strategies are more likely to have *internalizing disorders*, including anxiety disorders, PTSD, depression, postpartum depression, adjustment disorders, and borderline personality disorder. In contrast, insecure-avoidant and insecure-disorganized adults who tend toward avoidant relational strategies are more likely to have *externalizing disorders*, including substance abuse, alcoholism, intermittent explosive disorder, aggressive violence, and antisocial personality disorder.

If this chapter has helped you recognize yourself as insecurely attached, know that you're in the company of about half of the other people walking around the planet. *Indeed, the less awareness we bring to the programming we wired with our childhood attachment style, the more likely that we will continue its defensive and relational strategies in adulthood.* I believe such lack of awareness probably contributes to the fact that almost three quarters of adults maintain their childhood attachment style.

Yet your current attachment style and its default relational and defensive strategies don't have to be your destiny. Like anything else that we've wired through neuroplasticity, it can be changed over time. It's possible to rewire our ventral PSNS circuit and shift ourselves from an insecure to a secure attachment style—with more satisfying and supportive relationships, more capacity for self-regulation and recovery, and a wider window. As someone who started life with an insecure attachment style but has since rewired her neurobiology for secure attachment, I know this to be true.

Nonetheless, there isn't some pill or magic wand that can instantaneously undo our early wiring. Only through repeated experiences of fully recovering after stress arousal—as well as extending ourselves to show up more honestly and completely in our relationships, even when we're scared or angry—can we strengthen our ventral PSNS circuit over time.

As we widen our windows, it's important to remember that *our evolutionary heritage has incorporated its very own "do-over" capacity*—interactive repair—that magical ingredient that allows "good enough" parents to help their children wire secure attachment and wide windows. Even if we didn't get many positive experiences with interactive repair as children, we can certainly choose to learn and employ this skill now. The more we practice interactive repair, the more ingrained the implicit memory becomes—and the easier it is to employ again. Learning how to use interactive repair, and employing it after every attunement breach, is an important interpersonal skill we can practice to widen our windows, move ourselves toward or strengthen secure attachment, and build more satisfying and supportive relationships.

Before starting the next chapter, I encourage you to reflect in writing about your attachment style, both as a child and as an adult. With nonjudgmental curiosity, you might investigate *why* you developed this attachment style. We didn't get much say in how our neurobiology effortlessly adapted to our earliest social environment, but we can be grateful that it helped us survive until now. Even if you think you have a secure attachment style, it can be helpful to write down any insecure relational strategies that you tend to rely on when you're stressed, since these behaviors can help you recognize when you've moved outside your window.

As part of this exercise, you might also choose to speak with your parents, siblings, and other people who knew you as an infant and child. You might ask if you had a difficult and/or premature birth, colic, or medical treatments or surgeries as a child. If you don't already know, you might also ask about your parents' attachment styles and their own stress level during your early childhood, including any traumas, losses, or acute stressors with which they were coping. These individuals can help you remember your early behaviors that may help you identify your childhood attachment style.

CHAPTER SEVEN

Childhood Adversity

When I served in the Army, and for several years afterward, I thought of myself as a trauma magnet. "Trauma magnet" was a term first voiced by a guy I dated during that time—and his way of making sense of several events happening in my life. In retrospect, I know he cared for me a great deal but likely felt angry, afraid, and helpless, since there was little he could do to make things better for me.

From one perspective, he was correct: I *had* experienced a disproportionate number of serial traumatic events throughout my life. So the question naturally arose: *Why? Why did I have such a large lived body of extreme experiences, when many other people didn't?*

For those of us who started life with a narrow window due to insecure attachment style and/or developmental trauma, many aspects of our neurobiology—when left to their own unconscious devices—conspire to keep us that way. Individuals who experienced childhood trauma and adversity often endure additional traumatic events later in their lives. While we may appear to be "trauma magnets," it's way more complicated than that. Some distinct neurobiological processes explain how narrow windows from childhood tend to get narrower over time.

THREE PATHWAYS TO NARROWING THE WINDOW

Remember that when allostasis is functioning properly, we can mobilize stress activation to cope effectively with a threat or challenge and then, afterward, discharge any remaining stress activation to recover completely.

With chronic stress and trauma, however, our mind-body system doesn't complete a full recovery after the stressful or traumatic experience. Instead, it remains in an activated state, adds to our allostatic load, and leads us to develop an adversarial relationship between our thinking brain and survival brain. Over time, we build allostatic load, narrow our window, and eventually manifest symptoms of dysregulation.

There are three pathways to building allostatic load and narrowing our windows. This chapter will explore the first pathway—childhood adversity and developmental trauma. Chapter 8 will explore the second pathway, shock trauma in adulthood, while Chapter 9 will explore the third pathway, chronic stress and relational trauma in daily life.

Of course, not everyone experienced childhood adversity and developmental trauma. However, even if you believe you had a happy childhood—during which you wired a secure attachment style and a wide window—there's something important in this chapter for everyone, for two reasons.

First, many of the neurobiological adaptations that occur after childhood adversity can also occur after stressful and traumatic experiences later in life. Unless we fully recover afterward, *every* stressful or traumatic event we endure—every injury, infection, professional setback, spiritual crisis, and emotional upheaval—has a cumulative effect on our mind-body system. With childhood adversity, this accumulation starts much earlier and may even distort our neurobiological development. Nonetheless, the detrimental neurobiological adaptations in this chapter are just as likely after stress and trauma in adulthood.

Second, as my trauma magnet story suggests, our society has some deeply held assumptions about the causes and consequences of abuse, violence, obesity, addictions, and mental illness. Indeed, these beliefs and assumptions drive much of our culture's disowning of trauma, which I discussed in Chapter 2. These beliefs frequently shame and blame both perpetrators *and* victims for a lack of moral character or self-control—fueling stigmatization that blunts our collective understanding. These beliefs feed societal divisions and shape our policies on a range of issues, such as education, healthcare, welfare, law enforcement, incarceration, and affirmative action.

THE FIRST PATHWAY

Todd from Chapter 2, Julio from Chapter 5, and I all narrowed our windows through the first pathway, childhood adversity and developmental trauma. As our varied stories suggest, the first pathway is quite complex, involving transgenerational effects and several complex interactions between nature and nurture.

As Chapter 3 explained, environmental inputs and habits can lead to changes in our DNA or surrounding proteins, creating an *epigenetic change.* These epigenetic changes can turn gene expression on or off, with enduring effects in our mind-body system. Thus, while we may have a genetic tendency toward a particular trait, whether that trait actually manifests is strongly influenced by our environment and our habits. Childhood adversity has been linked with many detrimental epigenetic changes, which can then also be passed through offspring for several generations, as Chapter 3 explored.

Moreover, as Chapter 6 explained, stress arousal is contagious. Parents with insecure attachment styles and narrow(ed) windows are more likely to create the environmental conditions for their children to wire an insecure attachment style and develop a narrow window, too. Thus, chronic stress and trauma may have transgenerational effects via our early attachment bonds with our parents and other important caregivers.

Parents play a critical role in helping children develop the capacity for self-regulation and recovery after stress arousal and negative emotions. When this development is compromised by our earliest social environment, it leaves indelible patterns in our neurobiology and sets the stage for a narrow window.

For instance, mothers with an insecure attachment style—especially if they're also coping with family substance abuse or coercive and abusive behaviors from their partner—are more likely to develop postpartum depression. In turn, having a depressed mother—in utero and/or the first months of life—has been shown to dysregulate the baby's stress hormone system (HPA axis) toward hypersensitivity. In effect, this lowers the baby's threshold for distress and adversely affects the development of his ventral PSNS circuit.

When parents are coping with their own unresolved trauma or loss, it's especially hard for them to attune to their infants' and children's needs—increasing the likelihood that their children will develop an insecure attachment style and a narrow window. In effect, the parents' chronic stress and trauma echoes across the family.

One recent study with U.S. National Guard families highlights how parents' chronic stress and trauma exposure, narrow windows, and insecure attachment styles can reverberate intergenerationally via stress contagion to narrow their children's windows, too. Although this study examines combat exposure's effects, it's important to remember that *any* unresolved parental trauma or loss may have detrimental transgenerational impact.

In this study, researchers observed family members interacting several times over a two-year period after fathers returned from combat deployment. The researchers found significant links between the combat veteran father's PTSD, the mother's PTSD, and growing psychological symptoms and behavioral problems in their children over time. One year after the father returned from deployment, children were more likely to experience internalizing problems, such as depression and anxiety. After two years, however, the children were more likely to exhibit externalizing problems, such as aggression, lying, or breaking rules.

Nonetheless, *not all families saw these detrimental cascading effects—and parental attachment styles likely account for much of the difference.* In families where parents used behaviors associated with secure attachment—such as showing strong emotion regulation skills and positive interactions with each other and their children—the detrimental effects of the father's combat exposure and PTSD cascading to the children were dampened. Conversely, in families where both parents lacked adaptive emotion regulation skills, they were more likely to exhibit coercive behavior, toward each other and toward their children—exacerbating the cascading effects of their own trauma onto their children.

Other research echoes these transgenerational stress contagion effects that can result from parents' narrow(ed) windows. For instance, compared with adults whose parents did not experience the Holocaust, children of Holocaust-surviving parents were more likely, as adults, to experience dysregulation of their stress hormone system (HPA axis) and to develop PTSD, anxiety disorders, and depression. Likewise, having a parent who experienced either multiple military deployments or deployment-related PTSD significantly increases a child's risk for depression, anxiety, ADHD, and behavioral problems, compared to children whose parents don't have those experiences.

My own family history provides an example of these dynamics. My paternal grandfather, WD, was an infantry noncommissioned officer who saw combat in World War II and the Korean War, with duty in Germany's postwar occupation force in between. After his combat tours, he clearly had PTSD, although the diagnosis didn't formally exist yet; he also battled a gambling addiction. Both he and my grandmother Marie were heavy smokers and alcoholics. Although undiagnosed, Marie's behavior, as recounted to me later by her and my dad, suggests she also had untreated bipolar disorder, which led to wildly inconsistent parenting at best, and abusive and neglectful parenting at worst. As their only child, my father, Deane, bore the brunt of their dysregulation.

When my dad would talk to me and my sisters about his own childhood, he always characterized it as a "life with many adventures." When he was three, for instance, his parents sent him alone by train from North Texas, where WD was preparing to deploy to the Pacific, to rejoin relatives in Alabama, with a note pinned to his jacket so the conductors would know where to let him off. When he was six, he and Marie were on an early boatload of American dependents to Germany after World War II. There, he watched WD's dodgy entanglement in the postwar black market and rode to the local German school in a motorcycle sidecar with his bodyguard, a 6'5" Cherokee Indian incongruously named Tiny, who was assigned to protect him during the denazification campaign. When he was seven and their car broke down on a family vacation in France, an incompetent mechanic inadvertently set the car on fire—with my dad asleep in the backseat.

My father eventually attended West Point after narrowly avoiding juvenile detention in the 1950s for his gang activities in the Anacostia neighborhood of Washington, D.C. He served in Vietnam for nearly two years beginning in 1966, initially advising South Vietnamese combat forces and later, with a U.S. combat unit, conducting major search-and-destroy operations. In the fall of 1968, he started graduate school at Harvard, where he experienced hostility from faculty and fellow students and vandalism of his apartment. I believe those years at Harvard were psychologically more difficult for him than fighting in Vietnam. During this time, my parents married, and my mother got pregnant with me.

My mom, Cissie, had a challenging childhood, too. When my mom was five years old, she lost both her father and her younger sister during a six-month period. In fact, my mom was sitting right next to her three-year-old sister on a picnic bench when BettyLou unexpectedly stopped breathing, fell over, and died. (The family never understood exactly what happened.) Understandably, this was a massively traumatizing experience from which my mother never truly recovered. After the two deaths, my grandmother Louise went to work as a waitress—about the only work she could find, since she was functionally illiterate in English—to support my mom and her older brother, Dick. Cissie was the first person from her extended family to complete high school and attend college, on a field hockey scholarship.

In June 1970, my parents moved to West Point, where my father was slated to teach. Shortly before the move, in the fifth month of Cissie's pregnancy, her beloved dog was killed by a car. Never good with death, and losing one of her most important emotional bonds, she was reportedly inconsolable. She sank into a depression so deep that Louise and her doctors worried she'd lose the baby (i.e., me).

I was born jaundiced and soon developed colic. According to my parents, I would wail inconsolably for hours at a time, sometimes for most of the night. At the time, Deane was self-medicating his undiagnosed PTSD with hard work, alcohol, and tobacco, while Cissie struggled with postpartum depression and the simultaneous new challenges of motherhood, life as an Army spouse, and marriage to a traumatized combat veteran. With insecure attachment styles and narrow windows from their own childhoods, together they had few internal resources for coping with these many life stressors.

With this context in mind, is there any possible way that I could *not* have initially wired an insecure attachment style and a narrow window?

To be sure, such transgenerational effects are not limited to military families. Indeed, several studies about other kinds of family hardship corroborate their cascading effects. In this research, significant family transitions—including divorce, unemployment, homelessness, and economic hardship—are affected by the parent's own capacity for self-regulation and parenting skills, as well as by the quality of the parent-child relationship.

Family hardship and transitions are *always* challenging. Whether these events will have lingering detrimental effects on the children, however, depends greatly on how regulated the parents are. *The narrower the parents' windows, the worse the cascading effects on their children.*

ADVERSE CHILDHOOD EXPERIENCES

The final interrelated factor for transmitting stress and trauma intergenerationally comes from *adverse childhood experiences* (ACEs)—such as physical, sexual, and emotional abuse or physical and emotional neglect; being exposed to domestic violence; and having mentally ill, addicted, incarcerated, or separated/divorced parents. There are strong interrelationships between epigenetics, parents' insecure attachment styles, parental exposure to trauma and adversity throughout their lives, and ACE exposure for their offspring.

The first ACE study was a collaboration between the Centers for Disease Control and Prevention and Kaiser Permanente. It asked twenty-five thousand Kaiser patients in the San Diego area to answer ten questions about their experience with ACEs before their eighteenth birthday. Almost 17,500 patients agreed to participate; the researchers then compared the patients' ACE surveys to their detailed medical and mental health records.

This study was most important for highlighting just how common ACE exposure actually is. Study participants were middle class, well educated, and had Kaiser insurance, suggesting they were also financially secure and employed. They were also three quarters white. This is a demographic we might expect would not have much ACE exposure; however, only 36 percent reported no ACEs. Instead, more than a quarter reported having been repeatedly physically abused as a child. One in eight reported having observed domestic violence and abuse against their mothers. Almost 30 percent of women and 16 percent of men reported having been sexually molested by someone at least five years older than they were. One in eight reported an ACE score of 4 or more, meaning they'd experienced multiple categories of childhood adversity.

Since that study, there's been an explosion of empirical research looking at how ACE exposure affects children's learning and behavioral problems and adults' physical and mental health. Collectively, this research suggests that *the effects of ACEs are additive*: The more categories of adversity someone has experienced—the higher someone's ACE score, from 0 to 10—the more likely she'll experience learning and behavioral problems during childhood and physical and mental health problems in adulthood. Having an ACE score of 4 or more seems especially predictive of downstream negative effects.

Fewer studies have examined the impact of ACEs on low-income, urban, and minority populations; however, the evidence to date mirrors the research about insecure attachment styles in Chapter 6. Just as coming from a minority or lower socioeconomic background increases the likelihood of wiring an insecure attachment style, such backgrounds also increase the likelihood of experiencing childhood adversity. This makes sense, especially when we remember the data from Chapter 2 about the effects of relational trauma linked to poverty and racism.

One study tracked a cohort of minority children—93 percent African American and 7 percent Latino—born in 1979 or 1980 into underprivileged Chicago families. Of the 1,100 children in this study, only 15 percent had an ACE score of 0, compared with 36 percent in the original San Diego ACE study. In contrast, more than a third had an ACE score of 3 or more, compared with 22 percent in the San Diego study.

High-stress professions may also attract people with higher ACE exposure, compared to the general population, in part because of their recruiting patterns among minority and/or lower socioeconomic populations. For instance, U.S. military service-members during the all-volunteer force (AVF) era since 1973 are significantly more likely to have experienced ACEs than their civilian counterparts. One recent study examined more than sixty thousand Americans with and without military service, during the draft and AVF eras.

In the AVF era, men with military service had a disproportionately higher prevalence of experiencing all ACE categories than men without service. Only 27 percent of military men had an ACE score of 0 (compared with 42 percent of nonserving men), while 27 percent of military men had an ACE score of 4 or more (compared with 13 percent of nonserving men). In contrast, during the draft era, there was no significant difference in the spread of total ACE scores between men with and without service; the only difference was that military men were significantly less likely to have experienced household drug use than nonserving men.

Trends for women are different, since we couldn't be drafted. During *both* eras, therefore, military women had a disproportionately higher prevalence of experiencing several ACE categories. During the AVF era, 31 percent of military women had ACE scores of 0 (compared with 37 percent of nonserving women), while 28 percent had ACE scores of 4 or more (compared with 20 percent of nonserving women). A similar spread existed during the draft era.

These findings are consistent with earlier research suggesting that military populations may experience more ACEs, because they may enlist to escape violent, abusive, or dysfunctional home environments. Although other high-stress professions haven't received as much systematic study, the existing data suggest similar consequences later in life.

All of this empirical research may actually underestimate the true impact of ACE exposure. That's because many other types of childhood trauma are not captured in the ACE survey, such as bullying, poverty, racism, the death of a parent, homelessness, foster care, or surviving an accident, fire, or natural disaster. Furthermore, adults—especially men—tend to underreport their ACE experiences. This makes sense, given the wider disowning of trauma and mental health stigma that exists in our culture, as Chapter 2 explored. This underreporting tendency may be especially potent in high-stress professions, given understandable concerns about how reporting such experiences might hurt someone's chance of getting a security clearance.

For instance, although I experienced many ACEs, I never reported any of these experiences to the military's medical system while in uniform or to doctors after my service. At that time, I didn't even know what ACEs were—or what developmental effects they'd had on my neurobiology. Todd and Julio, who also have high ACE scores, were also completely oblivious to their effects until they were being trained in MMFT. I've observed this pattern frequently in hundreds of one-on-one conversations with MMFT participants. That's because, in our culture, our thinking brains are especially likely to disregard, disown, and devalue the patterns we wired in response to childhood stress and trauma. As Chapter 2 explained, our thinking brains don't want to focus on our past pain and the survival strategies we developed to cope with it; they much prefer to focus on the future, such as our life goals or external self-esteem boosters (such as fame, physical prowess, relationships, or possessions).

Nevertheless, these childhood patterns can continue to impair our ability to cope adaptively with stress and negative emotions as adults, especially when we aren't aware of them. Furthermore, because the treatment of stress disorders and trauma among adults often addresses recent events exclusively, it's easy for clinicians and individuals themselves to miss the importance of looking back to our earliest experiences to craft the best treatment plan today.

WHY DOES CHILDHOOD ADVERSITY HAVE SUCH ENDURING EFFECTS?

So what are the common neurobiological effects that come from experiencing childhood adversity or growing up with depressed, abusive, neglectful, addicted, or traumatized parents? Voluminous research points to several common adaptations.

Empirical research shows that early-life chronic stress leads to two structural changes in the developing brain. First, children with early-life chronic stress are more likely to develop larger and hyperreactive amygdalae, the survival brain region responsible for neuroception. Second, they are more likely to develop a smaller prefrontal cortex (PFC), the thinking brain region that controls executive functioning and assists with top-down regulation of stress and emotions.

In effect, these children build the survival brain's capacity for quickly appraising danger at the cost of thinking brain development. These adaptive shifts toward System 1 (survival brain functions) at the expense of System 2 (thinking brain functions) have *almost two decades* of conditioning before many thinking brain functions even develop. Not surprisingly, these children tend to show impairments in executive functioning, conscious memory and learning, emotion regulation, and impulse control.

It's just like that study in Chapter 3 of troops who deployed to combat in Iraq. After their deployment, these troops' brains showed greater capacity

for quick reaction time, which came at the expense of verbal ability, attention skills, and working memory. Think about it: If *one* year of hypervigilance can do that to an *adult* brain, imagine what happens to a still-developing
brain that marinates in childhood adversity for eighteen years.

Furthermore, the larger, hyperreactive amygdalae linked with childhood
adversity increase the risk of developing anxiety disorders and PTSD. If
someone first experiences an anxiety disorder, depression, or another mood
disorder during childhood or puberty, research shows that he's also more
likely to experience additional mental health problems later in life, compared to someone whose first episode occurs after she turns twenty.

Those of us who experienced ACEs are also more likely to have *faulty
neuroception*, such as perceiving a situation to be threatening when it's not—
or, conversely, perceiving a situation to be safe when it's actually quite dangerous. For instance, children who witness frequent conflict between their
parents are significantly more likely to read neutral facial expressions as
threatening. Our survival brains are especially sensitized to external or internal cues that unconsciously remind us of earlier stressors—especially if
those earlier stressors were chronic or repeated, or if we perceived ourselves
as helpless or lacking control. This survival brain conditioning also increases
our susceptibility to *kindling*, discussed in Chapter 5. With kindling, the
internal components of a memory capsule—such as a worried thought, racing heart, or other physical sensation—can trigger stress activation completely independent of any external cues. Each time kindling occurs, it
"ups the ante" and increases our internal sensitivity—so that, over time, the
survival brain neurocepts danger even when the external situation is actually safe.

In the nervous system, chronic stress and trauma during childhood compromise the development of our ventral PSNS circuit. That's because, as
Chapter 6 explained, the ventral PSNS continues to develop through adolescence. With a hypervigilant survival brain and an underdeveloped ventral
PSNS, we're more likely to "fall back" quickly to the second (fight-or-flight,
SNS) and third (freeze, dorsal PSNS) lines of defense. We're also less likely
to feel safe with others and to develop supportive, trusting, and satisfying
relationships. We're likely to have difficulty recovering completely after
stressful or traumatic situations, because the ventral PSNS also controls

the vagal brake and recovery functions. Thus, we tend to have lower thresholds for distress and less capacity for downregulating stress and negative emotions.

Third, childhood adversity impairs and dysregulates our *endocrine (hormone) system*—especially the HPA axis, which controls our stress hormones. With a dysregulated HPA axis, we produce abnormally high or low levels of stress hormones, a finding shown across many species after early-life stress and trauma. HPA axis dysregulation often ripples across the whole endocrine system, possibly upsetting metabolism and reproductive functioning. Diabetes, thyroid problems, and sexual dysfunction are all examples of endocrine system dysregulation.

Fourth, childhood adversity also adversely affects the *immune system*. As Chapter 3 explained, chronic stress and trauma during childhood can lead to detrimental epigenetic changes in the immune system that create chronic inflammation. Specifically, the macrophages—and microglia, specialized macrophages in the brain and nervous system—get programmed in a dysregulated way. Remember that macrophages are responsible for recognizing and destroying "bad guys," including infections, damage, and dead cells. After childhood stress, macrophages become extremely effective at turning inflammation on and less effective at turning inflammation off. These hyperactive macrophages also continue to release inflammatory cytokines to turn inflammation on, long after the infection, toxin exposure, injury, or physical trauma that triggered them is gone.

In turn, chronic inflammation can manifest in many different ways later in life, including chronic pain, fibromyalgia, chronic fatigue syndrome, arthritis, back pain, chronic headaches, migraines, eczema, psoriasis, cardiovascular disease, asthma, allergies, irritable bowel syndrome, and insulin resistance, a precursor to type 2 diabetes.

Chronic inflammation also underlies neurodegenerative diseases, including depression, anxiety disorders, PTSD, schizophrenia, Alzheimer's disease, and multiple sclerosis (MS) and other autoimmune diseases.

Chronic inflammation may be exacerbated by the dysregulated HPA axis, which may be overproducing or underproducing cortisol and other stress hormones. During acute stress, cortisol boosts immunity in the short term. However, chronic stress can suppress immunity—in part through dysregulated cortisol production. Cortisol *over*production has been linked with depression, type 2 diabetes, active alcoholism, anorexia, hyperthyroidism, panic disorder, and obsessive-compulsive disorder. Conversely, cortisol *under*production has been linked with PTSD, fibromyalgia, chronic fatigue syndrome, hypothyroidism, allergies, asthma, rheumatoid arthritis, and other autoimmune diseases.

Fifth, childhood adversity also alters the brain's *dopamine system*—increasing our vulnerability to addictive behavior: gambling, shopping, sex, and substances, especially alcohol, nicotine, and substance abuse. Dopamine is the one of the "feel-good" neurotransmitters. Not surprisingly, it plays a major role in motivation and goal-directed behavior, on the one hand, and procrastination, craving, and addiction, on the other.

The tendency toward addiction after childhood adversity likely comes from three related neurobiological adaptations. To begin with, the poorly developed PFC has less capacity for impulse control and top-down regulation of negative emotions, which may lead someone to look to addictive behaviors or substances to cope. That's because substances like alcohol, nicotine, and other drugs help dampen the chronic high arousal levels.

Another problem is that chronic stress produces high levels of stress hormones, which deplete dopamine. In turn, dopamine depletion increases drug, nicotine, and alcohol craving.

Finally, the dopamine system itself develops in an impaired manner, leading to relatively fewer dopamine receptors in the brain. And as Gabor Maté, a physician who treats and researches addictions, explains, "When our natural incentive-motivation system is impaired, addiction is one of the likely consequences."

When an addiction is just starting, the external stimulus—the addictive behavior or substance—will trigger and flood the brain with artificially high levels of dopamine. This helps the person fill his dopamine deficit and thereby mobilize motivation and energy. Over time, however, the brain's dopamine system becomes "lazy." Rather than functioning near full capacity, the dopamine system comes to rely on the artificial, external boosters. Substance abuse may also contribute to the loss of the brain's dopamine receptors, further impairing the dopamine system. Together, these dynamics help explain why someone will build up a "tolerance"—and thus require ever larger "hits" of the external substance or behavior to fill her dopamine deficit.

Childhood dysregulation of the dopamine system and the HPA axis may also contribute to depression in adulthood. In addition to being linked with chronic inflammation, depression is also linked with high levels of stress hormones and low levels of dopamine—both of which are more common after chronic stress without recovery. Dopamine's role in depression makes sense, because depression's defining symptoms are apathy and the inability to feel and pursue pleasure. Of course, these are also symptoms of dopamine depletion. Thus, it makes sense that depression and substance abuse, or other addictions, often occur together.

A sixth consequence of chronic stress and trauma in childhood can be *learned helplessness*, which actually helps explain the higher rates of chronic depression among people who suffered childhood adversity.

In animal studies, learned helplessness gets conditioned by exposing animals to "inescapable shock"—usually from being made to sit on an elec-trified metal grid. When jolted with electricity, initially the animal tries to escape but is blocked by a barrier. After being impeded from escaping dur-ing successive shocks, the animal makes fewer attempts to escape, eventu-ally "falling back" to freeze (dorsal PSNS). Then the traumatized animal simply accepts the shocks passively. The traumatized animal will remain immobile on the electrified grid without trying to escape, even when the researchers remove the barrier. In effect, the traumatized animal has learned to perceive that safety is not possible—even when it actually is.

Humans can also condition learned helplessness, especially when young children experience repeated terrifying events during which they have little control. Learned helplessness may contribute to depression in adulthood through distorted survival brain implicit learning. In effect, the trauma-tized survival brain generalizes from past experiences—when the person was truly powerless—to believe they are still powerless, helpless, and lack-ing agency now. These beliefs can contribute to apathy, passivity, "stuck-ness," and the inability to mobilize energy to change or adapt to the current situation—all symptoms often accompanying clinical depression.

Learned helplessness can also contribute to conditioning a default sur-vival strategy of freeze, submission, paralysis, or extreme procrastination in the face of *any* threat—even when standing one's ground, taking action, or getting away would be more effective. If learned helplessness is still being cued later in life, then the person has not yet fully recovered from those earliest events when this default programming was conditioned. Until the survival brain fully recovers, the person will tend to default to freeze during threatening situations in the future—further reinforcing the learned helplessness.

A final enduring consequence of childhood adversity is dysregulation of the *endogenous opioid system*, the endorphins that get released when we experience high stress arousal levels. With dysregulation of our endogenous opioid system, we're more likely to rely on *adrenaline-seeking behavior*, such as Greg's compulsive extramarital affairs and Todd's dangerous motorcycle driving in Chapter 2.

We're also more likely to experience *trauma reenactment*. With trauma reenactment, we may find ourselves mysteriously re-creating situations that echo aspects of old traumatic events. Trauma reenactment can partly be explained by faulty neuroception—inaccurate perceptions of danger and safety—and partly by distortions in the survival brain's implicit memory system.

Remember that the physical sensations, emotions, and body movements that we experienced during earlier traumatic events are stored as mostly unconscious cues in our implicit memory. These implicit memories may contribute to our choosing activities, relationships, or situations that allow us to reenact the previously encoded trauma-related experiences, emotions, body sensations, and beliefs. Such reenactments may occur in intimate relationships, work situations, repetitive accidents, bodily symptoms, psychosomatic diseases, and other seemingly random events. Importantly, someone engaged in trauma reenactment can play the role of either perpetrator or victim.

Why do we do this? That's still a matter of debate. Since Freud, some clinicians and researchers have suggested that we reenact traumatic events in an effort to heal them—unconsciously putting ourselves into similar, or symbolically similar, circumstances as the original event with the hope that *this* time, our mind-body system will finally respond in a way that successfully overcomes the danger. Because this survival brain process is unconscious, however, we're more likely to simply reenact the implicit memory and its patterned programming, emotions, and beliefs without healing them.

More recently, researchers have argued that trauma reenactment may also be linked with dysregulation of the endogenous opioid system. For example, many traumatized people seek out high-arousal activities and experiences that would repel most people. They also often complain "about a vague sense of emptiness and boredom when they are not angry, under duress, or involved in some dangerous activity." In fact, the rush of endorphins that accompanies these high-arousal states may make them addictive—hence the term *adrenaline junkies*—often leading dysregulated people to compulsively seek out activities that bring this rush.

Such adrenaline-seeking activities include extreme sports, skydiving, high-speed motorcycle riding like Todd enjoyed, extramarital affairs like the ones Greg enjoyed, and violent behavior toward others. Interestingly, they also include self-harming behaviors, such as head-banging, cutting, and self-starvation. All of these activities may trigger an endorphin rush and provide relief from anxiety. Even the recreational runner, biker, or surfer who feels compelled to get a daily endorphin "fix"—or else feel anxious and jittery—is actually self-medicating their underlying dysregulation. *If compulsion is involved, it's likely pointing to dysregulation.*

A similar pattern exists in abusive relationships, where tension gradually builds and culminates in an explosive, violent outburst—followed by the calm, loving respite of "making up." This cycle not only reinforces the traumatic bond between the abuser and the victim, it also triggers an endorphin release—and its associated physiological calm—for both parties. The tendency to reenact abusive relationships is more likely when trauma occurred in our early attachment relationships, perhaps because of the important role endorphins play in our attachment bonds.

CHILDHOOD ADVERSITY'S EFFECTS

Wow, huh? *So many different pathways* by which a mind-body system, adapted to chronic stress and trauma early in life, gets primed for hypervigilance, hypersensitivity, hyperreactivity, inflammation, and dysregulation.

Although any of these pathways could be activated after chronic stress and trauma at any age, the biggest and most enduring effects occur with chronic stress and trauma during childhood and adolescence, when our mind-body system is still developing. These neurobiological adaptations are wide-ranging, including structural changes in the brain; dysregulation of the autonomic nervous system and the endocrine, immune, dopamine, and endorphin systems; and cellular-level epigenetic changes that prime the body for chronic inflammation and disease.

All of these neurobiological adaptations ripple into our behavior. They make it more likely that someone will have difficulty paying attention, making effective decisions, controlling impulses and cravings, and regulating stress and negative emotions. They also create compulsive, unconscious tendencies toward addictions, trauma reenactment, and adrenaline-seeking, self-harming, or violent behavior.

Taken all together, it's not a pretty picture.

Now that you understand the underlying mechanisms by which childhood adversity changes the mind-body system, you won't be surprised by the empirical evidence of its effects. Like insecure attachment, childhood adversity sets the trajectory for a narrow window. We're more likely to build allostatic load, which can eventually manifest as physical and mental health problems.

Remember that ACEs have an additive effect—the more ACE categories someone experiences, the greater their likely dysregulation. For school-aged children, ACE exposure has been linked with obesity, ADHD, and other learning and behavioral problems. Indeed, *most* children with an ACE score of 4 or more experience learning and behavioral problems in school, compared with *just 3 percent* of kids of an ACE score of 0.

For adults, ACE exposure has been linked with a wide range of health problems. Compared with adults without ACEs, adults with ACE histories are more likely to be obese and have a shortened life span. They're more likely to develop diabetes, cancer, heart disease, high blood pressure, liver disease, bronchitis/emphysema, allergies, asthma, ulcers, and arthritis/rheumatism. They're more likely to contract sexually transmitted diseases and have unintended pregnancies. They're more likely to be tobacco users and abuse alcohol and other substances. They're more likely to experience depression, anxiety, and PTSD. They're also more likely to attempt suicide.

The additive effects of ACEs are especially poignant by adulthood. For instance, roughly one in eight people with no ACEs experience chronic depression. In contrast, two thirds of women and more than one third of men with an ACE score of 4 or more experience chronic depression. Likewise, compared to adults with no ACEs, adults with an ACE score of 4 or more are seven times as likely to be alcoholics and thirteen times more likely to attempt suicide.

Not surprisingly, many empirical studies show that people with ACE histories end up experiencing abuse and violence again—or abusing others—as adolescents and adults. Overall, roughly a third of people abused as children grow up to become abusers themselves.

Adults abused as children are also prone to enter violent relationships in which they're abused again—increasing the chance of a violent childhood for their own offspring. Indeed, being physically or sexually abused as a child—or witnessing domestic violence against a parent—more than doubles the chance of experiencing intimate partner violence as an adult. Compared with men who didn't witness domestic violence growing up, men who did are seven times as likely to abuse their own partners in adulthood.

Compared to nonviolent criminals, violent criminals are more likely to have witnessed violence as children. Among minority youth age nine to nineteen and living in high-poverty neighborhoods, those who were repeatedly exposed to violence during the five-year study period were *31.5 times as likely* to engage in violence themselves. Witnessing gun violence doubles the chance that teenagers will perpetrate serious violence themselves within the next two years. Thus, it's not surprising that childhood trauma increases the likelihood of arrest and recidivism, and inmates tend to have higher ACE scores than other adults.

These data are truly devastating in their wide-ranging effects.

COPING AFTER CHILDHOOD ADVERSITY

The final layer to add to this picture, which also contributes to narrowing the window, is this: Whenever we confront a stressful or traumatic event, we bring to that event all of the tools we have available at that time. Some of these tools are internal, such as our physical size and strength, our intellect and education, our capacity for self-regulation, or our faith. Some of these tools are external, such as our network of social support, financial resources, or power within institutions.

With this in mind, it should be clear that as we grow up, over time we gain access to more and more tools, internally and externally. Indeed, the tools that stressed infants and young children have at their disposal are extremely limited. They have not yet wired many internal tools—and they don't have access to many external tools, either.

Thus, for those of us who encountered a lot of childhood adversity, we wired powerful, deeply conditioned survival strategies *when we had the fewest internal and external resources available to us.* Since we were small, we couldn't really defend ourselves physically. With our immature thinking brains, we didn't yet have much capacity for top-down self-regulation to help us soothe ourselves and recover after stress and intense emotions. We also didn't yet have the reasoning and perspective-taking skills to understand that the neglectful or abusive behavior we experienced probably had very little to do with us. For some of us, turning to an adult wasn't an option, either, because that might run the risk that things would get worse.

So what could we *do* with all of that chronic stress activation? How does a child with so few internal or external resources survive?

Well, we could override how we were feeling by compartmentalizing, suppressing the physical and emotional pain, or disconnecting from our bodies and retreating into our heads. We could think that what was happening was all our fault, that we weren't (fill in the blank) enough. We could ruminate and worry.

We could try to assert control in the few ways available to us, such as controlling our food intake (too much or too little) or developing obsessive-compulsive habits, like pulling out our hair.

We could submit to what was happening and not make waves, by becoming compliant or pleasing people, with the hope that we might survive, as the first priority, and maybe even get what we need, as the second priority. We could become invisible or get sick.

We could check out, distract, dissociate, or numb ourselves—with television, books, video games, mobile devices, food, nicotine, alcohol, or other substances. We could suck our thumbs, rock our bodies, bang our heads, wet the bed, or hide under the covers.

We could act out, by throwing temper tantrums, bullying or fighting others, breaking rules, cheating, smoking, sleeping around, abusing drugs or alcohol, stealing, or engaging in violent behavior. Finally, we could hurt ourselves—by not eating, cutting, or attempting suicide.

For instance, to cope with the traumatic events I experienced early in my life, I used to bang my head against the wall. My parents tell me this head banging started when I was about six months old, as soon as I could sit up in the crib. Although I felt ashamed and repeatedly tried to break this habit from middle school onward, I continued it sporadically during stressful periods into early adulthood. Although it wasn't a conscious choice, I chose this habit as one of my primary coping strategies because it provided self-soothing, left no visible marks, and didn't involve any substances that might cloud my vigilance or my judgment.

From an adult perspective, these kinds of coping mechanisms may seem "childish"—and in some ways, they are. When we cope with stress using these kinds of strategies, it's a signal that we're relying on some of our *oldest* coping techniques, the ones we wired when we were young and didn't have all the intellectual, emotional, physical, spiritual, and relational resources we have available now.

Yet because we taught ourselves to cope with childhood stress and trauma using such methods, the pull to keep relying on them as adults remains extremely strong—especially when we're stressed, burned out, sleep-deprived, or emotionally unsettled. This, too, is neuroplasticity in action. That's why, without a deliberate intention to choose something different—something that takes advantage of the full range of internal and external resources we have today—we're still likely to default back to these oldest coping habits when we get stressed or triggered. *The Grand Canyons we wired during childhood chronic stress and trauma are the deepest canyons of them all.*

Unfortunately, these habits tend to keep our windows narrow. Why? The one thing all of these coping strategies share is that *none of them actually helps the mind-body system fully recover from stress and trauma.* While they can soothe our feelings of distress in the short term, none of them actually discharge stress activation and provide true recovery. As a result, our window stays narrow—further building allostatic load and further limiting the range of strategies we can draw on during stress in the future.

SURVIVORS WITH NARROW WINDOWS

Those of us who experienced childhood adversity are survivors. But we also wired some *very deep and enduring* neurobiological patterns during critical years of our mind-body system's development. This neurobiological profile— survivors with large allostatic loads—has some unique strengths and weaknesses.

Someone living this trajectory may appear extremely resilient and adaptive—almost superhuman in their capacity to weather adversity and perform challenging tasks. In fact, their mind-body system developed the capacity to function well, even thrive, during sustained high arousal states— *which is an appropriate response for extreme-stress environments.* This is a mind-body system that unconsciously craves crisis.

However, the resilience of someone living this trajectory is usually quite brittle, because their mind-body system's default high arousal state is mismatched for any circumstances where high arousal is not necessary, that is, most of daily life. Ironically, their suffering can be exacerbated by their self-sufficiency, which may isolate them and make it difficult for them to ask for support or help.

For people who don't come from such upbringings, it can be hard to understand why someone would repeatedly put themselves into harm's way for work, stay in abusive relationships, remain obese, destroy their life over an addiction, or commit violent crimes. Too conveniently, we judge their choices from a place of ignorance—not fully recognizing that each of us has the potential to exhibit the same behavior, given appropriate conditions. The more we collectively deny the understandable reasons for their behavior—and the more we condemn them for that behavior—the more we trap them into their current patterns and stifle their efforts to change.

The evidence in this chapter clearly demonstrates that childhood poverty, abuse, neglect, trauma, and violence have lifelong effects on the width of our window. Although childhood experiences don't determine destiny, our earliest social environments do deeply influence the trajectories of our windows. There are clear neurobiological reasons why someone from a challenging upbringing is more likely to have difficulty making good decisions, controlling impulses, regulating emotions, developing supportive relationships, and recovering after stress.

For those of us from such upbringings, understanding why we may be hurting and dysregulated today—connecting the dots from our earliest experiences until now—doesn't make the pain go away or heal our dysregulation. However, it can help us quiet our self-judgments, self-hatred, and shame about our physical and mental health conditions, addictions, and other coping behaviors. In fact, these limiting beliefs and emotions only perpetuate the hyperreactive and dysregulated programming we wired during early-life stress—and the coping mechanisms we adopted then that may no longer be serving us today. For instance, such shame and self-judgment fuels yo-yo dieting, binge-and-bust exercise regimens, extreme procrastination followed by binge work on deadline, and other stop-start attempts at coping with stress in new ways.

This is where the warrior qualities of wisdom and courage can help. Wisdom helps us understand how our neurobiology got wired in childhood and why it was so adaptive. It helps us appreciate that our neurobiological profile has unique strengths and weaknesses. Although early-life stress and trauma gave us a strength that many others may never develop, we also have less wiggle room with our lifestyle choices. Compared with people who wired wide windows in childhood, we'll always be more vulnerable to dysregulation, inflammation, and allostatic load. That's just how we got wired. Being aware of and accepting this truth about ourselves opens up the possibility for change. With this understanding, we can draw on courage to make deliberate efforts to rewire those neurobiological structures and widen our windows. We can experiment with new internal and external tools to access a wider range of adaptive responses during stress. We can learn to interrupt the survival brain's default programming and the coping habits we adopted in childhood.

Finally, we can remember that inherent in a narrow window today is also the possibility of a wide window tomorrow. The silver lining of childhood adversity is the potential for the widest windows of all.

Before starting the next chapter, please take some time to reflect on your own experiences related to the first pathway to narrowing the window. In your journal, you might reflect on your grandparents' and parents' life experiences, their attachment styles, and the width of their windows. You might reflect on the life stressors that your parents or other care providers were coping with during your early childhood, especially whether they were dealing with any unresolved trauma or loss.

Next, you could also take the ACE survey, which is available online, to calculate your ACE score. As noted, the ACE survey does not include all of the sources of childhood chronic stress and trauma that might have enduring effects on the mind-body system. With this in mind, Table 7.1 contains a longer list of potentially stressful and traumatic events that might have narrowed your window during your childhood, for you to consider in your reflection:

Table 7.1 Adverse Childhood Experiences That May Narrow the Window

Premature birth (especially before 30 weeks) and/or a difficult birth	Adoption
Childhood surgery, extensive hospitalization, or medical emergency (you or a sibling)	Foster care, juvenile detention, or incarceration
Frequent physical abuse	Witnessing a family member being abused, attacked, or killed
Frequent emotional abuse	Having a family member with mental illness, including depression, PTSD, and anxiety disorders
Bullying	Having a family member attempt and/or die by suicide
Sexual molestation, abuse, or rape	Having an alcoholic or addicted family member

Poverty and/or chronic hunger	Having a family member commit a serious crime or be incarcerated
Homelessness	Losing the family home or possessions through bankruptcy, parental joblessness, gambling, war, or natural disaster
Racism	Experiencing—or having a family member experience—a natural disaster, war, terrorist attack, genocide, mass shooting, accident, or another catastrophic event
Parental infidelity, separation, or divorce	Being forced to flee the family homeland or take political asylum
Domestic violence	Abandonment or forced extended separation from family members
Death of a parent, sibling, or other important care provider	

Finally, review the section of this chapter on the mechanisms by which childhood adversity has enduring effects and write down the ones that apply to you. You might also reflect on how you coped with stress and trauma as a child—and whether you still rely on any of those coping habits when you're stressed today.

CHAPTER EIGHT

Shock Trauma

"Martin," a dedicated first responder, came to an MMFT course because he was seeking new tools for chronic pain. In our one-on-one interview, I asked Martin to tell me when his pain had started. Since a "small fender bender" when his car was hit the previous year, he said, he'd been experiencing back pain and a stiff neck. Most mornings, he'd also awaken with a killer headache, which he thought came from grinding his teeth and clenching his jaw during the night. Usually, Martin told me, he could just "deal with it"— suck it up without taking anything. Recently, however, Martin felt like his pain was getting worse. Regular visits to the chiropractor weren't helping. Now, more days than not, he was popping ibuprofen like candy.

In the last few months, Martin had also started noticing new symptoms that would come and go—dizziness, ringing in his ears, and tingling in his arms. He was also sleeping poorly, waking up with nightmares. As a result, he was having trouble concentrating and remembering things. He was especially frustrated about missing appointments and misplacing his keys.

"Has anything shifted at work or at home recently?" I asked. "I'm wondering if there are any new stressors that started at the same time as the new symptoms."

"Not really," Martin replied. "Things are great at home. At work, we're preparing for a big inspection, but it's nothing we can't handle. I'm working slightly longer hours, maybe, but it's not really stressful."

"Well, even if your thinking brain doesn't find the inspection prep stressful, I'm wondering if your survival brain does. What does it involve?" I asked.

"Mostly it's a paperwork drill, but we also have to requalify on certain duties. Kind of like the predeployment training that the troops you've trained were doing."

"That makes sense," I replied. "So here's a weird question for you: Does any of that recertification involve driving?"

"Well, yes." Martin looked surprised and sat back in his chair. "How did you know?"

I smiled. "Another weird question for you: Do you remember what your nightmares are about? If so, would you be willing to tell me?"

Martin grinned. "Huh. The nightmares *have* been about accidents, but not necessarily car accidents. I only remember one where I was in a ten-car pileup on the highway. In the rest of them, I'm part of the team responding to big incidents, but either we get there too late or we can't stop what's happening. I wake up sweaty, with my heart racing. And then I can't get back to sleep for hours. But how does this fit with the upcoming inspection? And with that small fender bender last year? There's just no comparison between these things!"

"That's right, for your thinking brain, there *is* no comparison between these things," I replied. "But I believe that for your survival brain, these things *are* all related. You were the one hit in the fender bender, right?"

Martin nodded.

"So that was a stressful experience where you were not in control. And when the survival brain feels like it's not in control, that's when we experience trauma," I said. "It doesn't matter that it was just a fender bender. It was still traumatic for your survival brain and body. So while *you* aren't worried about this upcoming driving requalification, it appears that your survival brain *is*. It's creating all these new symptoms to tell you that. We just have to listen to its messages and help it finish recovering completely from the fender bender."

"But it really was no big deal." Martin frowned. "I'm over it."

"Again, *you* might be over it, but I assure you, your survival brain is not. Until we help it recover completely—at its own pace—you will keep having symptoms. You came to MMFT to learn new tools for your chronic pain, right? That requires letting your survival brain lead—and trusting that when you do, your survival brain knows exactly what it needs for complete recovery."

THE SECOND PATHWAY TO NARROWING
THE WINDOW

The second pathway to narrowing our window comes from *shock trauma*, which happens during acute or unexpected events that have a sudden, major effect on the mind-body system. During shock trauma, we experience too much stress arousal and emotional intensity too fast, thereby overwhelming our window of tolerance. Shock traumas include the things that we conventionally think of as "traumatic"—hurricanes, earthquakes, other natural disasters, terrorist attacks, mass shootings, combat, rape, kidnapping, and political captivity. Shock trauma's effects are magnified if the event is also experienced by significant others, including family members, friends, and coworkers.

As Chapter 2 explained, most of us are willing to grade the events in which shock trauma commonly occurs as "traumatic." Often these are unexpected events involving many individuals or the whole community, where many people are killed or wounded. These characteristics of the event make it more likely that *both* the thinking brain and the survival brain will perceive us not in control. With such events, often the best we can do is simply hang on while it washes through our lives and then pick up the pieces afterward.

Because these stressors are usually so big and extraordinary—Capital T traumas—our thinking brain is also more likely to consider how the event is affecting our mind-body system. It's less likely to devalue the source of our stress, so we're also likely to take active measures to reinforce recovery afterward. For instance, we might engage in meditation, prayer, or other spiritual practices. We might seek help from loved ones, a chaplain, a therapist, or a doctor. Or we might build in explicit time for physical and emotional recovery.

Importantly, however, as Martin's story suggests, we can also experience shock trauma during "smaller" events like fender benders or "minor" surgery. Like any trauma, *shock trauma is not in the event—it's in the mind-body system of the person experiencing the event.* Thus, even if the event is "no big deal" for our thinking brain, if the survival brain perceives us to be helpless, powerless, or lacking control, then we're still likely to experience too much stress arousal too fast, moving us outside our window. *This* is the defining feature of shock trauma—not any particular characteristics of the event that precipitates it.

Most of us don't experience shock trauma on a regular basis. If we have a narrowed window, however, we may experience shock trauma more frequently than our thinking brain believes. However, when our thinking brain dismisses a traumatic event as "no big deal," as Martin's did, we're also more likely to experience lingering detrimental effects afterward. Even with "small" shock trauma, unless the survival brain and body are able to completely recover afterward, we will still be building allostatic load and narrowing our window. Like Martin, we will eventually experience symptoms of dysregulation as a result.

SHOWING UP TO LIFE WITH THE MIND-BODY SYSTEM WE'VE WIRED UNTIL NOW

During crises and other events where shock trauma is possible, people naturally arrive to the event with different windows of tolerance. Thus, they bring different capacities for accessing and employing thinking brain functions and social engagement during the crisis. The width of the window we bring to a crisis depends on our allostatic load from early life experiences, as well as from later experiences of chronic stress and trauma without adequate recovery.

In other words, we show up to life with the mind-body system we've wired until now.

There's an excellent example of this dynamic among Israeli leaders' performance during the 1973 Yom Kippur War. To set the stage, in 1967 Israel had launched a stunning preemptive strike against its Arab neighbors, capturing significant territory from Egypt, Syria, and Jordan—including the Sinai Peninsula in the south and the Golan Heights in the north—which more than doubled the size of the territory Israel controlled. Understandably, the Arab states were humiliated by this military defeat and vowed to reclaim their lost territory.

Over the next several years, the Israeli public remained anxious about the possibility of an Arab attack, especially because 80 percent of the ground troops in the Israeli Defense Forces (IDF) are reservists who would have to be mobilized to defend the border without much warning. Nevertheless, based on analysis from the Israeli intelligence community, Israeli leaders dismissed several warnings in the spring and summer of 1973 that war was growing more likely.*

When Egypt and Syria finally attacked on October 6—on the first day of Yom Kippur, the holiest day in the Jewish calendar—Israeli leaders were caught by surprise. The invasion involved coordinated attacks in both the Sinai Peninsula in the south and the Golan Heights in the north. Over the next twenty-four hours, Egypt and Syria pushed deep into these territories with more than one hundred thousand troops, while Israel scrambled to respond. It took Israel two days to organize a successful counterattack against Syria and three days to halt the Egyptian advance. Eventually, Israel's forces gained the upper hand militarily and defeated Syria and Egypt. Fighting finally stopped in late October, with both sides having sustained heavy casualties.

As this background suggests, the invasion clearly was a shock trauma for Israeli civilian and military leaders. Nevertheless, they showed significant variation in their ability to respond during the crisis. On the one hand, Israeli Defense Minister Moshe Dayan—with more military and leadership experience, more positional authority and, as the recognized hero of the 1967 war, arguably more social influence—made remarkably poor decisions during the war. In contrast, the IDF chief of staff, Lt. Gen. David "Dado" Elazar, exhibited such level-headedness that he assumed a more powerful war decision-making role than his position would usually warrant. Why?

Interestingly, researchers have studied Israeli leaders' varying stress levels

---

* In his book *Smarter Faster Better*, Charles Duhigg explains why the Israeli intelligence community developed the blind spot that led it to dismiss credible warnings that war was likely in 1973. After the 1967 war, the leader of Israeli military intelligence, Eli Zeira, created what was known as "the concept" to assess Arab intentions for initiating war. First, Zeira assumed, Syria would not go to war unless Egypt joined them. In turn, he assumed, Egypt would not go to war until it received Soviet fighter-bombers (for neutralizing the Israeli Air Force) and Scud missiles (for holding Israeli cities hostage and thereby deterring attacks against Egyptian infrastructure). Because Egypt never received the Soviet fighter jets it wanted, Zeira and his team continued to dismiss warnings in 1973 that war was likely—even as Egypt acted in ways to further this misconception.

and decision-making capacity on the second day of the war, during the worst of the Arab attack. Although the researchers didn't frame their analysis in terms of the window, the leaders' decision-making behavior tracks with their windows of differing widths. While situational pressures on the Israeli leaders were similar, their decision-making behavior during the war showed considerable variance in their ability to cope with stress and to make strategically optimal decisions.

For instance, Defense Minister Dayan carried a large allostatic load—from his PTSD from previous combat experiences, as well as from a 1968 accident that left him with severe back pain, incessant headaches, and a single eye. Given these experiences, we would expect that Dayan's window was narrowed, so that during the war he likely would be aroused beyond his window, with survival brain processes—stress and emotions—predominating his decision making.

Dayan's behavior during the war demonstrates this. As the researchers note, Dayan's behavior "indicated that he was under an extreme level of stress and revealed signs of panic. . . . Dayan's form of expression, less than 24 hours after the war had begun, revealed an immense . . . fear that Israel was under an existential threat."

Not surprisingly, therefore, Dayan made poor decisions during the war, including what the researchers judged as "probably the single worst decision made throughout the war"—considered a "grave mistake" both at the time and in retrospect: Dayan aborted an air attack to damage Egyptian surface-to-air missiles in the south, despite its initial success, and diverted air assets to another air attack in the north. The northern attack subsequently failed because of its improvised execution without necessary intelligence support. In effect, Dayan's poor decision neutralized the Israeli Air Force, "render[ing] them unable to provide effective support to Israel's ground forces throughout the war."

Dayan's social engagement skills also showed strain during the war. Even before the war, Dayan was an "individualist" with "hardly anyone he respected enough to keep in confidence." His closest confidants were Prime Minister Golda Meir and her advisor, Minister without Portfolio Yisrael Galili. In fact, the three together formed Meir's "kitchen cabinet" for Israel's most important security decisions. Yet after Dayan's poor decisions during the first two days of the war, Meir and Galili "lost confidence in his judg-ment." Although he was the defense minister, they cut him out of war deci-sion making—choosing instead to let IDF chief of staff Elazar make the final decisions about the Israeli military's retreat.

In contrast, on the war's second day, Elazar "did not show any dramatic or notable signs of distress"—suggesting that he remained inside his win-dow. Not surprisingly, the researchers evaluate Elazar as having made "high-quality decisions" throughout the war. Indeed, Prime Minister Meir and many other observers noted that, unlike Dayan, Elazar remained "like a rock" and maintained excellent judgment throughout. Throughout the war, Meir "preferred his advice over Dayan's," which gave him an "informal sta-tus of seniority over the defense minister."

Elazar was also able to offer and receive social support. He had a small, informal group of generals who functioned as his unofficial staff; impor-tantly, this group was "not homogenous and did not provide Elazar with automatic agreement"—suggesting that he was capable of listening to and learning from diverse viewpoints, even during the extreme stress of the at-tack. In other words, Elazar kept social engagement online during the war—connecting with, advising, and gaining support from a wide network of people throughout.

To be clear, during shock trauma events—including unexpected or acute crises, existential conflicts, or situations where someone's life is literally on the line—*everyone* may initially experience stress arousal and emotional in-tensity that takes them outside their window.

However, after the initial shock passes, people with wider windows, like Elazar, are more likely to have the capacity to downregulate stress arousal and emotional intensity and get back inside their window. The wider their window, the more quickly the shock will pass—and the more quickly they will be able to access effective decision making and other thinking brain functions. They'll also have more capacity to connect with and support other people, using social engagement skills of the first line of defense.

In contrast, people with narrower windows, like Dayan, may be unable to downregulate their arousal and therefore may find optimal decision-making capacity beyond their reach. These people are likely to have stress and emotions driving their decisions for a longer period of time, perhaps even throughout the crisis. Because of their narrower window, they may also "fall back" to the second and third lines of defense, fight-or-flight and freeze. As a result, they're also more likely to experience interpersonal challenges during the crisis, such as withdrawing, snapping at others, or needing excessive reassurance and support.

In other words, events that we conventionally describe as "traumatic" will not affect everyone the same way.

When I first introduced this point in Chapter 2, I used the example of an infantry squad encountering an ambush. When thirteen infantrymen get ambushed, we can be sure that there'll be thirteen different mind-body responses. It can't be otherwise, since there'll be thirteen different windows encountering that ambush—each with its own default survival strategies.

Infantrymen with wide windows will be more likely to keep thinking brain functions online, even during the stressful ambush. They're more likely to assess the situation accurately, coordinate with other squad members, and then implement an adaptive response, such as mounting a counterattack. In other words, although they'll experience a rush of stress arousal, they're more likely to be able to access situational awareness and employ social engagement and fight-or-flight—the first and second lines of defense— as needed, until the group gets to safety.

In contrast, infantrymen with narrower windows are more likely to experience stress arousal levels that take them outside their window during the ambush. For these men, their thinking brain functions may operate in a degraded manner, impairing their situational awareness and decision-making skills. They're more likely to "fall back" automatically to the second and third lines of defense—fight, flight, and freeze—without effectively accessing situational awareness first, in order to discern if their default defensive strategy is appropriate right now. With their ventral PSNS unavailable, they're more likely to have difficulty orienting to their surroundings and coordinating with other squad members. They may feel dazed and confused or become overwhelmed with panic. They may also experience tunnel vision, sound distortions, a sense of time slowing down, or a sense of watching the scene from a distance, all symptoms of freeze.

After the ambush is over and the squad returns inside the wire, there'll also be thirteen different coping strategies afterward. For the infantrymen who came to the ambush with a wide window and allostasis functioning properly, no problem: Although they may remain activated and even experience acute stress issues afterward, eventually their systems are likely to recover and heal completely.

For the infantrymen who came to the ambush with a narrow(ed) window, however, by definition allostasis is *not* functioning properly. For these men, the dysregulating effects of the ambush will be much greater. Afterward, their ability to trust and lean on support from others will be lower. And with their existing allostatic load, it's going to be more difficult—if not impossible—for them to downregulate and fully recover afterward. Instead, the effects of the ambush are likely to linger on in their mind-body system without complete recovery—interacting with earlier, unresolved chronic stress and trauma and adding to their cumulative allostatic load.

Furthermore, for squad members who narrowed their windows through childhood adversity, they're likely to rely afterward on coping patterns they adopted during childhood stress and trauma—those "childish" coping habits I discussed in Chapter 7. For instance, they may self-medicate or numb themselves with nicotine, alcohol, comfort food, or other substances. Or they may use adrenaline-seeking activities—dangerous motorcycle riding, picking fights, sleeping around, or engaging in domestic violence—to get the rush of endorphins that soothes their anxiety. Or they may hurt themselves. Of course, none of these coping mechanisms helps their mind-body systems fully recover. Thus, while their thinking brains may feel more in control, these coping strategies will actually add to their allostatic load.

HOW PREVIOUS CHRONIC STRESS AND TRAUMA AFFECT OUR RESPONSE TO SHOCK TRAUMA TODAY

Developmental trauma, insecure attachment, and childhood adversity increase the likelihood that someone will experience shock trauma later in life. Several factors contribute to this elevated risk, as research in the last chapter explained. First, people coming from such upbringings are more likely to have faulty neuroception, so that they might appraise as "safe" situations that are actually dangerous, and vice versa. They may also have more difficulty making good decisions and controlling impulses. For instance, they may be coping with an addiction, which can fuel a range of high-risk choices.

Second, they're also more likely to be attracted, or even addicted, to high-stress-arousal environments, which feel familiar to their survival brain, nervous system, and body. Thus, they may compulsively seek out relationships, activities, and environments that bring that rush of adrenaline and endorphins that they need to feel calm and in control. Often these are mind-body systems that crave crises and emergencies. Thus, they may seek out a range of personal and professional environments in which feeding their "adrenaline junkie" addiction is socially sanctioned.

For instance, given the disproportionate ACE exposure among U.S. service-members in the AVF era, it's not surprising that recent research shows soldiers joined the U.S. Army with significantly higher rates of pre-enlistment PTSD, panic disorder, ADHD, and intermittent explosive disorder (IED) than those in the general population. Notably, 8 percent of soldiers entered the Army with IED—characterized by uncontrolled attacks of anger—a pre-enlistment prevalence *almost six times* the civilian rate. In addition, more than three quarters of soldiers who currently meet the criteria for a mental health disorder reported that they'd had at least one mental health disorder as teenagers before enlistment. Although some of these soldiers have may entered the service to escape abusive, violent, or dysfunctional home environments, others may have sought this work environment in which their high-arousal states were professionally adaptive.

A final reason why developmental trauma and childhood adversity increase the risk for later shock trauma is that these individuals may be caught in unconscious cycles of trauma reenactment, as I was. As Chapter 7 explained, people coming from such upbringings are more likely to experience abuse and violence again, or abuse others, as adults—increasing their risk of being in situations in which shock trauma occurs. For instance, the risk of being raped as an adult is *six* times as high for women with an ACE score of 4 or more, compared with women who experienced no ACEs. Likewise, compared with nonviolent criminals, violent criminals are more likely to have directly observed violence as children.

Importantly, however, developmental trauma and childhood adversity don't just increase the chance of experiencing shock trauma later in life. *They also increase the risk of developing PTSD or other stress-related physical and mental health problems after the shock trauma.*

One reason for this is that experiencing a freeze response during a traumatic event—what clinicians call *peritraumatic dissociation*—has been shown to be the single biggest predictor of developing PTSD and other stress spectrum disorders later on. And, as the examples in the previous section suggest, someone with a narrow(ed) window is significantly more likely to experience freeze during a shock trauma event than someone with a wide window.

For instance, not everyone who experiences a motor vehicle accident (MVA) will also develop whiplash syndrome. Some people who've experienced an MVA, like Martin, develop a group of symptoms, including spinal pain in the neck and back, ear and jaw pain, headaches, and numbness and tingling in the arms. They may also experience neurological symptoms, such as blurred vision, ringing in the ears, dizziness, and problems with memory and concentration. Finally, they may experience several PTSD symptoms, including nightmares, hypervigilance, excessive startle response, irritability, and anxiety or panic with driving or other accident-related cues. Often symptoms take days or weeks to develop and worsen over time. Indeed, many of Martin's symptoms are part of the stereotypical whiplash syndrome.

Research shows that whiplash symptoms have nothing to do with the speed of the accident, the violence of the crash, or damage to the cars. As Martin's "small" fender bender suggests, trauma is *not* in the event—it's in the mind-body system experiencing the event. For instance, race car and demolition derby drivers sometimes experience modest neck pain after crashes, but they typically show none of the other whiplash symptoms. This makes sense, since presumably they felt in control while they were driving, decreasing their likelihood of experiencing trauma.

Robert Scaer, a neurologist who has treated more than five thousand patients who experienced MVAs, has found a strong link between the people who eventually developed whiplash syndrome and their prior developmental and/or relational trauma. In his practice, the most powerful predictors of prolonged and/or severe postaccident whiplash symptoms include having experienced physical and sexual abuse, a difficult birth, intense medical treatment, or an alcoholic parent during childhood, or having experienced discrimination, harassment, or other relational trauma during adulthood.

I never asked Martin about his early-life experiences during our one-on-one discussion, since I wanted to focus our limited time together on teaching him how to support his survival brain in recovering completely. Given these data, however, it's very possible that Martin had a stressful childhood, too. Even without knowing his early-life history, Martin's symptoms certainly suggest that he experienced shock trauma during the fender bender.

Likewise, there's a growing body of empirical research about individuals who experienced childhood adversity who work in high-stress professions. In this research, not surprisingly, the more ACEs someone experienced, the more severe and enduring their symptoms after a shock trauma or stressful work events.

For instance, several studies with American, Canadian, and British troops and veterans diagnosed with PTSD found that most had histories of childhood adversity. ACE-exposed military service-members are also at greater risk for developing PTSD, depression, anxiety disorders, and alcohol or substance abuse after combat, deployment, or other stressful work-related events.

Several studies with paramedics and police show a similar pattern: Those who experienced childhood adversity showed significantly larger physical and emotional responses after threat exposure and/or traumatic incidents on the job. They also face a greater risk of developing PTSD symptoms later in life. One study even linked ACE exposure with police suicides.

Of course, because he was a first responder, Martin's story also provides an example of these findings with high-stress professions. Having narrowed his window through the car accident, Martin was now experiencing greater physical and emotional symptoms in anticipation of a job stressor, the upcoming inspection. With its corrupted implicit memory from the accident, his survival brain was especially sensitive to cues related to the upcoming driving requalification. In fact, many of Martin's newer symptoms—especially the nightmares—were evidence of kindling, growing physical and emotional symptoms possibly cued by anxiety about the upcoming driving test.

How severe and enduring the detrimental consequences will be after a shock trauma are intricately related to the internal and external resources that we have at our disposal for the recovery process. Just as having a wide window provides internal resources to help us recover after a traumatic event, having a large cash reserve, a steady income, institutional power, and/ or supportive relationships can provide external resources, too.

Conversely, the detrimental consequences after shock trauma tend to last longest for people lacking internal and external resources. I've already explored how narrowed windows contribute to this outcome, but it's also more likely when we lack social, institutional, and financial resources.

For instance, a year after Hurricane Harvey flooded Houston, the most vulnerable and impoverished neighborhoods showed the least signs of recovery; many residents still lived in mold-infested conditions, in temporary trailers, or with others while their homes were still being repaired. One year after the storm, compared with white Houstonians, twice as many blacks and almost three times as many Latinos reported that their homes were still unsafe to live in. Likewise, half of lower-income respondents said they weren't getting the help they needed, compared with a third of those with higher incomes.

SHOCK TRAUMA AS TIPPING POINT

I've observed one final feature of shock trauma in my own life and in teaching many people with narrowed windows: Shock trauma often occurs as a tipping point after a long series of assaults on someone's mind-body system.

Even with narrowed windows, people are often capable of functioning through chronic stress and trauma for long periods of time. For instance, someone may have survived a challenging upbringing. Now, as an adult, they may endure sleep deprivation, overwork, and relentless job deadlines, while juggling family responsibilities. Or they may regularly confront relational trauma through racism, sexism, heterosexism, poverty, discrimination, exclusion, or harassment. Living this way for years (or even decades), they may not be thriving exactly, but they're keeping it together and functioning, nonetheless.

Someone with this profile is actually living in a precarious balance, however, because with their narrowed window, they lack the internal resources to cope effectively with any new unexpected shocks or crises. They've been steadily accumulating allostatic load over several years. Thus, should a shock or crisis occur—a car accident, a terrorist attack, a fire, a hurricane, a school shooting, the unexpected loss of a loved one—they're already running on fumes. In turn, it's much more likely that their survival brain will feel powerless, helpless, or lacking control during the acute or unexpected event. It's a perfect setup for shock trauma.

Furthermore, for a mind-body system that's already teetering on the brink, this new shock can be the tipping point where symptoms of dysregulation first appear or, if symptoms were already present, multiply and intensify. To cope with these symptoms, given the person's narrowed window, they're also more likely to rely on short-term fixes that actually build allostatic load further. For instance, they may self-medicate with nicotine, alcohol, caffeine, comfort food, and other substances. Or they may double down on adrenaline-seeking, violent, or self-harming behaviors.

It's just like someone who's living paycheck to paycheck. Add an unexpected expense into the mix, such as big car repairs or emergency dental work, and their financial stability begins to evaporate. In turn, they may make short-term choices to cope with the unexpected expenses—such as maxing out credit cards with high interest rates—which actually add to their long-term financial instability.

For this reason, shock trauma is often the tipping point for symptoms of dysregulation. Fortunately, however, in my experience working with countless people, shock trauma is often the tipping point for recovery, too. Just as Martin's symptoms were the tipping point that brought him to MMFT, shock trauma often helps us access the motivation to experiment with new habits and approaches for coping with stress, negative emotions, and chronic pain. It doesn't matter what caused our window to narrow. We can always choose to engage consistently in habits that will rewire our mind-body system and thereby widen our window.

Before you continue to the next chapter, I'd like to invite you to reflect on your own experiences related to the second pathway to narrowing the window. In your journal, you might make a list of every event when you experienced shock trauma throughout your life. Remember that your survival brain and body may have experienced shock trauma during "small" or "minor" events, such as a fender bender, outpatient surgery, or the discovery that your partner was having an affair. Over the next few days, you may remember additional events to add to this list; keep writing them down so that you have as complete a list as possible. You might also ask your family members to help you recall shock traumas from when you were a child.

Once the list is complete, take some time to reflect in writing about whether you'd considered these events to be traumatic at the time—or whether your thinking brain had devalued, ignored, or denied the shock trauma. You might also reflect on what measures, if any, you took for deliberate recovery after these shock traumas. Finally, you might reflect on whether any of your experiences with shock trauma served as tipping points for symptoms of dysregulation manifesting initially or intensifying. Maybe this was the first time you even noticed signs of ongoing dysregulation. You might also explore the interaction between your symptom expression and new "minor" stressors long after a shock trauma occurred, just as Martin experienced new symptoms with his upcoming work inspection.

CHAPTER NINE

Everyday Life

Sometime during high school, I unconsciously learned that the best way to square the circle between too much to do and not enough time to do it was to get less sleep. Both my parents routinely traded sleep for work, community events, or household chores. At my parents' social events, I would also overhear guests joking about their overextended, sleep-deprived lives. "I'll sleep when I'm dead," they'd quip.

As a result, I came to understand that sleep was something optional. Excellent grades, tough classes, extracurricular activities, volunteering: These were essential for getting into a top college and thereby securing future success and happiness. But sleep? Not so much.

By the time I got to college, I took this sleep-is-strictly-optional calculus up a notch. My new norm was sleeping two to four hours each night, for weeks on end. This mostly worked, as long as I scheduled one weekend each month to crash with a cold.

My sleep deprivation went into overdrive in the Army, when pulling all-nighters for my job became routine. Right after my promotion to first lieutenant, I was serving in a major's billet, leading about forty-five soldiers who managed intelligence collection and analysis for the 1st Armored Division. We had just been fielded a new automated intelligence system—the first Army unit to receive it. After two months of training—and several weeks to test, debug, and repair software and interoperability glitches with the contractors—my group participated in a field training exercise (FTX).

My boss viewed this FTX as a chance to show off the new system to the rest of the division. As a result, *I* viewed the FTX as my report card—since the new system's success or failure would publicly demonstrate whether I'd led my soldiers effectively. Because I was a twenty-three-year-old doing a major's job, this felt like a lot of pressure.

Fielding a new technical system is always challenging, especially one that had been in the R&D pipeline so long that it arrived with obsolete software. Beyond learning how to operate the new system, we also had to make it work with our existing systems, write code so it could connect with four data streams coming into our headquarters, and create ad hoc software templates to access and manipulate these data. Not surprisingly, right after the war game started, new problems appeared.

In response, I simply chose to stay on shift for three days, until my boss finally ordered me to get some sleep. By then, I'd been awake for seventy-nine hours.

About forty hours into this marathon, I was stung on my neck by a hornet. As these stings swelled up painfully—my neck was almost even with my ear—my soldiers suggested using masticated chewing tobacco to pull out the toxin. After forty-six hours awake, this sounded like a reasonable idea. I waved off my soldiers' offers to chew the tobacco for me. Since I'm not a tobacco user, however, that nicotine entered my system fast! Paired with the bottomless cup of "ranger coffee" I'd been drinking—instant coffee mixed with the hot chocolate, sugar, and creamer packets from our MREs— soon I was talking faster than Alvin and the Chipmunks.

The more absurd the problems with our new system became, the more buzzed I got. Somewhere around the sixty-hour mark, it seemed like time for another nicotine hit—this time keeping it in my mouth. (From my well-rested perspective today, all I can think is: *Really?!*) My soldiers thought this was hilarious: *Check out the LT, dipping!* Those three days were one long punch-drunk blur of diagnosing problems, brainstorming and implementing solutions, and high-fiving until the next disaster cropped up—all the while collecting and analyzing intelligence for the war game.

I wasn't the only leader who acted this way. Trading sleep for work was pretty endemic to my entire chain of command. Indeed, my sleepless behavior at that FTX was publicly recognized afterward with an award; it also led to several superiors grooming me for early company command.

THE THIRD PATHWAY TO NARROWING THE WINDOW

The third pathway to narrowing our window comes from chronic stress and relational trauma in everyday life. As a result, we may experience chronic arousal for *too long* or *too often*, without adequate recovery. In turn, we slowly deplete our inner resources.

Unlike the last two chapters, which explored the downstream effects of events mostly outside our control, this chapter focuses on contributors to the window that are mostly up to us. We didn't get a vote on our families of origin—or the lifelong alterations in our mind-body system that resulted from our earliest social environment. Likewise, we don't have much control over the crises and other extreme events we experience throughout our life span, such as car accidents, natural disasters, or the loss of loved ones.

In contrast, we have a *remarkable* amount of agency and ability to influence our window through the dynamics in this chapter. Unlike stressful events outside our control, *lifestyle choices are almost entirely up to us*—especially when we're aware of them. Yet during Life on the Gerbil Wheel, many aspects of our culture foster, facilitate, and reward behavior that leads to chronic stress arousal. As Chapter 2 explained, in our culture, "being stressed" usually means we're busy, successful, powerful, and important.

Much of the third pathway's cumulative narrowing effect comes from many mundane lifestyle choices we make *every single day*. However, *exactly because* these are daily lifestyle choices, we're less likely to consider how they're affecting our window. In general, our thinking brains are more likely to devalue or ignore everyday stressors. Thus, we're less likely to take active measures to recover from them.

Many of us simply aren't paying attention to how these choices narrow our window. For instance, although I couldn't prevent or control the events that led to my PTSD, I certainly controlled how much I prioritized getting adequate sleep (i.e., not at all) and how many activities I chose to stack on my plate (i.e., too many).

THE SLEEP-DEPRIVED AMERICAN

Possibly the single most important choice we make in daily life affecting the width of our window is how much high-quality sleep we get on a regular basis.

Most Americans aren't getting enough sleep. The most nationally representative data come from more than 110,000 American adults in 2004 to 2007. In this survey, about 28 percent reported sleeping six or fewer hours per night; 31 percent reported sleeping seven hours; 33 percent, eight hours; and 8.5 percent, nine or more hours. This range correlates with a 2014 survey, when Americans averaged 7.5 hours per night.

According to the Department of Labor's annual time use surveys, work-related activity was the single biggest predictor of short sleep duration in the United States from the 1960s through the mid-2000s. Americans working at least fifty hours per week were the group most likely to get less than 6.5 hours of sleep each night. The second most likely predictor of short sleep duration was time spent traveling, commuting, and driving for errands and car pools.

Perhaps not surprisingly, most Americans under age sixty-five also sleep significantly longer on weekend nights—showing how many of us use weekend sleep binges to prop up weekday sleep deprivation. Nevertheless, this assumption is misguided—the data disprove that sleeping in on weekends can make up a weekday sleep deficit.

Short sleep duration is especially problematic among people working in high-stress professions. Many jobs in these environments require variable sleep schedules, extensive travel with jet lag, or shift work for round-the-clock operations. Shift workers comprise almost 15 percent of all full-time U.S. workers—and they're disproportionately found in safety-sensitive professions, including law enforcement, the military, firefighting, and healthcare. Several studies show that sleep disruptions among police, correctional officers, firefighters, and healthcare workers are linked with cognitive, emotional, and physical impairments—as well as impairments in decision making and risk assessment.

One recent study with five thousand U.S. and Canadian police found that more than 40 percent had at least one sleep disorder. In turn, police with sleep disorders were significantly more likely to report making a serious error or safety violation, missing work, falling asleep while driving, and receiving citizen complaints against them. Likewise, sleep-deprived health-care providers are significantly more likely to crash their cars, make medical errors, and cause injuries that break the skin.

Sleep deprivation is also extremely common in the military. Several studies show that service-members, both during deployment and after returning home, sleep on average 5.5 to 6.5 hours per night. For instance, six months after returning from Iraq, 72 *percent* of three thousand soldiers in a U.S. Army brigade combat team reported sleeping fewer than six hours each night. Even after controlling for combat exposure, this 72 percent were significantly more likely to experience depression, PTSD, tobacco and alcohol abuse, and suicide attempts.

So how much sleep do we actually need? Individuals differ in how much their cognitive and emotional performance declines after sleep deprivation. Differing responses to caffeine sensitivity, as well as genetic differences in metabolism and circadian rhythm regulation, may account for some of this variation. Nonetheless, the most recent data show that most humans require *at least eight hours of sleep each night* to protect against cognitive impairment and several other window-narrowing mechanisms.

Two different experiments subjected about one hundred healthy volunteers to differing lengths of sleep deprivation while monitoring their diet and sleep 24/7 in the lab. The first study randomized volunteers into four-, six-, or eight-hour sleep limits per night for fourteen days, while the other study randomized them into three-, five-, seven-, or nine-hour sleep limits for seven days. No one was allowed to nap. Both studies also included three baseline days and three recovery days, during which all volunteers slept eight hours each night.

In these studies, the only people who experienced no attentional lapses or declines in cognitive performance were in the eight- and nine-hour groups. *Every other group saw performance declines.*

What's more, the three-, five-, and seven-hour groups did *not* recover to baseline cognitive performance during the three recovery days, when they slept eight hours each night. This finding gives the lie to the assumption shared by many overworked Americans, that we can just sleep in on the weekend to make up for chronic weekday sleep deprivation.

In the even-numbered study, the four- and six-hour groups showed progressive worsening in cognitive performance over time, and neither group stabilized at a lower-than-baseline level. In fact, after fourteen days, both groups showed cognitive declines equivalent to those found among people awake for twenty-four or forty-eight hours!

Even a moderate sleep restriction of six hours each night—if sustained night after night—may impair cognitive performance to the levels seen when legally drunk. Other research equates the cognitive impairment after twenty-four hours awake with a blood alcohol concentration of 0.1 percent. In this study, the six-hour group's cognitive impairment was on par with that. For reference, the legal limit is 0.08 percent—that's why the risks of driving while drowsy are at least as great as driving while intoxicated.

Nonetheless, the four- and six-hour groups weren't aware of how much sleep deprivation harmed their performance. By the third day of sleep restriction, they reported feeling slightly sleepy—but insisted that sleepiness wasn't adversely affecting their cognitive performance. They maintained this subjective self-assessment for the rest of the experiment, even as their objective performance on the tests continued to plummet.

In other words, when we're sleep-deprived, we're not only lousy at judging how much sleep we need, *we're also oblivious to how much sleep deprivation is actually harming our performance.* Several studies have found this same result. To give this a real-world context: Of the roughly 2,200 Army brigade combat team soldiers who reported sleeping fewer than six hours each night, *only 16 percent* believed their job performance was impaired due to lack of sleep.

These data are clear. Sleep deprivation impairs executive functioning. Even mild sleep deprivation, when it becomes a chronic habit, affects our thinking brain's performance. We're more likely to make mistakes, misread situations, and make ineffective decisions—especially during ambiguous, fast-paced, volatile, or threatening situations. Sleepiness can also cause problems during routine activities at lower arousal levels, when the brain might easily engage in *microsleep*, which occurs when parts of the brain briefly shut down.

In 2005, 60 percent of adult drivers in the United States said they'd driven while drowsy in the previous year—and 37 percent said they'd actually fallen asleep at the wheel. From 2009 to 2013, U.S. drowsy drivers were responsible for one in five crashes in which someone was killed and one in eight in which someone was hospitalized.

Indeed, the two 2017 deadly collisions in the Pacific Ocean between high-tech U.S. Navy destroyers and slow-moving commercial cargo ships were at least partially attributed to sleep-deprived sailors. Numerous rail and truck crashes, including the 2013 derailment of a Metro-North passenger train in New York City that killed four and injured more than sixty people, have also been linked with sleep deprivation and sleep apnea. The Exxon Valdez disaster, the Three Mile Island meltdown, the Challenger shuttle explosion, the 1986 Chernobyl nuclear accident in the Soviet Union, and the 1984 Bhopal gas leak disaster in India have all been directly linked to sleepiness on the part of people involved.

When sleep deprivation impairs executive functioning, it's not just undermining our thinking brain functions. *It's also inhibiting our capacity for top-down regulation of the survival brain.* As a result, stress and emotions are likely to play a much larger role in our decision making and our choices when we're sleep-deprived. Fragmented sleep—from waking during the night because of anxiety or depression or as an effect of aging—may be at least as detrimental to cognitive and emotional functioning as short sleep duration.

Fatigue affects our perception. We have more trouble regulating stress arousal and negative emotions—and less flexibility for dealing with life's curve balls. Sleep deprivation also impairs our ability to recognize emotions in ourselves and in others—specifically the prosocial emotions, such as happiness and sadness. The ability to recognize survival-oriented emotions, such as fear and anger, seems to be unaffected.

Since emotions play a large role in social communication, sleep deprivation may also adversely affect our social engagement. We have less capacity for interacting effectively with other people, in both personal and professional relationships. Sleep deprivation also undermines our ability to make ethical decisions—it's been shown to decrease moral awareness, adversely affect moral judgment, and increase permissiveness for unethical or transgressive behavior.

One recent brain imaging study suggests *how* chronic sleep deprivation leads to greater stress and emotional reactivity. Chronic sleep deprivation is linked with three changes that make sleep-deprived brains jumpier and more anxious. First, the amygdala fires faster and more during neuroception. Second, there's *more* connectivity within the survival brain between the amygdala and brainstem, which facilitates faster stress and emotional arousal. Finally, there's *less* connectivity between the amygdala and the PFC, which impedes top-down regulation of stress and emotions.

For concrete examples, recall that study in which 40 percent of police officers had at least one sleep disorder. Compared with police without sleep disorders, these police were significantly more likely to report displaying uncontrolled anger toward a citizen or suspect—and to have citizen complaints filed against them. Similarly, firefighters with insomnia or nightmares were significantly more likely to report having problems regulating their negative emotions. In turn, they experienced more negative emotions overall and more depression symptoms. In fact, insomnia is the single strongest predictor of clinical depression—and virtually all depressed people have a sleep disorder.

Chronic sleep deprivation also has several other enduring dysregulating effects on our mind-body systems. In fact, just like transgenerational trauma and ACEs, *chronic sleep deprivation and shift work can lead to detrimental epigenetic changes*—in the genes regulating the circadian rhythm and those involved in the inflammatory, immune, and stress responses.

First, chronic sleep deprivation is correlated with *dysregulation of the nervous system—especially undermining the ventral PSNS*, which controls social engagement, recovery functions, and the vagal brake on the cardiovascular system. Specifically, sleep-deprived people are more likely to have low HRV, which means their vagal brake isn't functioning properly. When our vagal brake isn't functioning properly, we're more likely to develop high blood pressure and cardiovascular disease. Not surprisingly, therefore, chronic sleep deprivation and circadian misalignment—such as with shift work—are both linked with hypertension, atherosclerosis, and increased risk for heart disease, heart attacks, and stroke.

Second, chronic sleep deprivation results in *dysregulation of our endocrine (hormone) system*, especially the HPA axis. Sleep deprivation especially affects the stress hormone cortisol, because cortisol has a pronounced circadian rhythm—peaking in the early morning to mobilize energy for waking up and then tapering down across the day. Night-shift work shifts the timing of cortisol's rhythm, while fragmented sleep or rotating shifts may dampen the rhythm's amplitude. When we're sleep-deprived, we produce more cortisol, with profound effects on our stress reactivity, metabolism, and immune functioning.

Our circadian rhythm affects our mood, energy level, body temperature, alertness, and appetite, and the timed release of various hormones throughout our daily cycle. That's why when we mess with this internal clock—with extended or unusual work hours, or while jet-lagged—all the sleep in the world doesn't really make up for circadian misalignment.

When we're sleep-deprived, increased cortisol levels interact with dysregulated levels of two hormones that control appetite. Leptin (which helps us feel satiated) decreases, while grehlin (which makes us feel hungry) increases. Together, these three shifts increase our appetite beyond our true caloric needs. We tend to eat more, especially snacking more throughout the day and overeating at night. We also tend to crave calorie-dense, high-carb, and high-fat foods—which helps explain why we reach for fast food or sugar when we're exhausted.

Not surprising, chronic sleep deprivation and circadian misalignment are linked with abdominal weight gain, higher body mass index, obesity, metabolic syndrome, insulin resistance, and type 2 diabetes. In fact, prospective studies with both children and adults show that sleep deprivation today is significantly linked with more weight gain in the future. It's not a coincidence that the dramatic societal increases in obesity and diabetes occurred during the same period as trends toward shorter sleep duration and poorer sleep quality. As of 2010, 70 percent of Americans were overweight or obese. The takeaway: It is *extremely difficult* to maintain normal weight or lose weight when we're not getting adequate sleep.

Since cortisol also plays a critical role in immune functioning, chronic sleep deprivation has also been linked with *dysregulation in our immune system*. When we aren't getting enough sleep, we're more likely to get sick.

One study tracked healthy volunteers' sleep duration and "sleep efficiency"—the percentage of time in bed they were actually asleep—for fourteen days. Afterward, they were quarantined, given nasal drops with a rhinovirus, and monitored for five days. Compared with people sleeping at least eight hours each night, those who averaged seven or fewer hours were three times as likely to catch the cold. Moreover, people with fragmented sleep were five times as likely to catch the cold than people with at least 98 percent sleep efficiency. Other studies have tracked how chronic sleep deprivation harms our ability to create antibodies after vaccination. Compared with people sleeping at least eight hours, for example, sleep-deprived individuals produce about half the virus-specific antibodies.

Not getting enough sleep also increases our risk for chronic inflammation—and diseases linked with chronic inflammation, including chronic pain, depression, autoimmune diseases, and cardiovascular disease. Shift work, fragmented sleep, and short sleep duration are all linked with increases in several inflammatory markers, in both experimental research and epidemiological studies. Circadian misalignment can also suppress melatonin production—in turn, increasing our risk for cancer.

CHRONIC STRESS AND RELATIONAL TRAUMA IN THE WORKPLACE

Most American adults spend much of their time awake at work. Moreover, in numerous polls since 1989, 55 percent of American adults—and seven out of ten college graduates—say they get a sense of identity from their job. For both reasons, chronic stress and relational trauma in the workplace can have a tremendous impact on the width of our windows.

The United States has a well-deserved reputation for workaholism. For instance, most advanced industrialized states require their employers to offer at least twenty paid vacation days annually, in addition to paid holidays. In contrast, the United States doesn't have a legal minimum for either. Vacations are not considered a non-negotiable social right here as they are in other industrialized countries. Since the 1980s recession, many American employers have reduced the number of vacation days they offer during economic downturns.

On average, American workers get ten paid vacation days and six paid holidays. Yet almost a quarter of U.S. workers receives no paid days off. Furthermore, vacation benefits are not distributed equally: Compared with 90 percent of U.S. high-wage workers, about half of low-wage workers and a third of part-time workers get paid days off. On average, low-wage workers receive four days and high-wage workers fourteen days.

Even more, many Americans don't use the vacation days they *do* receive. In 1995, one third of U.S. workers took less than half of their vacation time, with 10 percent using none at all. And that was *before* American vacation habits started a rapid decline in 2000. Even when this downward trend eventually ended in 2016, 54 percent still didn't use all their paid vacation. Certain workers are especially likely to forfeit vacation days: low-wage workers, executives, millennials age eighteen to thirty-five, and people working more than fifty hours each week, the so-called workaholics.

Three perceived barriers to taking time off stood out in surveys. These barriers were especially prevalent among millennials and workaholics. First, almost half of workers were afraid of returning to a mountain of work. Second, a third believed that there wouldn't be anyone to cover their work. Finally, more than a quarter worried that a vacation would make them seem less dedicated or cost them a raise or promotion. Although their fears may be unfounded, the data suggest that organizational culture may be at least partly to blame. Two thirds of employees believed their company culture is "ambivalent, discouraging, or sends mixed messages about time off."

Compared with employees who used all their vacation, however, "work martyrs" were less likely in the last three years to have received promotions, bonuses, or raises. They also reported significantly more stress at work (74 to 68 percent) and at home (48 to 41 percent). This makes sense, since taking time off can alleviate burnout, create space for nurturing personal and community relationships, and allow for physical and mental rest, recovery, and relaxation.

American workaholics also put in long hours and work overtime—as well as during vacation and weekends. One in five U.S. workers say they routinely work at least fifty hours per week at their main job. Of these workaholics, 57 percent say their job has a bad impact on their stress level. Moreover, 64 percent of U.S. workers say they routinely work overtime and on weekends, compared with only 15 percent who say they never do. Almost a third of U.S. workers also complete a "significant amount" of work while on vacation. Working on vacation is especially prevalent among workaholics (52 percent) and those with high-paying jobs (43 percent).

For about a quarter of U.S. workers, job stress ripples adversely onto sleeping and eating habits. Shift workers, millennials, and people in low-wage and dangerous jobs are more likely to report such adverse effects. Not surprisingly, the worst effects are among fifty-hour-plus workaholics, with almost half reporting adverse ripple effects.

Finally, most U.S. workers say that when they're sick, they still go to work "always" or "most of the time." This includes two thirds of low-wage workers, half of restaurant workers, and more than half of workers in medical jobs. In another survey, millennials (76 percent) were far more likely than those older than 35 (56 percent) to have left the house the last time they were sick. Of course, by bringing their illnesses onto public transportation, into their workplaces, and into public eating establishments, people risk spreading their infections to everyone else.

To be sure, some of this workaholism is likely not by choice. About four in ten Americans—and half of millennials—work at least two jobs to make ends meet. For some, this helps to cover increasingly expensive housing costs, especially in urban areas, while others cope with significant college debt. And many families put up to 30 percent of their annual income toward childcare.

Many people consider being overworked and constantly stressed at work to be a normal condition over which they have little control. They also readily admit to spending much of the time they're at work inefficiently, such as procrastinating or surfing the web. So why don't they just work productively during a concentrated period and then leave to exercise, eat well, nurture relationships, enjoy hobbies, or sleep?

Asking this question usually elicits two answers, which point to wider cultural norms and practices that undergird how our society thinks about and organizes "work." The first answer is that someone works on the clock—they have to put in X number of hours each week, regardless of how long their assigned duties actually take. The second is that people believe they must be physically in their office or tied to an electronic leash 24/7 to demonstrate that they're "dedicated" and "productive."

Inherent in both answers is a sense of powerlessness. Of course, when our survival brain perceives us to be helpless, powerless, or lacking control, it's likely to trigger greater stress and possibly even traumatic stress. Thus, both answers can contribute to the collective narrowing of our windows—fueling a mismatch between our *work habits* and our *neurobiological needs* for self-regulation and recovery.

As Chapter 4 explained, when the survival brain perceives stressors to be novel, unpredictable, uncontrollable, or threatening to our survival, sense of identity, or ego, it will trigger greater stress arousal. With this in mind, it's easy to see how certain kinds of work may be linked with chronic stress and relational trauma—especially when we aren't explicitly engaging in true recovery.

Most obvious is working in physically dangerous or toxic environments. For instance, when we routinely work around loud noises, intense odors, environmental toxins, wounded or dead humans and animals, or threats to our physical safety, our survival brain is likely to turn stress on without ever turning it off. In these environments, we're also more likely to think about our own mortality or the injury or death of other people, which are called *mortality concerns*. Managing mortality concerns depletes our executive functioning capacity—and then when we're depleted, we're also more vulnerable to thoughts and feelings about death.

Several psychological stressors also may lead our survival brain to neurocept danger and mobilize stress activation—even as our thinking brain might dismiss the stressor as "no big deal." This category includes working in chaotic, disruptive, and uncertain environments. Working for a toxic boss, or in demeaning or exploitative conditions. Coping with workplace exclusion, discrimination, or harassment. Working in low-status jobs or at the bottom of a power hierarchy. Each of these factors can increase the likelihood that we'll experience chronic stress or relational trauma at work.

Several studies with the British civil service found an inverse relationship between a worker's place in the power hierarchy and stress-related illnesses. The lower their place in the hierarchy, the more hypertension and chronic bronchitis symptoms they experienced.

Likewise, research shows that people in low-status roles are more vigilant—spending more energy and effort monitoring themselves and others. In turn, this heightened vigilance depletes their executive functioning capacity. Thus, compared with bosses and other powerful people at work, people who perceive themselves to be in low-status, powerless, or subordinate positions are less likely to plan effectively. They also have greater difficulty inhibiting distractions and task-irrelevant information.

From the survival brain's perspective, perhaps the most stressful combination of work characteristics is when we perceive ourselves at work as coping with high performance standards and/or responsibility, paired with low decision-making latitude. This combination is common across industries and wages—such as waiters and short-order cooks, construction workers, IT specialists, lawyers, and corporate executives. This holds true even for people who work in high-paying jobs, hold leadership positions, or perceive their work to be meaningful and enjoyable.

Why is this combination so stressful? When we lack decision-making latitude, our survival brains may perceive the situation to be uncontrollable and unpredictable, increasing stress arousal. Importantly, however, decision-making latitude doesn't just pertain to the *content* of our work, such as whether we control the necessary resources, funds, and people to get the job done. It also includes *decisions related to our own work autonomy*—such as where, when, and how we complete our work.

For example, in one study, 95 percent of workers said working privately, in a discrete space with a door, was important, but only 41 percent said they could do so—and almost a third reported having to leave the office to get work done. Another study linked working in open office spaces with higher stress levels, higher blood pressure, high staff turnover, and greater workplace conflict. Moreover, contrary to conventional wisdom, open-space offices correlate with significantly less face-to-face interaction among employees. Without the capacity to close a door, employees are more likely to withdraw socially from their cubicle mates and interact instead via electronic means.

We may also perceive low decision-making latitude when we have to manage the expression of our emotions at work, known as *emotional labor*. Emotional labor involves modifying our emotional expression—our speech, facial expressions, and body language—to satisfy organizational goals and requirements. For instance, we may need to outwardly express an emotion we aren't actually feeling inside. Or we may need to *suppress* an emotion we're feeling, because it isn't considered appropriate at work.

Emotional labor is common with jobs that require face-to-face or voice-to-voice contact with the public, such as politicians, or require the worker to provoke an emotional state in others, such as teachers, chaplains, therapists, or sex workers. Emotional labor is also likely when workers must express emotions at work in particular ways—for instance, to create a cohesive organizational culture or standardize the employer's brand or reputation (e.g., flight attendants or workers in tourism).

Regardless of the workplace, emotional labor can be especially challenging for underrepresented individuals, who may—for racial, gender, ethnic, religious, or sexual orientation reasons—feel disconnected from and unsupported by their organization's dominant norms and practices. They may feel isolated and lack mentors. Their employer may also turn to them frequently to meet particular quotas, such as requiring their disproportionate service on committees. Exacerbating matters, they may also be subject to racism, sexism, and heterosexism at work, as Chapter 2 explored. With this in mind, not surprisingly, one study found that female managers were more likely than male managers to report engaging in emotional labor.

Emotional labor is also common in many high-stress professions, such as law enforcement, firefighting, the military, disaster work, and emergency medicine. In these environments, the work itself tends to elicit negative emotions, and these professions socialize a strong cultural expectation of emotion suppression and self-reliance. By socializing suck it up and drive on and powering through as the appropriate response to stress, pain, injury, or adversity, however, these professions also condition other emotional and behavioral responses to be *not* appropriate. For instance, as Chapter 2 explained, male-dominated environments usually consider anger—but not fear or sadness—to be appropriate responses to stressful situations, thereby socializing irritability, rage, and violence as culturally acceptable. Thus, emotional labor in high-stress professions can exacerbate the likelihood of individuals—especially men—disowning and externalizing their emotional pain and trauma. Not surprising, these professions are also particularly prone to expressing distress via somatization, since members may perceive less stigma for seeking help for physical rather than emotional concerns.

Recent research suggests that emotional labor increases psychological strain and depletes mental resources, especially when *surface acting* is involved. Surface acting is faking an outward emotional expression while overriding internal feelings. Emotional labor has been linked with cognitive declines and errors in work performance—as well as with emotional exhaustion, burnout, and lower levels of job satisfaction.

CHRONIC STRESS AND RELATIONAL TRAUMA OUTSIDE WORK

We can also experience chronic stress and trauma in our relationships. We may have to interact with a stressed, traumatized, addicted, or abusive family member. Or we may be responsible for caring for a family member with a disability or a chronic physical or mental illness. Or we might be juggling responsibilities at work and at home. For instance, empirical research shows that women consistently report higher stress levels than men, which some researchers attribute to women averaging nearly three times as much unpaid domestic work as men. Of course, whether we experience trauma in these situations depends on whether our survival brain perceives us to be powerless, helpless, or lacking control.

Numerous studies show how partners in a romantic relationship form one neurobiological unit. Because of stress contagion in their attachment bond, they regulate each other's stress arousal and emotional well-being. Of course, this connection works both ways. When partners feel satisfied in the relationship, for example, physical contact and proximity between them helps reduce their anxiety and stress during stressful situations. However, when partners are *not* satisfied with the relationship, *physical contact and proximity between them has actually been shown to increase stress arousal levels.* Thus, when we're in a romantic partnership with someone who doesn't provide a "secure home base" and meet our attachment needs, our physical and mental health can be undermined.

One study asked middle-aged women about their marriages twice, about eleven years apart. Compared with women satisfied with their marriage on at least one occasion, women who reported being dissatisfied both times were three times as likely to have metabolic syndrome. Metabolic syndrome includes high blood pressure, high blood sugar, high cholesterol, and excess fat in the abdominal region—all risk factors for chronic illness.

We can also experience chronic stress and relational trauma in our schools and communities. Chapter 2 explored how poverty, sexism, hetero-sexism, and racism can all contribute to chronic stress and relational trauma in our everyday lives. Of course, discrimination, prejudice, harassment, and bullying don't have to actually be experienced to create toxic effects in our mind-body systems. It's possible to experience a surge of stress arousal while reading or watching the news about events where other members of our identity group are being marginalized. We can also experience this surge while remembering or anticipating events when we are marginalized our-selves.

Finally, as I'll explore further in Chapter 17, we can experience chronic stress and relational trauma due to loneliness and/or social isolation. Accord-ing to a large recent survey, nearly half of Americans say they sometimes or always feel alone and "left out." Empirical research shows that people who are chronically lacking in social contacts are more likely to experience ele-vated levels of stress hormones and chronic inflammation. Having few or low-quality social ties has also been linked with the development of cardio-vascular disease, high blood pressure, repeated heart attacks, autoimmune disorders, cancer, and slowed wound healing.

THE STRESS REACTION CYCLE

Ironically, the final contributor to narrowing our window through the third pathway comes from how we cope with our chronic stress. Some habits help us feel better—or more in control—in the short term, while actually build-ing allostatic load. Coping with stress in this way creates a vicious feedback loop, which Jon Kabat-Zinn calls the *stress reaction cycle*. And, as neuroplas-ticity shows us, each time we consciously or unconsciously repeat a habit, it's not only easier to repeat the habit again in the future but also *that much harder* to interrupt it and choose something else.

For example, we may engage in chronic workaholism and excessive busy-ness with low-priority tasks. Or we may engage in chronic procrastination and avoidance. We may do any of the following: Rely on chemicals—including caffeine, sugar, nicotine, alcohol, and illegal, over-the-counter, or prescription drugs—to increase or decrease our arousal and mask stress symptoms. Overeat, skip meals, or choose unhealthy or fast food, especially to help us soothe distress, fill emotional emptiness, or feel more in control. Use television, social media, mobile devices, the Internet, or video games to numb or distract ourselves. Stay up late, sleep irregularly, trade sleep for other activities, or lie awake at night to worry about our relationships or work. Deprive our bodies of the movement, exercise, and time in nature they require—instead consistently choosing sedentary, indoor activities. Compulsively engage in self-harming, high-risk, or adrenaline-seeking behaviors: cutting, extreme sports, gambling, aggressive driving, or infidelity.

We can also rely on thinking brain habits that perpetuate the stress reaction cycle. For instance, we may do any of the following: Engage in chronic worrying, planning, ruminating, or catastrophizing about worst-case scenarios. Compartmentalize or pretend that there's nothing wrong. Devalue our stress by using positive self-talk or reframing techniques that subtly disown aspects of the situation. Compare our situation to that of the friend who's got it much worse than we do. Feel self-critical and ashamed about being stressed, such as thinking that "this really is not a big deal, and why can't I just handle it already?"

Sometimes these habits are extremely effective strategies for short-term coping, especially during life-or-death situations. And, to be clear, even people with wide windows occasionally eat junk food, skip their workout to meet a deadline, enjoy a glass of wine at the end of a crappy day, or drink strong coffee to jump-start their day after a sleepless night.

The problems arise when these choices become *habitual and compulsive*, when we feel like we *need* them to function and cope in our daily lives. That's when they perpetuate the stress reaction cycle by suppressing, denying, distracting from, compartmentalizing, or self-medicating the stress arousal we're experiencing, without supporting a complete recovery. That's when these behaviors add to our allostatic load—until eventually the mind-body system's dysregulation develops a life of its own.

With most of these choices, we usually feel better in the short term—which is exactly why we're drawn to them when we're stressed! In fact, when we feel their pull, it's a clue that our mind-body system is activated and needs some recovery. This information can cue us to build in recovery activities soon, before we become even more dysregulated.

Interestingly, while these coping habits are individual choices, several cultural norms and practices incentivize us to default to them. Why? Many of these habits mask stress so that we can just keep pushing—*and this choice to keep pushing frequently aligns with the norms, goals, beliefs, and practices of our workplaces, education systems, and social institutions.* Because of these social norms, we're likely to drift unconsciously toward dysregulation. It's like entropy—it's where our Paleolithic wiring ends up when it's left to its own unconscious devices in Life on the Gerbil Wheel.

With our widespread collective tendency toward overworked, sleep-deprived, and stressed lives, our whole society tends toward a dysregulated baseline. Why does this matter? Well, when most members of an organization are overworked, sleep-deprived, and stressed, everyone moves toward degraded thinking brain functioning. We set unrealistic deadlines and expectations, adding self-imposed time pressure onto our duties. Then, everyone feels "behind," so we scurry to "catch up"—working long hours and weekends, guzzling caffeine to force our depleted thinking brains to concentrate, and skipping workouts and healthy meals to snack from the vending machine.

Furthermore, once these habits become a workplace norm, they reinforce those perceptions mentioned earlier in this chapter—that taking time off will make us seem not "dedicated" or "productive." As a result, work deadlines start to preempt self-care. Over time, we can end up sedentary, poorly nourished, disconnected from our relationships, and overreliant on artificial energy mobilization—from caffeine, nicotine, sugar, and other stimulants; electronic stimulation like loud music, action video games and movies, and time on our devices; or excessive cardiovascular exercise that masks deeper depletion and exhaustion. After too many months (or years) of this, about all we can do when we get home is numb out in front of the television with alcohol or comfort food. All the while, our allostatic load builds.

Of course, thinking brain degradation from chronic stress isn't just impairing our work performance—it's also undermining our ability to access top-down regulation of stress, cravings, and negative emotions.

Say you've experienced several weeks of sleeping only six hours each night because of pressing work and family demands. Your executive functioning is seriously impaired at this point, even if you don't know it. Since you're still working long hours, you're also spending most of what's left of your already-depleted executive functioning credit bank at work.

With this in mind, *of course* you're crankier, more impatient, or more anxious about things—there's nothing left for regulating these negative emotions. *Of course* you're more likely to cheat on your diet or your spouse, skip your workout, or procrastinate on an important long-term goal—there's nothing left for curbing cravings and accessing willpower. *Of course* you're more likely to overindulge in alcohol or distract yourself with social media, video games, and television—there's nothing left for managing stress in a healthier manner. And each time you make these choices—even if you did it on autopilot—you're also making it easier to default this way again in the future. Is it any wonder that allostatic load starts to build?

As our window narrows from chronic stress over time, it creates several vicious cycles. First, chronic stress—and the stress reaction cycle habits that mask it—disconnects us from what our mind-body system *truly needs* for self-regulation and recovery. When we're in this place, it becomes almost impossible to choose anything except what feels soothing *right now*. We get that wired-but-tired feeling, where it's hard to motivate ourselves to do the things that will truly help us recover.

Second, with impaired executive functioning, we get distracted more easily. It's harder to resist that little dopamine hit that comes from constantly checking our phones or social media feeds. In one study, for example, for every hour of interrupted sleep the previous night, people wasted 8.4 minutes in online puttering. Paradoxically, this is also when we're more likely to try multitasking—which, as Chapter 3 explained, further impairs executive functioning and amplifies stress arousal. Together, these dynamics dissipate our energy, rather than helping us harness and focus our energy for meeting concrete goals.

Third, we may swing back and forth on the procrastination-overwork pendulum. Initially feeling overwhelmed, exhausted, and burned out, we procrastinate. In fact, being "too tired" is the most common reason that people give for procrastination. Then, feeling behind, we pull an all-nighter or work through the weekend to catch up. Rinse and repeat.

Fourth, chronic stress narrows our attentional focus to "putting out fires" and handling what's immediately in front of us. Or we might spin our wheels with "workcrastination"—handling less important or less challenging tasks, like the email queue—while putting off tasks for achieving long-term personal and professional goals. Our lifestyle choices start to reflect this narrowed focus and misplaced prioritization of the urgent over the truly important.

Those of us who started life with narrow windows have less wiggle room to escape these vicious cycles before we experience physical or mental illness. Yet, as Chapter 7 explained, many of us from such upbringings are unconsciously attracted to high-stress occupations. Indeed, *stress reaction cycle coping habits are so common in these work environments as to be one of their defining features.* These environments socialize and culturally normalize artificial energy mobilization—inadvertently encouraging people to rely on nicotine, caffeine, sugar, electronic stimuli, multitasking, adrenaline-seeking behaviors, and even vigorous cardiovascular exercise that masks underlying depletion.

This is how, for instance, someone spends three days on shift during a field training exercise—propped up by sugar, massive quantities of caffeine, and the questionable choice to introduce nicotine into her system—and then gets professionally rewarded for it.

I want to acknowledge the momentum—and inertia—created by these vicious cycles. When we're caught in them, it can feel extremely daunting to try to break them. *Nevertheless, it's also important to own our agency in perpetuating these cycles.* These lifestyle choices *are* up to us.

It takes consistent, intentional effort to prioritize and practice lifestyle choices that feed self-regulation and recovery. Our neurobiology actually works best when we alternate periods of genuinely productive worktime, mobilizing energy and giving tasks our full attention—followed by time away from work, when we can renew our cognitive, emotional, and physical resources. Such renewal comes from healthy food, hydration, exercise, sleep, time in nature, leisure activities, and supportive relationships. Although it may be challenging to develop these new habits, over time they can develop inertia, too—as *virtuous* cycles—helping us access joy, creativity, connection, health, and well-being.

Before starting the next chapter, please reflect on your own experiences related to the third pathway to narrowing the window. In your journal, you might make a list of experiences at work, school, and home that contribute to chronic stress and relational trauma in your daily life. You should include chronic stressors that you're currently facing, as well as chronic stressors from the past from which your mind-body system may not have completely recovered. Over the next few days, you might remember additional stressors to add to this list; keep writing them down so that you have as complete a list as possible.

With nonjudgmental curiosity, you might also investigate the ways that you habitually cope with stress and relational trauma in daily life. First, make a list of any stress reaction cycle coping habits that you tend to rely on. If your coping habits have changed over time, it's useful to reflect on when and why your coping habits changed (in both beneficial and detrimental directions). It can also be helpful to reflect honestly and nonjudgmentally about how much sleep and exercise you're routinely getting. Especially if you're sleeping less than eight hours each night or exercising fewer than three times each week, you might ask what's getting in the way of getting adequate sleep and exercise. Finally, you might reflect on your consumption habits, conceptualized broadly—not just the food, beverages, and other substances you typically consume, but also your consumption of news and information, social media, and entertainment media (including music, movies, television, and the Internet), and the quantity, quality, and content of routine social interactions with friends, neighbors, work colleagues, and family. Each of these consumption streams can have significant effects on our mind-body system.

CHAPTER TEN

Stuck on High/
Stuck on Low

This final chapter in Part II pulls together the science behind the window. To get the most out of this chapter—and to best prepare yourself for the window-widening tools in Part III—pull out your notes from the exercises in earlier chapters. If you didn't complete those exercises as you were reading along, I encourage you to go back and complete them now. You'll get much more out of this chapter if you do.

At the end of this chapter, I'll ask you to review your notes from those previous exercises and complete one more. My hope is that by the time you finish that reflection, your thinking brain will have a much richer appreciation for all that your mind-body system has endured—as well as a greater understanding of why your survival brain, nervous system, and body behave as they do. This understanding is actually the first step in creating an allied relationship between your thinking brain and survival brain—and widening your window.

THE THREE PATHWAYS TO NARROWING
OUR WINDOW

As a reminder, the first pathway to narrowing our window is from chronic stress and developmental trauma during childhood. The second pathway is shock trauma, when we experience *too much* stress activation *too fast*. And the third pathway is a challenge that goes on *too long* or comes up *too often*—leading to chronic stress and/or relational trauma. It doesn't matter whether we experience all three pathways or just one—without adequate recovery, we turn stress arousal on without turning it off, and allostasis stops functioning properly. In turn, the mind-body system gets focused on the immediate needs of survival and stops attending to long-term needs—especially recovery, repair, healing, and growth. The end result is allostatic load.

Although it may take time for allostatic load to manifest symptoms, eventually the allostatic load *itself* can narrow our window. For instance, coping with a chronic physical or mental health problem—such as chronic pain, diabetes, obesity, insomnia, depression, anxiety, or PTSD—can add more stress arousal to our system. For instance, not only is chronic pain debilitating and depleting, but it also fuels anticipatory stress—such as when we worry it will never get better. In turn, anticipatory stress about pain is linked with greater pain intensity, more fatigue with pain, and a greater likelihood that pain becomes chronic.

Perhaps more importantly, whenever we experience any kind of chronic physical or emotional pain, we often feel helpless, powerless, and lacking control—*all features that increase the likelihood we'll experience traumatic stress on top of managing the chronic disease*. This exacerbates the adversarial relationship between the thinking brain and survival brain. Thus, over time, the allostatic load itself accelerates the narrowing of our window—leaving us with ever-dwindling inner resources for coping with our ever-growing allostatic load.

FINDING OURSELVES OUTSIDE OUR WINDOW

Chronic stress tends to slowly deplete our resources and reserves, while acute or traumatic stress may overwhelm our mind-body system through too much stress arousal, too quickly. In either case, if we don't reset and recover, over time our window narrows.

When this occurs, even small things can push us over our *stress capacity threshold*. The stress capacity threshold is another way of thinking about having moved outside our window. Obviously, someone who has narrowed their window will have a lower stress capacity threshold than someone with a wide window. Both shock trauma and chronic stress without adequate recovery can push someone over their stress capacity threshold and outside their window.

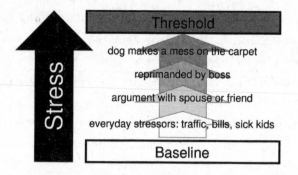

Figure 10.1: The Stress Capacity Threshold

The Stress Capacity Threshold depicts how we can move outside our window when we experience chronic stress and trauma without adequate recovery. The width of our window is the space between our baseline and our threshold; someone with a wider window can tolerate more stress arousal before they exceed their stress capacity threshold. However, when our window narrows, we are no longer operating at our neurobiological baseline but in a chronically activated state. From this place, even a small stressor can push us over our stress capacity threshold and outside of our window.

Exceeding our stress capacity threshold has four common consequences. First, *we're more likely to behave in ways that work against our values and goals.* That's because, as Chapter 5 explained, when we're outside our window, our thinking brain functions are degraded—undermining our ability to concentrate, see the bigger picture, think clearly, solve problems creatively, and make effective and emotionally intelligent decisions. We're also more likely to have difficulty with top-down regulation of our stress arousal and negative emotions. It'll be harder to access willpower and control impulses—increasing the likelihood that we'll give in to cravings, temptations, and unethical or violent behavior.

A quintessential example of this is coming home at the end of a stressful day to find that the dog made a mess on the carpet—and then "lose it" by shrieking or lashing out at the animal (see Figure 10.1). While we know it's not really the dog's fault, the dog becomes the focal point for all of our unresolved anxiety and frustration—and we act out.

After living in a state of chronic stress arousal for weeks, months, or years at a time, our mind-body system no longer recovers back to the base-line. Instead, we start living consistently up near the threshold, hovering just below our stress capacity threshold all the time. From this place, it doesn't take much to push us over—leading us to overreact in the face of small hassles, like an unexpected traffic delay.

Second, when we've exceeded our threshold, in the face of *any* stressor, even a "minor" one, *we're more likely to "fall back" to the third line of defense (dorsal PSNS, freeze).* Outside our window is when we're most likely to get overwhelmed and paralyzed or experience a freeze response. At the same time, it's also when we're most likely to see our performance suffer. For instance, we might "choke up" and not be able to perform at all. Alternatively, we might succumb to extreme procrastination. In effect, when we exceed our stress capacity threshold, we find ourselves way out on the right end of the Yerkes-Dodson curve (from Chapter 5). And, as you can see in Figure 10.2, at high distress levels, performance drops off completely.

Third, when we've exceeded our threshold, *we move into the zone of an adversarial relationship between our thinking brain and survival brain.* This means

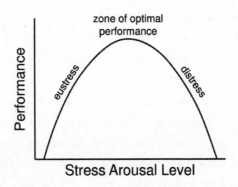

Figure 10.2: The Yerkes-Dodson Curve

The Yerkes-Dodson Curve posits an inverted U–shaped relationship between perceived stress levels and performance. Performance on a task improves as someone approaches moderate stress arousal, but steadily decreases past this point until it drops off completely with freeze. Eustress ("good" stress) describes low stress arousal levels, while distress describes high stress arousal levels.

that the survival brain will be more likely to engage in faulty neuroception, perceiving danger in situations that are actually safe, and vice versa. In addition, thinking brain functions, such as executive functioning, explicit memory, and effective and ethical decision making, will be compromised. Degraded thinking brain functions may manifest as inaccurate situational awareness, memory problems, distractedness, waves of anxious planning thoughts, or defensive reasoning.

Social engagement skills will also be impaired—decreasing our ability to interact cooperatively with others and offer and receive social support. Even with a secure attachment style, in this place we're more likely to exhibit insecure relational strategies, such as withdrawing, seeking excessive reassurance, or relying on coercive or violent interpersonal behaviors.

We're also more likely to experience *survival brain hijacking*, in which emotions and stress arousal bias our perceptions, absorb most of our attention, and drive our decision making and behavior. Here, the survival brain and nervous system are likely to cue default programming, even if it's not the most appropriate choice for the situation at hand. While this programming plays out unconsciously, our thinking brain may try to blame others. Or the thinking brain may take the programming personally, getting caught up in self-judgment and shame.

Conversely, we're also more likely to experience *thinking brain override*, in which we "live in our heads"—disconnected from survival brain signals, including our emotions and physical sensations in the body. We know that thinking brain override is occurring when we find ourselves engaging in suppression, denial, compartmentalization, and suck it up and drive on.

Finally, when we've exceeded our threshold, *we tend to rely heavily on stress reaction cycle habits.* After living with chronic stress arousal for weeks, months, or years at a time, our mind-body system no longer recovers back to the baseline in Figure 10.1. Instead, we begin to perceive our chronically activated state as *the new normal.* We normalize the allostatic load we're carrying around, often without even knowing it.

In this place, we're more likely to make *maladaptive coping choices*, such as eating poorly, skipping our workout, and pulling an all-nighter to finish the project on deadline, having procrastinated on our social media feeds and Netflix all week. Or relying heavily on nicotine, sugar, caffeine, alcohol, or other substances. Or engaging in adrenaline-seeking, violent, or self-harming behaviors. Of course, we're more likely to rely on these habits when we're on autopilot—denying, ignoring, distracting from, suppressing, self-medicating, masking, compartmentalizing, or avoiding our stress arousal. Although stress reaction cycle habits usually help us feel better or more in control in the short term, they actually build allostatic load—narrowing our window further and increasing our dysregulation.

DYSREGULATION

Over time, with repeated surges of stress hormones—and a lack of healing, repair, growth, and recovery—allostatic load produces wear and tear on major body systems. You might remember that allostasis happens through communication between (1) the brain; (2) the autonomic nervous system; (3) the immune system; and (4) the endocrine system, especially the HPA axis, which controls the stress hormones. Thus, allostatic load can manifest as imbalances or malfunctioning in any of these four systems.

Think about that. So many aspects of our mind-body system can be impaired through the dysregulation of these four systems—including neuroception, executive functioning, memory, willpower, creativity, emotions, metabolism, appetite, weight, body temperature, respiration, blood pressure, sexual drive, sleep, inflammation, healing capacity, and our default defensive and relational strategies.

Moreover, since these four systems connect to and direct the other organs, dysregulation can also manifest as imbalances or malfunctioning in other body systems, including our cardiovascular and respiratory systems, reproductive functioning, digestion and elimination systems, and the functioning of our cells, skin, fascia, muscles, and skeleton.

In other words, dysregulation can affect *all aspects of our mind-body system*—and thus may manifest as a range of cognitive, emotional, physiological, spiritual, and behavioral symptoms.

Symptoms of dysregulation often occur at points of weakness, such as where we've previously experienced an injury, toxin, infection, or physical trauma. Alternatively, our points of weakness may relate to a genetic predisposition. As Chapter 3 explained, a genetic tendency is not destiny—however, accumulating allostatic load is often what initiates the epigenetic changes that "turn on" a genetic tendency. To say that differently, our genome is not static. It's constantly shifting through epigenetic changes—*and many detrimental epigenetic changes result from chronic stress arousal.* That's one reason why many manifestations of dysregulation get grouped together as "stress-related illnesses."

Over the years teaching in high-stress environments, I've worked with people who've stockpiled six or seven different "unrelated" diagnoses—and at least that many different prescriptions—for the collection of symptoms they currently manage. (As you've probably surmised, at one point I was one of these people, too.) For instance, "Bob" had been diagnosed with ADHD in middle school; since then, he'd been taking Ritalin for attention problems and restless hyperactivity. A few years later, another doctor focused on Bob's insomnia, prescribing sleeping pills. After years of pushing his mind-body system through challenging work, a third clinician zeroed in on Bob's chronic pain and gave him a narcotic. Shortly before I met him, a mental health provider had diagnosed major depression and had also started him on an SSRI antidepressant. He also took a beta-blocker for hypertension.

At core, stress-spectrum illnesses, both physical and mental, share a common origin. (See Table 10.1 for a list of some common stress-related diseases and disorders.) The high rates of *comorbidity* among these diseases—meaning that these disorders often occur together—should make it clear that each exemplifies different aspects of dysregulation. It's only modern Western medicine that artificially divides them into discrete diagnoses. That's not to suggest that these diseases are always and only stress-related; however, especially if someone has several concurrently, then stress is likely an underlying driver of their incidence.

Table 10.1: Some Common Stress-Related Diseases and Disorders

PTSD	Insomnia
Acute stress disorder	Sleep apnea
Anxiety disorders	Hypertension (high blood pressure)

Attention deficit hyperactivity disorder (ADHD)	Cardiovascular disease
Depression	Metabolic syndrome
Substance abuse	Diabetes
Alcoholism	Migraines
Eating disorders	Asthma
Chronic pain	Allergies
Fibromyalgia	Psoriasis
Irritable bowel syndrome	Eczema
Ulcers	Rosacea

Dysregulation usually manifests in three patterns: (1) *hyperarousal*, or "stuck on high;" (2) *hypoarousal*, or "stuck on low;" or (3) oscillating between hyperarousal and hypoarousal.

I have yet to find empirical research about the relative prevalence of these three patterns. I've often heard trauma clinicians say that about two thirds of the population tend toward "stuck on high" when dysregulated, while one third tends toward "stuck on low." Although I haven't found empirical evidence to support this claim, my anecdotal experience from teaching also suggests that most people tend toward "stuck on high," at least initially.

With "stuck on high," the mind-body system's default response becomes too much SNS arousal, or *hyper*arousal. With this pattern, the survival brain is hypersensitive and reactive—constantly scanning the internal and external environment with hypervigilance. With hypersensitive neuroception, the mind-body system (over)mobilizes stress arousal. The HPA axis tends to produce elevated levels of adrenaline and the other stress hormones made by the adrenal glands (catecholamines). In effect, the mind-body system is hyperactively trying to cope, using some variant of fight-or-flight (the SNS defensive strategies).

With that in mind, symptoms associated with "stuck on high" include hypervigilance, excessive startle response, hypersensitivity to light or sound, hyperactivity, restlessness (such as fidgeting and bouncing knees), muscle tension, chronic pain, insomnia, nightmares, flashbacks, panic attacks, prolonged anxiety, rage attacks, violent outbursts, prolonged irritation and irritability, and emotional reactivity. Kindling from internal cues is also more common when we're "stuck on high."

Behaviors associated with "stuck on high" include adrenaline-seeking activities in search of an endorphin rush, such as engaging in extreme sports, excessive cardiovascular exercise, infidelity, gambling, high-speed motorcycle riding, dangerous driving, playing violent video games, and feeling a pull toward action and horror movies. There's also a tendency to rely heavily on nicotine, caffeine, sugar, and other stimulants.

As you might imagine, "stuck on high" is the most common pattern of dysregulation in high-stress professions—at least initially, as I will explain in a moment. Not only do these work environments tend to attract adrenaline junkies, but they also reward the workaholic, hard-charging, sleep-deprived behavioral choices linked with "stuck on high." Or, to say that differently, it's difficult to function effectively in such environments when you're "stuck on low."

With "stuck on low," the mind-body system's default response is not enough SNS arousal—hence, *hypo*arousal—and too much dorsal PSNS in defensive mode. Thus, "stuck on low" behavioral responses fall along the freeze spectrum, including learned helplessness, extreme procrastination, overwhelm, apathy, and dissociation. Dissociation occurs when we subjectively experience ourselves as numb, separate, or distanced from our body, emotions, and distress. With this pattern, the HPA axis produces too much cortisol and other glucocorticoids. And, as Chapter 5 explained, high glucocorticoid levels are linked with depression, memory problems, impaired executive functioning, metabolism problems, and weight gain.

Thus, while "stuck on high" is associated with *hyperactive* coping, "stuck on low" is associated with *underactive* coping. Someone "stuck on low" may feel apathetic or depressed, without feeling motivation or pleasure. "Stuck on low" may also manifest as not trying to mount any coping response at all.

Symptoms associated with "stuck on low" include depression, forgetfulness, fogginess, mental spaciness, clumsiness, chronic fatigue, low physical energy, exhaustion, sleeping excessively and still feeling tired, loss of libido, emotional numbness, overwhelm, and feelings of alienation. Someone who is stuck on low may also experience physical numbness, which can lead to chronic pain and/or inflammation in the dissociated or numb parts of the body.

Behaviors associated with "stuck on low" include sedentary behavior, television binging, overeating, alcohol and drug abuse, isolation, diminished sexual activity, procrastination, denial, and avoidance. In my anecdotal experience, a pure "stuck on low" is quite rare in high-stress environments; instead, people in these environments are more likely to follow the third pattern.

The third pattern *oscillates between "stuck on high" and "stuck on low,"* with corresponding swings in the predominance of SNS and dorsal PSNS survival strategies (and their related symptoms). With this pattern, for instance, someone might experience weeks of "stuck on high" hyperactivity, workaholism, and insomnia, followed by a "stuck on low" crash brought about by exhaustion, a migraine, or a cold. Then they may sleep excessively and hide out at home for several days, feeling catatonic, depressed, and numb.

Alternatively, someone may have a condition in which they oscillate between symptoms that seem diametrically opposed to each other, at opposite ends of autonomic functioning. For instance, someone with irritable bowel syndrome may experience bouts of constipation during the "stuck on high" phase, followed by bouts of diarrhea during the "stuck on low" phase. Likewise, someone with a sleep disorder may experience insomnia during the "stuck on high" phase, followed by chronic fatigue and sleep binges during the "stuck on low" phase. At its extreme, this third pattern may even resemble the highs and lows of bipolar disorder, alternating between a frenetic, manic hyperactivity and a depressed, apathetic collapse.

Trauma researcher Robert Scaer suggests that this third pattern is especially common with the passage of time among people who've survived childhood trauma or experienced many shock traumas in succession. It's also common among people who've experienced both childhood trauma and later shock traumas, a condition sometimes called *complex trauma*. For example, I've often seen this third pattern among combat veterans after multiple combat tours, especially when they also have a high ACE score. Of course, I experienced this pattern eventually myself.

For those of us with histories of complex trauma, or multiple shock traumas in succession, often we first manifest the classic "stuck on high" PTSD symptoms—hypervigilance, heightened stress reactivity, nightmares, flashbacks, insomnia, and chronic hyperarousal. With the passage of time, however, these "stuck on high" symptoms tend to be replaced with a secondary group of symptoms or diseases linked with "stuck on low." While periods of hyperarousal still occur, they tend to become more intermittent. At the same time, symptoms of dissociation and freeze become increasingly common.

In other words, without adequate recovery or treatment, over time the "stuck on high" phases likely get shorter and more intense, while the "stuck on low" phases likely last longer. As someone on this trajectory approaches midlife, they become more likely to experience freeze than earlier in their life—especially during unavoidable or irresolvable conflict, such as often occurs in personal or professional relationships.

One classic indicator of this particular trajectory of dysregulation is persistent low levels of cortisol. As Chapter 7 explained, diseases linked with cortisol *under*production include PTSD, fibromyalgia, chronic fatigue syndrome, hypothyroidism, allergies, multiple chemical sensitivities, asthma, rheumatoid arthritis, and other autoimmune diseases. Not coincidentally, these are also diseases linked with "stuck on low."

Regardless of which pattern, symptoms of dysregulation often present *subclinically* at first—meaning they don't require clinical treatment yet. Mild dysregulation serves as a yellow flag, a cue from our mind-body system that we need recovery. Having one symptom occasionally is not a problem. However, we need to pay attention when one or more symptoms become chronic or disabling.

One helpful threshold to keep in mind: Mild or moderate dysregulation is almost certainly present whenever we find ourselves pulled toward stress reaction cycle habits. If we don't prioritize behaviors that support true recovery soon—and instead give in to the short-term relief that these coping habits bring—we end up masking, self-medicating, or suppressing the underlying dysregulation. Of course, this merely exacerbates the dysregulation.

Over time, without adequate recovery, our symptoms of dysregulation will continue to worsen, in part so that we can unconsciously manage the excess stress arousal coursing through our mind-body system. For instance, when someone is restless and fidgeting, such as bouncing their knees or flipping a pen, their mind-body system is trying to cope with excess stress arousal and anxious energy. In addition, symptoms increase over time because the imbalances and dysfunctions from the growing allostatic load multiply.

The more dysregulation worsens, the more extreme our coping behaviors become to manage our intensifying symptoms. This is why extreme behavior is a hallmark of extreme dysregulation. Indeed, deliberate self-harm, violent behavior, substance abuse, and other addictive behaviors are all maladaptive attempts to self-medicate our symptoms of dysregulation—while actually exacerbating the dysregulation.

Because dysregulation can affect *all* aspects of our mind-body system, I have grouped the symptoms into five categories: Table 10.2 includes physiological symptoms; Table 10.3, cognitive symptoms; Table 10.4, emotional symptoms; Table 10.5, spiritual symptoms; and Table 10.6, behavioral symptoms. These symptoms highlight how dysregulation is not "mental weakness"—it's an indication that the mind-body system is no longer functioning the way it was wired to do. The more symptoms we're currently experiencing, the worse our current level of dysregulation.

Table 10.2: Physiological Symptoms of Dysregulation

Hypervigilance/feeling on guard	Exaggerated startle response
Muscle tension/neck and back problems	Dizziness
Racing or pounding heart	Changes in appetite (too much, too little)
Chronic pain/fibromyalgia	Weight loss or gain/metabolism problems

Chronic inflammation or inflammatory diseases (e.g., allergies, asthma, autoimmune diseases)	Gastrointestinal symptoms (constipation, diarrhea, spastic colon, heartburn, ulcers)
Sleep problems (difficulty falling/ staying asleep/sleep apnea/ oversleeping)	Restlessness/fidgeting/bouncing knees/unable to settle down
Nightmares/night terrors	Hypersensitivity to sound or light
Jerks in the body while sleeping	Skipped menstrual periods
Physical numbness/lack of feeling in body parts/deadened sensations	Severe premenstrual syndrome (PMS) symptoms
Headaches/migraines	Changes in sexual drive/loss of libido
Nausea/upset stomach/vomiting	Erectile dysfunction/premature ejaculation
Hyperactivity	Increased urinary frequency
Feeling outside of/disconnected from/unable to sense your body	Temperature shifts (chills, hot flashes, night sweats)
Chronic fatigue/low physical energy/lethargy	Hormonal imbalances (e.g., thyroid problems, diabetes)
Feeling weak or collapsed in the body	

Table 10.3: Cognitive Symptoms of Dysregulation

Memory problems	Impaired decision making
Forgetfulness/amnesia	Inability to make and keep commitments

Missing appointments or misplacing/losing things (e.g., keys, glasses)	Catastrophizing/worst-case-scenario planning
Difficulty focusing/attention problems	Self-critical or self-blaming thoughts
Reduced ability to formulate plans	Intrusive or obsessive thoughts
Decreased concentration/easily distracted	Suicidal thoughts
Disorientation about time, location, or direction	Ruminating
Mental spaciness/fog/confusion	Excessive worrying

Table 10.4: Emotional Symptoms of Dysregulation

Abrupt or extreme mood swings	Depression
Emotional numbness	Easily and frequently feeling "stressed out"
Exaggerated emotional responses/ emotional flooding (unable to control emotions)	Rage attacks, temper tantrums, and anger
Heightened emotional reactivity/ overreacting to things	Irritability
Panic attacks, anxiety, or phobias	Sadness/grief
Avoidance (avoiding situations that bring up certain emotions)	Feelings of impending doom/fear of being followed/dying/going crazy
Withdrawal, isolation, and feelings of alienation (e.g., "no one can understand")	Apathy/loss of interest in life

Despair/overwhelm/frequent crying	Feelings of helplessness/powerlessness
Overcautiousness	Shame or feelings of inadequacy
Fear of being alone	Fear of being with others or leaving home

Of course, as humans we all experience waves of emotion as a natural part of the survival brain's response to stimuli. Dysregulation occurs when an emotional symptom is *chronic or disabling*, such as when it impedes our ability to do our work or interact effectively with others. Emotion dysregulation is also likely when a particular emotional state becomes our prevailing mood for several weeks.

Table 10.5: Spiritual Symptoms of Dysregulation

Loss of belief/meaning	Feelings of hopelessness
Sense of meaninglessness	Crippling doubt
Extremist/radical beliefs	Survivor guilt
Dogmatic or rigid beliefs/black-and-white thinking	Feelings of alienation
Existential crisis	Loss of identity

As with emotions, someone may experience spiritual symptoms for reasons other than dysregulation. However, when other forms of dysregulation are present, dysregulation is usually contributing to spiritual symptoms.

Table 10.6: Behavioral Symptoms of Dysregulation

Inability to love, support, nurture, or bond with other people, especially loved ones	Lack of boundaries in relationships/excessive clinging or people-pleasing behavior

Bonding with others through trauma	Disrupted relationships
Avoidance behavior (certain people, places, or things)	Starting many projects and not completing them, or difficulty starting projects
Attraction to dangerous or high-risk situations	Extreme procrastination
Extreme sports and/or excessive exercise	Temper tantrums/uncontrolled temper
Adrenaline-seeking behavior	"Acting out" (screaming, throwing things, shouting, hitting, kicking, punching the wall)
Alcohol or substance abuse	Domestic violence or other violent behavior
Excessive use of nicotine and/or caffeine	Lack of sexual interest
Other addictions (work, shopping, gambling, sex, porn)	Compulsive masturbation
Disordered eating (e.g., anorexia/undereating, binging, purging, overeating)	Infidelity/affairs/sexual promiscuity
Being accident-prone or bumping into things	Self-destructive behaviors (e.g., extreme fasting, cutting)
Relying on stress reaction cycle habits to feel better	Deliberate self-harm/suicidal behavior
Overly rigid boundaries in relationships/holding people at a distance	Obsessive-compulsive behaviors (e.g., compulsively rechecking things or counting)

WHEN PEOPLE ARE LIKE SALMON

Nature's most striking example of allostatic load comes from salmon, who trek back upriver to spawn the next generation. Depending on the species, salmon may travel for nine months and as many as one thousand miles upstream—fighting the current, leaping over rocks, pushing to return to their breeding ground. During this migration, they even stop eating.

By the time they lay and fertilize their eggs, however, their chronically high levels of stress hormones have exhausted their energy stores and devastated their immune systems. As a result, after breeding, the salmon die. In other words, while their stress hormones help them mobilize enormous amounts of energy for their trek, chronic exposure to toxic stress levels eventually kills them.

As the last several chapters have made clear, we humans have many pathways for enacting our own version of the salmon trajectory. From our earliest experiences in the womb and our early attachment bonds, to childhood adversity and developmental trauma, to accidents, illnesses, injuries, losses, and other shock traumas over the years, to our many stress reaction coping habits—it's somewhat miraculous that we're still functioning.

We've all observed people like the salmon. These are people who perform brilliantly through periods of prolonged stress and adversity, only to see their lives eventually "go off the rails" in a spectacular fashion. By "go off the rails," I mean things like suicide, a nervous breakdown, or other debilitating psychological injuries. A heart attack, cancer, or other life-threatening medical diagnoses. Allegations of domestic violence, sex scandals, DUIs, or other violent, criminal, or unethical behavior. Indeed, in the #MeToo era, examples of dysregulated individuals finally being called to account for such off-the-rail behavior has become quite common.

I am not pointing fingers here. By 2004, after decades of revving my mind-body system without recovery, I was like the salmon myself, having ravaged my immune system and lost my eyesight.

Unlike the salmon, however, we humans are *not* destined to die after remarkable accomplishments. Stress arousal allows us to mobilize immense amounts of energy to accomplish incredible feats or survive horrific events—but allostasis doesn't necessarily have to lead to allostatic load. *We have an extraordinary amount of agency in this process.* Although we can't choose our family of origin or control the stressful or traumatic experiences we've endured, we can always choose how we direct our attention, care for our mind-body systems, and nurture our relationships and communities.

STRESS EQUATION REDUX: PEOPLE DON'T HAVE TO BE LIKE SALMON

To end this chapter, I invite you to engage in a deeper reflection about the current state of your window. I recommend that you pull out your journal. If you completed the exercises in the earlier chapters, take some time to go back and read through all that you wrote in response to those exercises.

If you still haven't engaged in those reflective writing exercises from the earlier chapters, I strongly urge you to get a notebook and go back and complete them now. Many of the tools in Part III are best accomplished in tandem with reflective writing, so it's time for you to create one place to collect data and gather your thoughts.

At a *minimum*, you need to determine your attachment style, both as a child and as an adult, and make a tally of your life experiences along the three pathways to narrowing the window. *The goal is to help your thinking brain truly appreciate all that your mind-body system has endured, so that it will be less likely to devalue your stress and your dysregulation.*

Take some time to make an exhaustive list—and be sure to include events that your thinking brain might have devalued as "not that bad" or "minor." Remember, whether those events were actually stressful or traumatic was not up to your thinking brain, because neuroception is a survival brain job. In your exhaustive list, I want you to include everything that your *survival brain* might have neurocepted as significantly stressful or traumatic. Keep in mind the level of internal and external resources you had available to you at the time you encountered the stressful or traumatic events. For example, watching our parents scream at each other was extremely stressful when we were young children, without much internal capacity for down-regulating stress. In contrast, as adults, we're much less likely to feel threatened by their screaming behavior or take their argument personally.

After you review your responses to the exercises from Chapters 4 through 9, the second reflective task is to review the five tables with symptoms of dysregulation in this chapter (Table 10.2 through Table 10.6) and make a list of your current symptoms. As you do so, note whether your symptoms indicate that you're "stuck on high," "stuck on low," or oscillating between these two poles. If you cannot discern one of these three patterns, you might investigate whether there are any other patterns. For example, you might explore whether particular circumstances, locations, relationships, or activities trigger certain symptoms of dysregulation. *The goal is to identify how your mind-body system experiences stress—and the ways that you commonly cope with it.*

Of course, since many behavioral symptoms of dysregulation are stress reaction cycle coping habits, doing this task will also help you compile a list of ways that you commonly cope with stress. Feel free to add other coping strategies that are not in the five tables.

Next, review again the notes you jotted down during the exercises in Chapters 4 and 5. For instance, you can review any insights you had about how an adversarial relationship typically manifests between your thinking brain and survival brain. How do those insights fit together with your current symptoms of dysregulation? Likewise, you might compare the symptoms of activation from the visualization exercise in Chapter 4 with your current symptoms of dysregulation.

The third reflective task is to review the list of stressors you wrote at the beginning of Chapter 4. Begin by adding additional stressors that you've identified since you first wrote the list. Once the list feels complete, look at each stressor on the list and give it some additional labels. Is it chronic or acute? Internal or external? Is it a stressor of anticipation? Is it a sign of kindling?

With all of this additional information about your current sources of stress, now I want to reprise the Stress Equation from Chapter 4 (see Figure 10.3).

As earlier chapters explained, where we find ourselves on the continuum between stress and trauma is the result of how our survival brain neurocepts a threat or a challenge. When it perceives the stressor as novel, unpredictable, uncontrollable, or threatening to our survival (or our ego), it will neurocept greater danger—and thus trigger more stress arousal. Likewise, when it perceives that we do not have the internal and external resources to deal with the stressor, it will also neurocept greater danger and trigger more

$$\text{Stressor} \quad + \quad \begin{matrix}\text{Perception}\\\text{of threat}\end{matrix} \quad \longrightarrow \quad \text{Stress}$$

$$\left(\begin{matrix}\text{internal or external}\\\text{event}\end{matrix}\right) \qquad \left(\begin{matrix}\text{survival brain's}\\\text{neuroception}\end{matrix}\right) \qquad \left(\begin{matrix}\text{activation in the}\\\text{mind-body system}\end{matrix}\right)$$

Figure 10.3: The Stress Equation

The Stress Equation explains how the mind-body system produces stress activation/stress arousal. Whenever we experience a (1) *stressor,* an internal or external event that (2) the survival brain perceives as threatening or challenging, then our mind-body system turns on (3) stress arousal, which is physiological activation in the body and mind.

stress arousal. *Thus, whether we can access agency in the face of the stressor plays a tremendous role in whether (and how much) we will experience stress or trauma.*

With this in mind, the final reflective task is investigating your list of current stressors in light of the Stress Equation. *The goal is to understand where and how you can influence your stress.*

First, we generally have little influence over our stressors. However, if we can eliminate, change, or influence a stressor without jeopardizing our goals or values, we probably should. With that in mind, you might review your list of stressors and see if there are any that you *could* influence.

For instance, if one of your stressors is sleep deprivation, would it be possible to make some different choices to get more sleep? Or if you work for a toxic boss, would it be possible to change jobs? If sitting in traffic is on your list, could you change when or how you commute to work? You get the idea. While we generally don't have a lot of control over most internal or external circumstances, if you discover points of leverage where you could reduce or eliminate stressors, by all means, do! Not only could this boost your sense of agency, but it will also free up bandwidth for working creatively with the remaining stressors that you can't influence.

Part III will address how to find agency with the second and third components of the Stress Equation—changing how we perceive the threat as well as work with stress activation once it's been triggered. Here, I will note that *over the longer term,* we may develop the most influence over the equation's second component, how we appraise stressors. It's possible to shift our perception of stressors over time by changing the way that we relate to them. However, since this appraisal process happens in the survival brain, just telling yourself "this is not a big deal" isn't actually going to change the perception! Such thinking brain assessments may dismiss, devalue, or criticize the survival brain's neuroception process—in the process, actually making our stress arousal worse.

Instead, *it takes time to teach our survival brain and body how to shift their relationship with particular stressors.* I'll explain more about this in Part III, but for now, you can rely on the warrior qualities of wisdom and courage to facilitate this eventual shift. For instance, you can draw on wisdom to recognize when your survival brain has neurocepted something as threatening or challenging. You can also draw on courage to build your capacity to tolerate the stress arousal in your mind-body system—allowing the stress, negative emotions, and distressed thoughts to be there, without needing them to be different.

In terms of the third component, we probably gain the most leverage in the short term from working with stress activation once it has arisen. When stress arousal gets triggered, the first thing you can do is not take it personally. This is how we mobilize energy to respond to a threat or a challenge, so once the survival brain neurocepts danger, we can't somehow magically avoid stress arousal. It's just not neurobiologically possible!

In Part III, I'll teach you some exercises for downregulating stress arousal and building an allied relationship between the thinking brain and survival brain. For now, you might redirect your attention away from any stressed thinking brain habits that amp up stress arousal—such as rationalizing the distress away, judging ourselves for being stressed, ruminating about the stressor, catastrophizing about what-if worst-case scenarios, or comparing our experience to someone else's. For instance, you might direct your attention instead to pleasant sounds or attractive colors in your surroundings. Or you might notice how your body is in contact with and supported by your surroundings, such as a chair, a bed, or the grass outside.

You can also reflect on all of the stress reaction cycle habits you've developed to distract from, deny, self-medicate, mask, or avoid experiencing the discomfort of stress activation in your mind-body system. This is another place where wisdom and courage come in handy. Wisdom helps us see these patterns clearly, while courage helps us admit and take responsibility for our unskillful choices that send us repeatedly around the merry-go-round. As we become more intimately acquainted with our own particular repertoire of stress habits, we can make conscious choices to interrupt them and replace them with other coping strategies that help us widen our window—such as sleep, exercise, healthy food, social support, and other habits to facilitate self-regulation. I'll talk more about this in Part III, as well.

Developing more adaptive coping strategies is especially important for dealing with chronic stressors, which are a routine part of our daily lives. If we can better learn to manage stress from the chronic stressors—and regularly reset back to our baseline, without inadvertently adding to our allostatic load—we'll have more capacity available for accessing agency, even during acute stressors.

Part III

WIDENING
THE WINDOW

CHAPTER ELEVEN

The Warrior Traditions

As Part II explained, where we find ourselves on the continuum between stress and trauma has everything to do with how our mind-body system, consciously *and* unconsciously, perceives our current situation—and especially whether we feel like we have agency. The capacity to find agency during challenging events plays a critical role in whether we'll experience stress or trauma. *The less agency we perceive we have, the more traumatic the experience will likely be for our mind-body system.*

MMFT's goal is to build our capacity to find agency and access choice in *every* situation, no matter how challenging, stressful, or traumatic. Regardless of the circumstances, if we can find agency, we're more likely to remain inside our window, with an allied relationship between the thinking brain and survival brain. With agency, we're also more likely to function well during stressful events and recover completely afterward—two facets of improved performance. With this in mind, Part III begins with a survey of different approaches for finding agency during stress.

STRESS INOCULATION TRAINING

High-stress organizations are in the business of training and preparing their people for unexpected and unwanted events. In the organizational behavior literature, these are called *high-reliability organizations* (HROs)—because they're capable of functioning reliably, safely, and effectively during widely variable circumstances. The corporate world has been especially keen to learn from HROs about training and maintaining organizational resilience during instability, turbulence, and chaos. Although they don't put it this way, what they actually want is to learn how to help their people find agency during stress.

In late 1995, my Army unit prepared to deploy to Bosnia, as part of the NATO Implementation Force (IFOR) responsible for peace enforcement after the three-year civil war. Before we deployed, we conducted intensive predeployment training using the most common HRO approach for training individuals to find agency during stress, called *stress inoculation training* (SIT).

As part of this process, we each carried spreadsheets—wrinkled, muddy, and often held together with duct tape—that were covered with initials that "certified" that each of us was ready to deploy. Each certification represented some interval of training and preparation, ranging from one-hour briefings to multiday events. Vaccinations and a cavity-free mouth. Will and powers of attorney. Recertification on mission-essential tasks. Marksmanship qualifying scores on all assigned weapons. Refresher training on nuclear, biological, and chemical weapon procedures and protective gear. You get the idea.

Those spreadsheets were the physical "proof" that we were ready to respond to any unwanted events we might encounter in Bosnia. While most HROs no longer use hard-copy spreadsheets, their digital equivalent still remains the U.S. government's methodology for systematically training and certifying soldiers, diplomats, and other first responders for deployment. Anyone who's deployed anywhere since the 1990s has completed a similar training and certification process.

More generally, SIT is how most organizations train and prepare their members to do their jobs during stress. Crisis simulations, emergency preparedness drills, field training exercises—even a typical fire drill—are all examples of SIT.

SIT is based on the principle that we experience greater stress arousal when we perceive something to be novel, unpredictable, uncontrollable, and threatening to our survival, sense of identity, or ego. You may remember these stressor characteristics from Chapter 4. Building on this principle, SIT has two goals.

SIT's first goal is to expose people to the specific types of stressors they might possibly encounter in the "real world" so that those stressors will seem more familiar, predictable, and controllable—in theory, reducing stress arousal when they encounter those stressors again in the future. In other words, SIT attempts to modulate how the survival brain neurocepts these stressors.

SIT's second goal is to practice individual and group tasks during simulated scenarios to prevent performance degradation during stress. By repeatedly practicing basic skills and standard operating procedures during stressful situations, individuals condition themselves to perform these tasks automatically. In other words, SIT also attempts to modulate how the thinking brain appraises the stressors, to increase the sense of self-efficacy and agency during stress.

You can see both of these goals in the training my unit completed before our Bosnia deployment. Since we knew that Bosnia had been heavily mined during the war, for instance, we conducted SIT scenarios that involved locating, avoiding, and defusing mines. The idea was that when we eventually encountered mines in Bosnia, we would know what to do—and thus feel more in control.

SIT remains extremely popular for training and preparing people to perform well under pressure. By reducing stressors' novelty and habituating people to the stressors' effects, SIT is expected to improve mission performance in challenging environments.

Despite SIT's popularity, however, it has several downsides. First, while SIT may reduce the perception of novelty, unpredictability, and uncontrollability in specific simulated scenarios, *its benefits are often limited to the trained scripts and contexts*. For instance, in a study of civilian firefighters conducting training during actual fires, the more that firefighters repeated a training scenario, the less anxiety and fewer cognitive difficulties they experienced with that scenario. In new scenarios, however, they experienced anxiety and cognitive difficulties equal to—or even above—what they'd experienced before the first scenario. In other words, the firefighters were unable to translate and extend the emotional and cognitive skills they developed from repeating the first scenario to other scenarios, even when the new scenario scripts were only slightly different.

Second, even when SIT improves performance on repeated scenarios, it may undermine individuals' ability to retain—and later access and employ—the trained skills. That's because SIT scenarios are designed to expose individuals to the "sights, sounds, and smells" of real-world stressors, making them as realistic and as stressful as possible. Indeed, a common belief is that SIT should be *as* stressful as—if not intentionally *more* stressful than—actual combat or other real-world operations. The more challenging the preparatory tests, the better prepared we are—or so the theory goes.

Paradoxically, however, by making the training as stressful as possible, individuals often exceed moderate stress arousal levels during SIT scenarios. When this happens, thinking brain functions—including executive functioning and explicit learning and memory—will be degraded. As Chapter 5 explained, the thinking brain's *explicit* learning and memory system works best at *moderate* arousal levels.

Certainly, we still get something out of training at high arousal levels—if only the self-knowledge that we can function during high stress levels. And of course, the survival brain's *implicit* learning and memory system is still taking it all in. That's why SIT is designed to help individuals overlearn certain tasks and make them automatic.

Nonetheless, with thinking brain functions degraded, the ability to *consciously retain, apply, and generalize* the learning from SIT to other (similar or dissimilar) real-world scenarios will likely be compromised. That's likely why those firefighters experienced cognitive difficulties during new, slightly different scenarios.

The way that most HROs compensate for this particular downside is to create and practice many different scenarios, each one slightly different. Of course, this compensation creates a third downside: *Having to train and prepare for so many slightly different scenarios can be extremely time-consuming.* For instance, from the time the Dayton Peace Accords were signed until my unit deployed, we mostly worked sixteen-hour days, seven days a week, to fit in all of the different SIT scenarios that we were required to have practiced and certified.

The fourth downside flows directly from the third. With our calendar this overscheduled, we also experienced *a massive narrowing of our windows from the chronic stress of such time-intensive SIT and its corresponding sleep deprivation.* Because of SIT's window-narrowing effects, everyone's allostatic load increased—and most of us experienced symptoms of dysregulation—*before* we even deployed. I've observed this same dynamic among the hundreds of troops I've trained before their deployments to Iraq and Afghanistan.

Resilience is the ability to function well before and during stressful circumstances *and* to recover back to baseline afterward. This is what keeps our window wide. However, most SIT focuses only on task performance during stress—not on recovery—meaning it addresses only half of what resilience requires.

Thus, while SIT might help someone function well during stress *in the short term*, it can have some detrimental window-narrowing consequences over the longer term. In this, SIT shows its cultural affinity with grittiness, suck it up and drive on, and powering through. Recovery is generally devalued in our society—so we shouldn't be surprised to see how it's also devalued and ignored in HRO training regimens. Nonetheless, without recovery, allostasis stops working properly and we build allostatic load.

There's ample empirical evidence from HROs about how SIT is linked with cognitive degradation, anxiety, mood disturbances, and higher levels of perceived stress—all signs of narrowed windows. In addition to that firefighter study, research with military SIT programs—including field training exercises, survival training, and predeployment training—shows how SIT-related stress exposure is significantly linked to declines in thinking brain functions, including symptoms of dissociation, declines in attention and problem-solving skills, inaccuracies in pattern recognition, and declines in working memory capacity. Furthermore, since working memory capacity is needed for top-down regulation of negative emotions and stress arousal, it's not surprising that these studies also show increases in anxiety, other negative emotions, and perceived stress levels.

To be clear, training at high stress arousal levels—outside one's window—is not a problem. Doing so can help us feel more in control when we find ourselves in high-demand or extreme-stress situations in the future. By habituating us to function during high arousal levels, this *does* provide a certain kind of agency.

The problem comes from training at high arousal levels without a conscious focus on recovery. This is what makes SIT such an imbalanced and incomplete approach for finding agency during stress. Without an explicit focus on recovery, SIT encourages people to turn stress on without ever turning it off—in the process, building allostatic load, narrowing their windows, and leading to symptoms of dysregulation.

As a result, individuals often end up with narrower windows after SIT—narrower windows that they take with them when they finally encounter the "real world." In this place, they're more likely to exceed their stress capacity threshold—and end up *outside* their window—with less capacity to recall and apply lessons from their training.

Perhaps more important, with its relentless focus on practicing as many templated scenarios as possible, SIT often falls short in training people how to improvise when reality inconsiderately refuses to match the rehearsed scripts. By disproportionately emphasizing anticipation, SIT often creates a training deficiency in adaptability, improvisation, and recovery.

DOMAIN-GENERAL TRAINING: A BIGGER BANG FOR YOUR BUCK

There's another way to access agency during stress that's built on different principles. It's called *domain-general training*. Here, the goal is to build generalizable qualities and skills through consistent practice—and then trust individuals to apply these qualities and skills to particular situations, regardless of the circumstances.

Most skills-training paradigms—including SIT—rely on the principles of *domain-specific learning*, meaning that the learning doesn't transfer to new tasks or contexts. Indeed, the more we practice a particular task, the more efficient we become at performing it—as significant evidence shows, across virtually all perceptual, cognitive, or motor tasks in which people may be trained. Just as the firefighter study suggested, however, skills training is usually highly specific: People improve only on the trained task, with little or no transfer of learning to other tasks, even very similar ones.

In contrast, a handful of training paradigms confer *domain-general learning*. Domain-general learning occurs when an individual not only improves on the trained task(s) but can also transfer that learning to new tasks and to other domains. Thus, domain-general learning is *not* specific to trained stimuli, tasks, or contexts. The evolutionary purpose behind domain-general learning was for humans to be capable of altering their behavior adaptively to novel, nonrecurrent environmental cues.

Domain-general learning paradigms are typically more complex than domain-specific learning paradigms. To date, research has identified empirical benefits from four domain-general training regimens—action video games, musical training, athletic training, and specific forms of mental training.

For instance, action video-gamers have shown a variety of improvements in certain aspects of attention, eye-hand coordination skills, and even skill transfer to real-world tasks, like piloting procedures or laparoscopic maneuvers. In contrast, players of puzzle, fantasy, or role-playing games don't show these same effects. The distinction makes sense, perhaps, because action video games are fast-paced and unpredictable—characteristics that may be more similar to real-life contexts than other game types. Of course, action video games have other potential downsides, in terms of increasing stress arousal, when we aren't being careful to mitigate them.

Likewise, research has shown domain-general learning effects from musical training. Children who learn how to play a musical instrument see improvements in their time-space reasoning skills, IQ scores, mathematical ability, and verbal memory.

In the athletic domain, expert players of several sports show improvements in perceptual, cognitive, and motor skills important for their given sport but also transferrable to other domains—including selective attention skills, eye-hand coordination, spatial orienting ability, and faster visual reaction times.

More generally, *aerobic exercise* has been linked with a range of *cognitive* improvements in children and adults. Compared with sedentary adults, older adults who regularly perform aerobic activity show improvements in cognitive performance on a range of measures. Likewise, regular physical exercise in children and teenagers has been linked with better perceptual skills, memory, IQ scores, verbal tests, reading comprehension, mathematics tests, and academic readiness and achievement.

Certain forms of mental training constitute the fourth domain-general training paradigm. To be clear, however, there's little empirical evidence of the beneficial effects promoted by various "brain training" programs. In most cases, these "brain training" regimens have deliberately parsed out different cognitive processes for training—such as memorizing lists to enhance semantic memory, or training pattern identification skills to improve visual recognition. When someone is asked to memorize a list, however, they're not training other aspects of executive functioning, such as inhibition control or attention skills. Thus, not surprisingly, while the evidence may show improvements on specific trained tasks, it doesn't show domain-general learning effects.

These studies with "brain training" regimens highlight a trade-off inherent in how we learn. Breaking larger tasks into smaller cognitive subcomponents, as many of these games do, does allow for faster learning during skill *acquisition*. However, this task disaggregation can have negative effects on skill *retention* and *employment* later. While musical and athletic training also break skills down into subcomponents initially, over time the training becomes more integrated, complex, and holistic, ensuring that skill retention and employment is generalizable.

The first form of mental training that *does* provide domain-general learning is visualization of a physical skill, such as visualizing yourself running a race, performing surgery, or playing piano. Mental practice of the skill doesn't just improve muscle memory, as physical practice does. It also strengthens a more generalized understanding of the physical skill, which then promotes transferring that skill to other contexts. Incredibly, this research shows that *mental* practice of a *physical* skill—or practicing the skill both in the mind and with the body—makes skill transfer to other contexts easier than after only practicing the skill physically.

In addition, two forms of mindfulness meditation—what researchers call *focused attention* (FA) and *open monitoring* (OM)—have also been shown to confer domain-general learning. Importantly, domain-general learning has been documented for these two forms of mindfulness meditation, not for other forms of "meditation."

Focused attention techniques involve picking a target object, like sounds or the sensations of contact between the body and its surroundings, and beaming attention steadily on this target. If the mind wanders off, the meditator notes what has happened and returns their attention to the target. FA techniques build attentional control.

In contrast, open monitoring techniques involve nonjudgmentally observing the entire field of awareness, moment by moment, without focusing on any particular target object. Someone using OM techniques would simply observe the flow of stimuli—sights, sounds, tastes, smells, touches, other physical sensations, thoughts, and emotions. In practice, it's almost impossible to employ OM techniques successfully without first having developed robust attentional control with FA techniques. MMFT includes both FA and OM techniques.

Different domain-general learning effects are available after different amounts of FA and OM practice. Studies show beneficial effects after shorter periods of practice, as little as several weeks. However, more robust changes in certain facets of attention, perceptual discrimination, emotion regulation, immune functioning, and telomerase levels—an enzyme that affects cellular aging and cancer risk—have been documented only after intensive training for several months or years.

All domain-general training regimens share three features.

First, they vary stimuli and/or tasks, forcing us to learn at more abstract levels and understand how to use the skill in different settings. This contextual variation protects us from only memorizing specific templates and scripts (e.g., "*if this happens, then I do that*").

For instance, remember those London taxi drivers from Chapter 3, where the longer a cabbie drove a taxi, the bigger their hippocampus became? (The hippocampus is the brain region that allows us to consolidate explicit memory.) Here's an interesting twist about domain-general learning: The cabbies' hippocampal changes have *not* been observed in London bus drivers on established bus routes. In other words, because cabbies must vary their routes based on their passengers' needs, they constantly experience variation—thereby promoting domain-general learning. In contrast, bus drivers stick to prescribed routes—missing out on this variation and its corresponding domain-general learning effects.

Second, these training regimens involve *embodied* learning. They tap multiple processes at the same time, in both the mind and body. The action video-gamer, athlete, and musician engage and develop several perceptual, cognitive, and motor skills in tandem as they practice their activity—selective attention skills, eye-hand coordination, spatial orienting ability, working memory, and faster visual reaction time, to name just a few. Even the mindfulness practitioner and the athlete mentally rehearsing his upcoming race are tapping multiple cognitive and physical systems in parallel.

That's because where we direct our attention has profound ripple effects through the survival brain, nervous system, and body. Therefore, as we visualize or pay attention, we are also activating physical processes—even if we are sitting still.

Third, these training regimens aim for *moderate* stress arousal, the zone of optimal performance. Motivation is generally highest and learning most efficient when tasks are *slightly* outside our comfort zone—when we believe the task is challenging, yet still doable. Thus, these regimens usually involve task progression—advancing incrementally from one level of difficulty to the next—so that practice remains engaging but not overwhelming. We observe this principle when a video-gamer advances to the game's next level or the musician learns to play ever-more-challenging pieces.

In contrast, skills training that will *not* lead to domain-general learning often occurs at *low* arousal levels, which predicts low skill retention. Think about all those annual training requirements your workplace requires, such as online training modules about your company's sexual harassment policy or cybersecurity procedures. Be honest: How much have you really retained from the last time you completed such training?

Moreover, skills training at *high* stress arousal levels, including SIT, *also* predicts low skill retention. We simply don't retain well skills that we learn when we're outside our window—even if the training was intended to promote domain-general learning. For this reason, mind fitness training comes with an extra benefit: By teaching you to identify and then modulate your stress activation, you'll learn how to deliberately shift your arousal upward or downward to remain at moderate arousal levels. By modulating your stress to stay inside your window, you'll set the conditions where domain-general learning is most likely.

Earlier, I mentioned that domain-general learning has an evolutionary basis—so that humans could alter their behavior adaptively to novel, nonrecurrent environmental cues. Thus, the *original* domain-general training regimen is actually *instrumental parenting*. Parents teach their children domain-general skills through many instrumental processes. For instance, they direct their child's attention in particular ways ("*look at the clouds!*"), which teaches the child's brain attention-directing processes. They soothe their child when she's distressed, so that her survival brain and nervous system begin to associate distress with the possibility of self-regulation. They help the child with motor and cognitive functions for which she is not yet capable—in effect providing surrogate capacity until her own comes online.

Instrumental parenting epitomizes the three features of domain-general training. Parents vary stimuli and tasks, forcing the child to develop abstract skills that can be applied in many settings. Such learning is, by definition, embodied—tapping many physical and cognitive processes at once. Parents also gradually increase task difficulty over time, so that the child can stay at moderate stress arousal, the arousal level best associated with skill retention. For instance, parents might help with portions of an activity that are still beyond the child's ability, while simultaneously providing a chance for the child to learn through imitation.

As these many examples suggest, domain-general learning is very *fungible*, since it's applicable to many different settings and situations.

Say, for instance, a diplomat or soldier is preparing to deploy to another country with an unfamiliar language and culture. Ostensibly, they could receive domain-general or domain-specific training. With domain-general training, they might learn emotional intelligence skills, such as how to read emotions in themselves and others, regulate their negative emotions, and see situations from other people's point of view.

In contrast, with domain-specific training, they might be given a list of foreign language phrases and specific cultural rules for that country, such as "don't point your feet at others" or "women must cover their hair and walk a few steps behind the men." This was the type of training that we received before our Bosnia deployment.

Comparing these two hypothetical training paradigms, it's clear that the diplomat and soldier who receive domain-general training in emotional intelligence would be able to use those same skills while interacting with work colleagues and family members, too. In contrast, the list of rules specific to the other culture is less helpful for these other contexts.

There are some important reasons why organizations prefer domain-specific training, with its checklists, rules, scripts, templates, and guidelines for particular situations. Efficiency and speed are highly prized in our culture, as Chapter 2 explored—and domain-specific training fits with these cultural values. It's relatively quick, sometimes involving only a few hours for a briefing or a few minutes for an online training module. It's scalable and relatively easy to deliver via technology.

It's also easily assessed. Even with live training, most instructors simply memorize a script and then teach to "task, condition, and standard." After receiving the training, either individuals can perform the task to standard in the assigned conditions, or they cannot. If not, it's then relatively straightforward to provide remedial training to bring someone up to standard.

Although these training regimens have their place, they aren't actually training domain-general skills. They aren't improving core cognitive processes, strengthening mental agility, or widening the window with more capacity to regulate stress and negative emotions. Since 9/11, there's been an insatiable desire for training individuals in resilience, adaptability, and intuitive decision making. *And yet most programs that HROs and other organizations have funded and widely implemented with the goal of training domain-general capacities are actually domain-specific training programs!*

One definition of insanity is trying to grow oranges by planting apple trees.

THE WARRIOR TRADITIONS

I designed MMFT specifically to help people working in high-stress environments access the powerful benefits available from domain-general training. In order to do this, I anchored MMFT in the one training lineage that successfully accomplishes this goal in a high-stress context.

While most domain-general training is *not* intended to improve performance in high-stress situations, one form of domain-general training specifically was—the warrior traditions. The warrior traditions include structured teachings and techniques that have historically prepared warriors to engage in heroic action on behalf of others. Across the board, the warrior traditions trained individuals to wield latent violence *ethically* and *effectively* in the service of defending their communities. They focused on teaching fungible skills, applicable for any situation, and then trusted warriors to access and apply them whenever and however the particular circumstances required. These skills supported warriors in keeping their windows wide—so that they could access awareness, self-regulation, and ethical action, regardless of the situation.

The Warrior—or the essential characteristics that most people association with the term *warrior*—is one of the most enduring and heroic archetypes in human consciousness. In its positive form, the Warrior embodies *a path of service*, not self-interest—the willingness to risk even one's own life to defend others. Of course, like all archetypes, the Warrior also has a negative manifestation: unbridled competition to gain power and/or control over others, unrelated to any altruistic intent, larger social purpose, or higher ideals. The warrior traditions explicitly aimed to cultivate the archetype's positive qualities, while protecting against its negative ones.

The warrior traditions were established for training young men to defend their communities from threat, because of their innate strength, stamina, and health. These men were trained to hone certain qualities of mind and body, such as self-discipline and mental agility. In the process, they also learned to tame aspects of masculine adolescence, such as the tendency toward quick temper or macho behavior. Such domain-general training became the foundation for the warrior traditions.

Right now, you may neither think of yourself as a warrior nor feel particularly connected to this archetype, but anytime you speak out against an injustice, protect someone else from harm, risk your life or your livelihood to stand up for a principle, you are calling on the Warrior. For instance, the Warrior is present when a firefighter helps someone out of a burning house, a teacher stops a bully on the playground, a neighbor organizes a community sit-in to draw attention to an environmental problem, a whistle-blower comes forward to expose workplace fraud, a mother protects her child, or a driver stops to help someone change a flat tire.

Warrior traditions throughout the ages, from the Tibetan warriors and Japanese samurai in the East to the Spartans and Native American tribes in the West, have offered different practices to train the mind-body system to embody the qualities of wisdom and courage with a wide window. Although the list of specific warrior qualities varies somewhat by tradition, wisdom and courage show up consistently as the most important.

As the warrior traditions have traditionally defined these qualities, *wisdom* is the ability to see clearly how things are right now—not how we want them or expect them to be—and then to use that information to make the most effective choice in the moment. *Courage* is the ability to stay present with any experience, even an extremely difficult one, without needing it to be different. Together, these two qualities are a pathway toward agency and effective action in any sphere but especially in high-stress environments.

By strengthening the mind-body systems of individual warriors—and widening their windows—the warrior traditions aimed to nurture the *warrior ethos*. Even older than the just-war tradition or laws of warfare, the warrior ethos constitutes the "ethics of last resort" for warfare and other in extremis situations. This ethos can shape human behavior in situations of chaos, violence, stress, and ambiguity much more effectively than rules, laws, orders, or economic incentives. In other words, the warrior ethos was intended to provide warriors with internal self-control so that their incredible capacity for violence is effectively harnessed.

In perhaps the oldest Western description of the profession of arms, Plato was very clear that the path to the guardians' godlike behavior lay through the cultivation of wisdom and bravery, because "the soul that is bravest and wisest will be least confused or disturbed by external influences." Warriors must be able to embody integrity, even in our morally complex world where few things are absolutely wrong or right. With wisdom and courage, a powerful warrior can see clearly, tolerate what's happening rather than get jerked around by it, and then choose the most effective course of action.

CONSISTENT PRACTICE ON THE PATH TO MASTERY

All warrior traditions share three characteristics. First, while they teach different techniques—some focused more on the body, others on the mind—all warrior traditions follow a path of consistent practice to cultivate self-discipline and mastery.

The best-known warrior body practices include the martial arts. Initially, martial arts focused on sword fighting, archery, and hand-to-hand combat with weapons. Over time, however, being able to perform certain physical movements without weapons became acknowledged as separate skills—the beginning of the unarmed martial arts that still continue today. Other warrior body-oriented practices include breath control techniques, physical exercise, horsemanship, and yoga. With body-focused practices, warriors cultivate physical qualities such as stamina, strength, speed, and agility.

In contrast, some warrior practices exist primarily to train the mind. For instance, the Lakota Sioux's vision quest and the Aborigines' walkabout involved warriors retreating into nature for extended periods to "enter the silence," brave the elements, and confront their fears. Spartan warriors in ancient Greece entered *agoge* training when they were just seven years old, learning to endure harsh environments, sleep deprivation, and pain with equanimity. Stoic warriors in ancient Greece and Rome engaged in daily contemplative practices, such as reviewing each day's behavior, learning from their mistakes, and developing freedom from their emotions. Chinese warriors cultivated awareness through the mindful movement practice of tai chi, while Japanese samurai were trained in *zazen*—a kind of mindfulness training—and *koans*. *Koan* practice (pronounced "koh-an") involves reflecting on a puzzle not solvable through logical reasoning, to teach the mind to surrender old constructs and intuit new insights.

Regardless of the technique, all warrior traditions shared a common understanding of the *goal* of practice—to follow the path consistently and thereby cultivate self-mastery. Whatever his chosen practice, the warrior can't achieve mastery simply by learning a technique, understanding it intellectually, or reading about it. Instead, he must exert himself through consistent effort to embody that technique fully.

In *The Book of Five Rings*, for instance, the sixteenth-century swordfighting (*kendo*) master Miyamoto Musashi writes that "in order to master the ways by which one can be victorious over others. . . . You must walk down the path of a thousand miles step by step." Practicing one thousand days is said to be "discipline," while practicing ten thousand days is said to be "refining." His comment echoes recent laboratory research on skill acquisition, which demonstrates that we can acquire expert skill only through extended periods of deliberate practice.

Yet the path isn't about "making progress" or striving to get somewhere. Such striving can actually work *against* cultivating warrior qualities. In fact, the more compulsively a warrior struggles for a particular achievement, such as winning martial arts belts or attaining particular mind states, the more attached her ego becomes to that outcome—and the less likely she can access wisdom and courage.

In this situation, we can lose sight of the deeper purpose, which is honing the qualities of mastery. As Aristotle noted, and recent neuroscience research corroborates, we develop character traits through repetition: "These virtues are formed in man by his doing the actions." Excellence, self-mastery, and character are not onetime events. They are habits. The very act of practicing consistently—beginning again when we falter, learning from our mistakes, treading the path sincerely—is the objective.

TRAINING FUNGIBLE QUALITIES

Another goal of the warrior traditions is to cultivate foundational qualities in the mind-body system. This makes sense intuitively to anyone who exercises. A body strengthened through weight training, for instance, has more capacity to carry a heavy rucksack for long distances or push a vehicle stuck in the mud back onto pavement. In other words, the strength from weight training is *fungible*, which means it can be employed in every facet of our lives.

In the Japanese warrior traditions, for instance, this is known as *ken-zen ichi nyo* ("body and mind, together"). Ideally, we can embody these qualities every moment, regardless of what's happening. Ultimately, warriors practice consistently to widen their windows, which then allows them to navigate any challenge with flexibility, presence, and sense of humor intact. As samurai Musashi put it, "The true path . . . is such that it applies at any time and in any situation."

Warriors aim to cultivate fungible qualities without getting attached to any particular outcome. Warriors may *prefer* a particular outcome, but that's not what's most important; instead, they should focus on embodying their training, regardless of outcome. As the Roman Stoic Cicero put it, "One's ultimate aim is to do all in one's power to shoot straight"—however, "to actually hit the target" is "to be selected but not sought."

And yet—here's the paradox—when the warriors' focus is on embodying their skills, they're more likely to create circumstances that set the conditions for success. In *The Art of War*, the collective wisdom of a lineage of Chinese warriors written about 2,300 years ago, Sun Tzu explained it this way:

Invincibility lies in oneself.

Vincibility lies in the enemy.

Thus the skilled can make themselves invincible.

They cannot cause the enemy's vincibility.

Thus it is said, "Victory can be known. It cannot be made."

The warrior recognizes that he can't control the enemy or environment around him. He can't "make" victory happen. However, by training to become invincible—widening his window and honing what *is* up to him—the warrior becomes skillful at recognizing and taking advantage of the moment of the enemy's vincibility. In this way, victory "can be known," but not made.

WISDOM AND COURAGE WORKING TOGETHER

All warrior traditions use consistent practice to cultivate wisdom and courage, which show up consistently as the most important qualities across all warrior traditions.

Wisdom is the ability to see things clearly—not how we want or expect them to be, but *how they actually are*—and then to use this information to make the most effective choice in this moment.

We usually think of wisdom as a static quality acquired through life experience, and many warrior traditions traditionally trained this component of wisdom through contemplative practices about death and impermanence. Yet wisdom also involves a clear, objective apprehension of what's happening right now. *This second component of wisdom cannot be planned—it's available only when we pay attention, moment to moment.* With training, we can learn to bring impartial observation and nonjudgmental curiosity to each moment—without judging, evaluating, comparing, or expecting this moment to be like a previous one.

However, as many warrior traditions point out, we can't simply muscle the thinking brain into setting aside its expectations, comparisons, opinions, and judgments. We can only train the mind by *directing the attention* so that, over time, it builds the capacity to set these things aside naturally.

If we try to circumvent that training, and *force* such internal commentary aside, it usually backfires on us. A tightly *controlled* mind may superficially appear similar to a well-*trained* mind, but it can't achieve the same clarity, flexibility, and freedom. That's because *tight mental control consumes a tremendous amount of energy to keep the internal chatter at bay.* This not only depletes executive functioning but leaves little attention available to see and respond fluidly to the current situation.

Skillful action informed by wisdom only emerges from an accurate, unvarnished appraisal of the unique contours of *this* situation in *this* moment. As the Sun Tzu lineage of warriors put it, "These are the victories of the military lineage. They cannot be transmitted in advance."

Wisdom requires trusting that when we fully arrive in the present moment and see it clearly, from this awareness will emerge the most perfectly appropriate response for this exact situation. When we operate from this place, it can even feel like we're not doing anything at all.

In contrast, courage is the ability to stay present with any experience, even an extremely difficult one, without needing it to be different. Indeed, courage is so foundational for warriors that the Tibetan word for warrior, *pawo*, literally means "one who is brave."

We usually think of courage as the ability to move forward physically or metaphorically, despite obstacles or challenges, in order to make things different from the way they are. It's not surprising that the quintessential image of courage in our culture is the dirty, hungry, exhausted, and terrified soldier still moving forward as bullets fly around him.

To be clear, the warrior traditions' definition of courage is not incompatible with this more typical one. However, their definition of courage acknowledges that before we can successfully act to make reality different, first we must willingly submit to seeing clearly and allowing reality to be exactly as it is. Otherwise, we may waste effort and energy trying to change aspects of reality that are actually beyond our control. Thus, to access agency most skillfully, first we must be able to see and tolerate the situation.

Courage is more than just bravery during the physical hardships of battle and other adversity. Courage helps us stay present with and tolerate all discomfort, vulnerability, and pain—physical, psychological, emotional, intellectual, and spiritual—without denying it, avoiding it, or pulling away.

Courage also involves honesty: being honest with ourselves about what's happening, taking responsibility for our vulnerabilities and mistakes, and connecting truthfully with others from the heart.

Warriors cultivate courage to accept the truth of our human condition—that we're mortal, that we'll inevitably experience things we don't want to experience, and that life isn't under our control. Courage allows us to face these truths directly.

For instance, the Spartan king Agesilaus was reputedly once asked to name the supreme warrior virtue, from which all other virtues derived. His reply: "Contempt for death." Yet courage is not simply a contempt for death as much as a willingness to accept that death can come at any moment—and continue to take action anyway. When we cultivate this willingness, we can live fully, authentically, and fearlessly in the present moment. This moment right now is all that we have.

While all this makes sense in theory, it can be difficult to live in practice—especially during challenging circumstances. Nonetheless, regardless of what's happening, we can choose to keep showing up and making the most of the situation. This choice, over and over, is central to the Stoic understanding of courage. As Epictetus explains, *a warrior's job is to find his agency*—and to keep finding it again and again, even when his agency is constrained and he's feeling vulnerable. For the Stoics, the way to do this is to keep concentrating on the factors that are *"up to us"*—internal qualities in the mind-body system and our internal reactions to external events.

Perhaps the reason why all warrior traditions have practices to cultivate both wisdom and courage is that these two qualities support and reinforce each other. *We need both to fully access either one.* Wisdom requires seeing the truth—even when truth is unpleasant or uncomfortable—but it's courage that allows us to do this, without giving in to denial or dishonesty.

It takes courage to see and take responsibility for our weaknesses, limitations, vulnerabilities, and unskillful choices. This is especially difficult in our society, because modern warrior culture tends to perpetuate the delusion that we're invulnerable. Yet for us to reach our full potential, we need to see our shadows clearly and then choose to learn, grow, and change. Who we'll become in the future always begins with the total awareness and acceptance of who we are right now. In turn, we can stop wasting energy denying what's already here—freeing us to see clearly what's happening and then respond effectively.

Courage also helps us take responsibility for and not second-guess previous choices. If wisdom *was* present, courage supports us in trusting that it was the right choice—regardless of how things turned out later. And if wisdom *wasn't* present when we made the initial choice, courage helps us learn from the situation so we can make wiser choices going forward.

Although there may be many courageous paths we could follow, wisdom helps us discern the best one for the current circumstances. Some situations require charging forward despite the whizzing bullets, while other situations require sitting this round out so we can fight another day. Only wisdom allows us to discern. We can't rely on a formula or script, because there's only one best choice for any particular situation.

Truly courageous people sometimes choose to yield, withdraw, or surrender. As Part I explored, American culture in general—and modern warrior culture in particular—has strong conditioning to "play hurt" or "power through," even when doing so isn't required in a given moment. However, although it may look like "warrior behavior" to keep going, wisdom helps us understand how getting injured or burned out can actually undermine our long-term effectiveness. Gritty persistence is not always the most appropriate path. Sometimes our goals are best achieved by pausing, regrouping, and gathering our internal and external resources, so we can move forward again at a later time.

Ultimately, wisdom and courage together open the possibility for masterful action. When we have the wisdom to see things clearly, and the courage to confront the truth about ourselves and our environment, we become truly available to serve in the world.

TRAINING THE WARRIOR ETHOS

In our society, we usually consider ethical behavior to result from moral character—a disposition that either we have or we don't. In the process, we've lost touch with a truth from the warrior traditions: Character isn't fixed and immutable—*it comes from repeatedly cultivating virtuous qualities, widening our windows, and making skillful choices.* In other words, character can be strengthened through training and consistent practice.

Throughout the ages, warrior traditions focused on cultivating *both* wisdom and courage, because they understood that *we need both qualities to fully access either one.* Together, these two qualities help us access agency.

Yet modern warriors don't cultivate both qualities equally. Modern HRO training regimens mostly ignore embodied practices for cultivating wisdom. This imbalance likely contributes to the increases in impulsive decision making, poor judgment calls, unethical and violent behavior, and moral injury we've observed in the last two decades. The warrior ethos constitutes the "ethics of last resort" in even the direst of circumstances—but only if both wisdom and courage have been developed equally.

Building on millennia of warrior traditions with the same aspiration, MMFT cultivates wisdom and courage—two qualities necessary for finding agency, functioning adaptively during stress, and recovering afterward. Indeed, MMFT's first module explicitly focuses on the warrior traditions, to situate its skills training within this ethical framework.

Specifically, MMFT cultivates two fungible skills, attentional control and tolerance for challenging experience, which are actually moment-by-moment micromanifestations of wisdom and courage. Furthermore, by deliberately pairing mindfulness skills with nervous system self-regulation skills, MMFT teaches individuals to widen their windows *and thereby increase the likelihood that their thinking brain capacity for ethical and effective decision making will remain online during stress.* It also increases the likelihood that individuals can access agency even during extreme events, decreasing the likelihood that they'll experience trauma.

We can't predict the future, but we can cultivate qualities that can be applied in any possible future, especially wisdom and courage. To rouse wisdom in any moment, we only need ask: *What's happening right now?* To rouse courage, we only need ask: *Can I be with this experience, just as it is, without needing it to be different?*

While not all of us will experience combat, we can each cultivate our inner Warriors through domain-general training practices from the warrior traditions. As we widen our windows, we can learn to embody wisdom and courage—and access agency—in every aspect of our lives. We can trust that we have the capacity to show up with flexibility, resilience, and resourceful-ness, wherever we find ourselves on the stress-trauma continuum. This starts with teaching our thinking brain and survival brain to work together as allies, which is where we're headed next.

CHAPTER TWELVE

The Thinking Brain and Survival Brain as Allies

Building an allied relationship between the thinking brain and survival brain requires training our attention in a systematic way. That's because where we direct our attention has profound ripple effects through the survival brain, the nervous system, and the body.

Since the survival brain is not verbal, the way that it communicates with us is through emotions and physical sensations. Whether we receive the survival brain's transmission correctly, however, depends on our capacity to notice, tolerate, and *accurately* interpret the message being conveyed by emotions and physical sensations. This capacity is called *interoceptive awareness*. Interoceptive awareness is one aspect of *interoception*, which includes the ability to recognize bodily sensations, be aware of emotional states, and regulate physiological processes to keep the mind-body system functioning, especially allostasis.

This chapter builds toward instructions for beginning your own mind fitness practice. To orient you, I want to preview this chapter's progression: Interoception is important for building an allied relationship between the thinking brain and the survival brain. Although mindfulness can help us cultivate interoceptive awareness, mindfulness alone may backfire for people with narrow(ed) windows. MMFT was explicitly designed to address this challenge.

Because this book is *not* the MMFT course but also includes information not directly covered in MMFT, this chapter will (finally) explain what MMFT is—and how and why this book differs from it. Finally, I'll conclude with some practical guidance for mind fitness practice and instructions for the first MMFT exercise.

245

BUILDING INTEROCEPTIVE AWARENESS

As Chapter 1 explained, awareness belongs to neither the thinking brain nor the survival brain. Thus, interoceptive awareness lets the thinking brain gather the information it needs from the survival brain and body in order to diagnose our current stress level. If the current situation doesn't require—and may actually be hindered—by our current stress level, the thinking brain can then direct our attention in ways that interrupt our default programming and instead facilitate the survival brain to initiate downregulation and recovery functions.

Two thinking brain regions—the *insula cortex* and the *anterior cingulate cortex (ACC)*—play a critical role in interoception. Neuroscientists argue that the insula and the ACC together provide top-down control of the survival brain processes that regulate stress and emotions. The insula and ACC are also involved with accessing the ventral PSNS circuit—including the social engagement and attachment system, the vagal brake on the cardiovascular system, and recovery functions. Simply put, when these two brain regions operate efficiently, they support our rapidly and effectively switching between thinking brain and survival brain functions.

The insula plays a major role in allostasis; it helps the brain become aware of imbalances in, or conflicts between, internal systems. The ACC orchestrates emotions and plays a role in emotion regulation and impulse control. Specific regions of the insula and ACC also serve as the brain's *pain distress network* for both physical and social/emotional pain.

Since the insula and the ACC together provide top-down control of survival brain processes regulating stress and emotions, *we can improve the functioning of this regulatory loop by cultivating interoceptive awareness.* We do this by building our capacity to pay attention to physical sensations and sensory stimuli, such as sights, sounds, or smells. In turn, better interoceptive awareness—especially the ability to sustain attention on unpleasant stimuli—helps us improve our survival brain's functioning during stress arousal and intense emotions.

Many practices can help us cultivate interoceptive awareness, including mindfulness meditation, tai chi, and some forms of yoga and martial arts. *Mindfulness* is the ability to pay attention and notice what's happening while it's happening—without getting caught in judgments, comparisons, narratives, or emotional reactivity. When we observe our experience with mindful awareness, it's similar to the way that a scientist observes an object under a microscope—clearly and exactly as it is, without any preconceived notions, agendas, or expectations. When we engage in mindfulness of physical sensations and sensory stimuli, we cultivate interoceptive awareness.

Interoceptive awareness works differently from executive functioning—*which means it can be available to us even when thinking brain functions are degraded during stress.* As Chapter 5 explained, executive functioning and explicit memory are usually degraded during chronic stress and trauma. In this depleted state, thinking brain skills that rely on executive functioning—including planning, problem solving, decision making, willpower, and self-control—can also suffer.

But because it acts as a bridge, efficient interoception has been shown to improve thinking brain functions and performance during stress. In brain imaging studies, for instance, military and civilian "elite performers"—including special operations forces, Navy SEALs, elite athletes, and adventure racers—have demonstrated insula and ACC activation patterns consistent with more efficient interoceptive functioning during stress. Importantly, this research captures static snapshots comparing elite performers with healthy, non-elite-performing people. For this reason, it's unclear whether the elite performers showed more efficient interoceptive functioning during stress because of selection effects or their training.

When we're dysregulated, our interoceptive functioning may be compromised. For instance, infants and children who experience difficulty developing their ventral PSNS circuit are likely to develop compromised interoceptive functioning. Compromised interoceptive functioning plays a critical role among adults with depression, anxiety disorders, PTSD, hypochondrias, and addictions. Each of these issues exemplifies an adversarial relationship between the thinking brain and the survival brain.

Nonetheless, efficient interoceptive functioning isn't simply a trait that we have or we don't. As the principles of neuroplasticity would suggest, it's something that we can deliberately train and cultivate. In fact, *MMFT research has shown that developing more efficient interoceptive functioning is possible in just two months.*

For instance, compared with Marines in a control group, MMFT Marines showed significant changes in interoceptive functioning during two stressful tasks in the brain scanner—cataloging emotional faces and enduring a restricted breathing challenge. In fact, after ten weeks of stressful predeployment training and the twenty-hour MMFT course, their insula and ACC activation patterns had shifted to match the pattern observed among the "elite performers" in the earlier studies.

IT TAKES MORE THAN MINDFULNESS

But building interoceptive awareness can feel challenging. That's because narrow(ed) windows are often associated with compromised interoceptive functioning. In this situation, when we bring our attention to physical sensations and emotions, the survival brain may neurocept danger and then produce *more* stress activation.

Unfortunately, the explosion of popular interest in mindfulness has created an expectation that there's some quick and easy path to peak performance and bliss. Indeed, there are books, magazines, blogs, apps, and podcasts for applying mindfulness in many realms, including education, business, leadership, parenting, nursing and care providing, and politics. Over time, these texts and tools have promised ever-better benefits with ever-less commitment required: *Mindfulness for Dummies. 10 Mindful Minutes. One-Minute Mindfulness.*

As a recent *New York Times* article put it, "Meditation is a simple practice available to all, which can reduce stress, increase calmness and clarity and promote happiness. Learning how to meditate is straightforward, *and the benefits come quickly.* Here, we offer basic tips to get you started on a path toward greater equanimity, acceptance and joy. Take a deep breath, and get ready to relax."

Clearly, we Americans like our silver bullets. Why not have a mindful one, too?

In fact, mindful awareness does take place nearly instantaneously— moment by moment, as we observe our experience unfolding. As a result, one minute's mindfulness can actually have a profound effect: It can help us choose composure instead of reactivity, conscious intention instead of autopilot, connection instead of withdrawal. Nevertheless, we won't be able to access mindful awareness—*especially when we're stressed*—if we've only been practicing one minute each day.

Beyond the misguided expectation of a mindful silver bullet, it can be a major setup for suffering when someone tries to square a dysregulated mind-body system with the hype promised by the "Mindfulness Revolution." What happens when someone reads all these media promises, tries to practice, but then doesn't see these touted benefits? Chances are, they're likely to think that they're doing something wrong—or even worse, that something is "wrong" with them.

Unfortunately, for people with narrow(ed) windows, *mindfulness practice by itself has the potential to make their dysregulation worse.*

In particular, a mindfulness-only training regimen increases the risk that someone with a narrowed window will become more aware of their dysregulation but not understand how to work with it effectively. I know—not only from personal experience but also from observing this trend among the first group of Marines I trained in 2008.

Through these experiences, it became clear to me that it's ethically imperative, in order not to cause harm, that the introduction of mindfulness practices in any high-stress environment—and among people with narrow(ed) windows—needs to be paired with skills for nervous system self-regulation.

Though some exceptions exist, most mindfulness-based interventions (MBIs) were not designed to accommodate and reregulate deep-seated mind-body dysregulation. For instance, mindfulness-only training gives someone the tools to *be aware of* and *accept* the racing thoughts, intense emotions, rapid heart rate, shallow breathing, and butterflies in the stomach that come with stress activation.

In contrast, MMFT *also* enables that person to direct their attention in particular ways in order to *downregulate* and *reduce* that stress arousal—by completing a discharge of the energy and hormones mobilized with stress, thereby bringing their mind-body system back into balance.

Mindfulness alone, without skills to reregulate the nervous system, may actually flood our mind-body system with heightened attention on the stress response, which often worsens our ability to self-regulate and exacerbates symptoms. That is, if you're extremely aware of your mind-body system and you're feeling stressed, all you may be able to do is focus on the stress, which could actually amplify the stress arousal and its cognitive, emotional, and physiological effects.

With this in mind, the findings of a recent study evaluating the efficacy of mindfulness training with more than three hundred middle and high school students, from diverse socioeconomic backgrounds, makes sense. Students in the mindfulness group completed eight weeks of training in the .b ("dot be") mindfulness in schools curriculum, which is based on the mindfulness-based stress reduction (MBSR) and mindfulness-based cognitive therapy (MBCT) programs for adults.

In this study, the mindfulness group showed no benefits at testing sessions either immediately after the training or three months later. In fact, three months later, *anxiety was higher* specifically among male teens in the mindfulness group relative to males in the control group. Likewise, among participants (both genders) who'd started the study reporting relatively fewer depressive symptoms and fewer weight/eating concerns, anxiety levels also increased over time.

Dr. Willoughby Britton, a clinician and researcher at Brown University who spearheads the Dark Night Project, argues that Western scientific research about mindfulness has been biased toward overrepresenting positive results and examining potential benefits, without adequate attention to potential harms or risks—because our culture assumes that mindfulness practices mostly exist for stress reduction and enhanced performance.

To counter this bias, her team interviewed experienced practitioners/teachers with many thousands of hours of practice who've had challenging meditation experiences: Almost three-quarters indicated moderate to severe impairment with symptoms. Those with psychiatric histories (32 percent) and trauma histories (43 percent) reported more severe and/or longer-enduring symptoms. Her team is currently investigating similar experiences among novices who completed an eight-week MBL. Other recent research suggests similar results.

And indeed, some mindfulness-based interventions note their contraindication for individuals actively experiencing post-traumatic stress or trauma. For instance, the University of Massachusetts's Center for Mindfulness—where

Jon Kabat-Zinn created MBSR—states that MBSR is not advised during active PTSD or other mental illness. Instead, they suggest that individuals seek other training or treatment if they have "a history of substance or alcohol abuse with less than a year of being clean or sober, thoughts or attempts of suicide, [or] recent or unresolved trauma," or if they are "in the middle of major life changes."

Importantly, *all of these criteria are quite common in high-stress environments,* where dysregulation from chronic stress or trauma is almost certainly present. Mindfulness training alone is likely not sufficient to address this constellation of challenges.

In the detachment I trained during the 2008 MMFT pilot study, about two thirds of the Marines already had one to three prior combat deployments. One Marine had also deployed six times as a private military contractor. Since this was a reservist unit, they hailed from a range of civilian backgrounds, but by far the largest group, 40 percent, worked in high-stress civilian occupations—including SWAT team members, narcotics detectives, firefighters, and EMTs.

During their individual practice interviews with me, I soon learned that six out of ten Marines were actively experiencing multiple symptoms of dysregulation. For many, this distress was getting worse as they prepared to deploy. Their symptoms covered the range of cognitive, emotional, physical, spiritual, and behavioral symptoms of dysregulation I outlined in Chapter 10.

Those with symptoms of dysregulation included two thirds of the Marines with prior combat deployments and half of the Marines in high-stress civilian occupations but without combat experience. As I grew to know them, I learned how their windows had narrowed through stressful and traumatic experiences in their pasts—during prior combat deployments, their high-stress civilian occupations, and early-life adversity.

Over the years, as I've worked with many other individuals and organizations, I've learned that this degree of symptom expression is pretty typical in high-stress contexts—despite what formal statistics might say. For instance, I routinely see this level of symptom expression among my Georgetown students.

When I was teaching those Marines, I'd not yet created MMFT's progressive sequence of mind fitness exercises, specifically designed to move someone from dysregulation to regulation. Instead, I'd started by teaching them shorter versions of two practices common to many mindfulness-based interventions: the "body scan," during which someone directs their attention sequentially to observe sensations at each part of their body, and "awareness of breathing," one of the classic forms of mindfulness meditation.

What's supremely ironic is that by choosing to teach these two exercises, I was consciously disregarding important lessons from my own early experience with mindfulness training, back when I'd experienced PTSD—because I'd assumed that my own experience was an anomaly. None of my mindfulness teachers taught about dysregulation, so it never occurred to me that *my own initial reaction to practice was actually quite common among people with dysregulation*!

Although I never learned the body scan during my own early years of practice, my initial exposure to awareness of breathing had not been pretty. In fact, when I first learned this practice in fall 2002, I practiced only ten minutes each day, simply because that was all I could tolerate—and only then by gritting my teeth.

As with many beginners, those ten minutes seemed *interminable*. I was first ashamed and then amazed at the constant chatter of my mind. It seemed almost impossible to just sit there, feeling all that restlessness coursing through my mind-body system.

Yet sometimes as I observed my breathing, I would flip into panic. I'd find myself suddenly gasping for air, unable to breathe—my body reliving again, at a cellular level, several traumatic events from my life. Whenever this happened, afterward I'd be panicked for days, with a fresh surge of flashbacks, claustrophobia, nightmares, insomnia, nausea, and hypervigilance. I didn't understand why this was happening until my own clinical trauma training many years later. In effect, paying attention to my breath triggered unresolved memory capsules stored in my survival brain via kindling. In the process, this bottom-up flooding actually retraumatized my mind-body system further.

At that time, I figured I must be practicing incorrectly or didn't have enough discipline to sit with the discomfort. As a result, I'd grit my teeth and practice some more—*all the while, only reinforcing my suck it up and drive on override conditioning further.*

It was only when I started observing *almost all* of my Marines having similar reactions to the body scan and awareness of breathing that I finally connected the dots. Even though I taught ten-to-twenty-minute versions of the exercises—as opposed to forty-five minutes, as in other MBIs—it was still too much. These two exercises were simply beyond what their mind-body systems could tolerate.

Even among the Marines most sincere about practice, within five to ten minutes of starting an exercise, they reported feeling like they were going to jump out of their skin or needed to punch the nearest wall. They reported pounding hearts, shallow breathing, nausea, flashbacks, intolerable anxiety, irritation, rage, or restlessness. Some even reported feeling dizzy, confused, checked out, or spacey. In other words, their reports mirrored my own early experience.

So why did we all have these reactions to practice?

When we experience prolonged stress or trauma without adequate recovery, the integration of information between our thinking brain and our survival brain can become disconnected or disordered, which contributes to our symptoms of dysregulation. Thus, to fully recover and thereby widen our window, we need to engage in and complete the processing of stressful and traumatic experiences at all three levels—cognitively, emotionally, and somatically (in the body). This happens only with the integration of top-down (thinking brain) *and* bottom-up (survival brain) processing.

As Chapter 13 will explore in more detail, effective bottom-up processing can occur anytime the survival brain neurocepts safety—and we allow the mind-body system to discharge the stress activation. Not surprisingly, therefore, efficient interoceptive functioning is necessary for bottom-up processing to occur.

Most MBIs begin with awareness of breathing, which they consider to be "relatively neutral sensory stimuli." Yet for many people—including those of us with histories of asthma, near-drowning, or any other traumatic event where we experienced a freeze response which, by definition, involved air constriction and/or oxygen conservation—*breathing sensations are almost certainly not neutral, especially when we're stressed.* When someone with such a history directs their full attention to breathing, breathing sensations may be extremely activating and can trigger more stress arousal and/or panic.

As a result of these insights from the 2008 pilot study, I restructured MMFT to make it more stress- and trauma-sensitive. For instance, I stopped teaching the body scan exercise entirely, and I started waiting at least a month to introduce awareness of breathing.

As a result, with MMFT, you can learn to direct your attention inward to assess your stress arousal level but also toward attentional cues that can help your survival brain and body feel stable. MMFT's carefully progressive sequence of body-based exercises focuses on *grounding*, so that the survival brain can neurocept safety and then initiate recovery. Over time, this enables our mind-body system to recover and widen the window gradually— without flooding or retraumatization—while also building greater capacity for sustained attention to body sensations.

In fact, experienced mindfulness practitioners who participate in MMFT are often surprised to discover how dysregulated their mind-body systems are, despite their decades of intensive mindfulness practice. "I had no idea that there was a way to work with these intense symptoms besides simply observing and accepting them!" they tell me. "This is the missing piece I've been looking for."

In sum, to prevent retraumatizing the mind-body system and reinforcing thinking brain habits that override the survival brain—all of which could exacerbate dysregulation further—*it's important to build our capacity for interoceptive awareness gradually.* Otherwise, we'll only further entrench the vicious cycles associated with an adversarial relationship between them. It's especially easy for those of us who cope with stress via thinking brain override to reinforce our override conditioning. Therefore, *slower is actually faster.* The introduction and sequencing of the target objects of attention in the exercises really matters.

WHAT IS MIND FITNESS TRAINING?

I designed MMFT with two overarching goals in mind: to help individuals widen their windows and to do so in a stress- and trauma-sensitive manner. To achieve these tailored goals, MMFT draws from two lineages: mindfulness training and body-based trauma therapies for reregulating the nervous system and survival brain after trauma, such as sensorimotor psychotherapy, Somatic Experiencing, and the Trauma Resilience Model.

MMFT has three components: (1) mindfulness skills training; (2) an understanding of our neurobiology and body-based self-regulation skills training to regulate the nervous system; and (3) concrete applications of both types of skills to participants' personal and professional lives. This blend of mindfulness skills training with body-based self-regulation skills training is crucial for widening the window, increasing resilience, and enhancing performance in high-stress situations.

A major focus of MMFT is improving self-regulation, at both the micro and macro levels. When we're regulated, we're more likely to find agency and access choice in *every* situation, no matter how challenging, stressful, or traumatic it might be.

When MMFT is taught over eight weeks, the first four two-hour sessions occur in the first two weeks, to front-load the neurobiology context for the skills taught in the course. These sessions focus on the scientific foundations of the neurobiology of stress and resilience. They also introduce the basic exercises for self-regulation. The other four two-hour sessions are taught in the fourth, fifth, seventh, and eighth weeks. These sessions teach content about habitual reactions, decision making, emotions, interpersonal interactions, and conflict. They also introduce more advanced exercises for self-regulation interpersonally. During the third week, participants have an individual practice interview, and during the sixth week, they complete a four-hour practicum to refine mindfulness and self-regulation skills.

In addition to the eight-week format, MMFT has also been taught as a weeklong intensive course or as introductory workshops. In these formats, participants learn some of the intellectual context intensively and then complete the eight-week exercise sequence afterward, on their own.

MMFT participants are asked to complete at least thirty minutes of mindfulness and self-regulation skills exercises daily, outside class sessions. Daily practice can be divided into several practice periods throughout the day. MMFT's exercises range from five to thirty minutes—which is notably, and deliberately, shorter than the forty-five-minute exercises included in many other MBIs. Participants initially use audio tracks to guide the exercises, but over time they can practice without audio support. Some exercises are conducted while sitting quietly or lying down, some while stretching, and some are designed to be integrated into daily-life tasks.

Through these exercises, MMFT aims to cultivate two core domain-general skills: *attentional control* and *tolerance for challenging experience*. Because these skills undergird other competencies needed for effective decision making and interpersonal interactions—such as situational awareness, emotional intelligence, and mental agility—cultivating these skills can provide a big return on training investment. Attentional control is mostly cultivated through focused attention (FA) techniques, while tolerance for challenging experience is cultivated through both FA and open monitoring (OM) techniques.

Attentional control is the ability to direct and sustain attention deliberately on a chosen target over time. Attentional control leads to improved concentration, more capacity to inhibit distractions, and more capacity to remember and update relevant information.

Tolerance for challenging experience is the ability to pay attention to, track, and stay present with such experience without needing it to be different. Such challenging experiences can be external (e.g., harsh environmental conditions or difficult people) or internal (e.g., physical pain, stress activation, intense emotions, distressing thoughts, nightmares, or flashbacks). Without training our capacity to tolerate challenging experience, many of us default to checking out, distracting ourselves, or trying to "fix" the discomfort with stress reaction cycle habits and other impulsive or reactive behavior. Importantly, tolerance for challenging experience is *not* the same thing as "sucking it up." That's because sucking it up is actually an aversive form of thinking brain override—which, by definition, means we're not fully present to *all* of the information available in the challenging experience.

As I've said, this book is not the MMFT course—it covers additional topics not directly addressed in MMFT, but by necessity it also can't replicate all of MMFT's experiential practices.

To help you *safely* begin a mind fitness training regimen, in this book I'm including only the two MMFT exercises that best facilitate our survival brains to neurocept safety and recover from stress activation—the Contact Points Exercise (in this chapter) and the Ground and Release Exercise (G&R,

in the next chapter). These two exercises also allow us to cultivate attentional control and tolerance for challenging experience, the two domain-general skills that support wisdom and courage. In this book, I've deliberately chosen *not* to include any MMFT exercises that sometimes lead to increased stress activation among participants with narrow(ed) windows.

To be clear, if you're confronting profound dysregulation, I believe working with a therapist trained in body-based trauma techniques (such as sensorimotor psychotherapy or Somatic Experiencing) is essential. They can help you pace your survival brain's bottom-up processing, so that it happens gradually and safely. I *strongly* recommend you seek out a trained professional to help you navigate your way through the process, so that you don't inadvertently flood your system, retraumatize your survival brain, and exacerbate your dysregulation.

This book provides a broader and deeper look at the intellectual content presented in MMFT. Chapters 14–18 also provide additional strategies that are not typically taught to everyone in the MMFT classroom—some of which build on the two MMFT exercises in this book. Usually, a trainer offers these strategies contingently, such as in response to a group's questions or during individual practice interviews. After you finish reading the book, the appendix provides a suggested window-widening sequence and summary of the book's exercises and strategies, for easy reference in the future.

MMFT RESEARCH FINDINGS

MMFT has been tested through rigorous neuroscience and stress physiology research—through four studies, funded by the U.S. Department of Defense and other foundations—with results published in top-tier peer-reviewed scientific journals.

I've already shared the brain imaging research that shows how MMFT Marines saw shifts in their insula and ACC activation patterns, linked with more efficient interoceptive functioning during stress.

Other MMFT research suggests additional facets of a wider window. For instance, U.S. combat troops preparing to deploy to Iraq and Afghanistan who received variants of the eight-week MMFT course showed significant benefits on several outcome measures—including improved cognitive performance, better regulation of negative emotions, and better physiological self-regulation and resilience. These findings are notable because predeployment training has previously been linked with declines in cognitive performance and mood, as well as increases in anxiety and perceived stress levels.

First, MMFT participants showed *improved cognitive performance—with more capacity to keep thinking brain functions online during stress*—during stressful predeployment training. Troops trained in MMFT saw significant improvements in sustained attention and working memory capacity, as well as protection against working memory degradation—all objective indicators of executive functioning. Troops who improved their working memory capacity also reported experiencing fewer negative emotions. MMFT troops who reported increases in mindfulness also reported reductions in their perceived stress levels.

Second, in terms of physiological self-regulation, MMFT participants showed significantly *more efficient stress arousal before and during combat drills, followed by a more complete recovery back to baseline afterward.* Compared to control Marines, MMFT Marines experienced faster activation to higher stress arousal levels, but they recovered from those higher peaks significantly faster and more completely. Stress arousal and recovery were measured in two ways—through heart rate and breathing rate, gathered via bioharnesses the Marines wore before, during, and after the combat drills, and through blood drawn at different time points during the study.

Similarly, compared to control Marines, MMFT Marines showed *lower concentrations of neuropeptide Y* (NPY) in their blood after the combat drills, indicating a faster return to baseline after stress arousal. NPY is thought to enhance cognition during stressful conditions. Because it's released at the same time as adrenaline but has a longer half-life in blood plasma, it's considered to be more conducive to stable measurement and analysis. Thus, blood concentrations of NPY are a good indicator of how much stress arousal is still present after stressful events—and NPY is a key biomarker of resilience in the scientific literature.

The MMFT and control groups showed no resting state differences in NPY levels at initial testing and two months later. However, there were significant differences between the groups after the combat scenarios, when the Marines were actively experiencing stress. After the drills, the MMFT Marines showed significantly lower NPY levels, suggesting a quicker recovery back to baseline. Of course, the MMFT Marines had actually experienced *greater* stress activation, since they showed higher peaks in heart rate and breathing rate during the drills. For this reason, their lower NPY levels afterward demonstrate how their mind-body systems recovered more efficiently and more completely. This pattern—faster arousal to higher peaks, followed by faster recovery—is the hallmark of a wider window.

Of course, with recovery back to baseline, our mind-body system can better access PSNS recovery functions and attend to those "long-term projects" put on hold during stress arousal—sleep, digestion, elimination, healing, growth, tissue repair, and reduction of inflammation.

With this in mind, it's not surprising that the MMFT Marines also experienced *significant improvements in sleep quality*—including longer sleep duration and decreased use of over-the-counter and prescription sleep aids. During the predeployment training period, the MMFT Marines reported sleeping, on average, one hour more each night after MMFT, as well as using fewer sleep aids. In contrast, the control Marines reported sleeping, on average, forty-five minutes fewer each night and relying more heavily on sleep aids.

Finally, the MMFT Marines also experienced significantly *higher concentrations of insulin-like growth factor (IGF-1)* in their blood after the combat drills. IGF-1 facilitates tissue repair, and it's produced when we get restful sleep. Thus, paired with the Marines' self-reported sleep results, IGF-1 is an objective indicator of improved sleep quality. IGF-1 has been linked with better immune functioning and better health outcomes—both indicative of a reduced allostatic load.

Taken all together, this research suggests that MMFT may provide greater resources for widening the window. Mind-body skills training to improve interoceptive functioning—like MMFT—can facilitate better responses to both stress and emotions, even in high-stress environments characterized by depleted executive functioning and narrow(ed) windows.

With better interoceptive functioning, we develop a more adaptive stress response and more efficient recovery afterward. More capacity to regulate stress and negative emotions. More capacity to keep thinking brain functions online, as well as to interrupt impulsive, reactive behavior, even during stressful situations. Most importantly, all of this can be trained—simply by building our capacity for interoceptive awareness.

PREPARING YOURSELF FOR MIND FITNESS TRAINING

When we start practicing mind fitness exercises, we may notice autopilot and mind-wandering. We may also notice other thinking brain habits, such as judging, comparing, narrating, planning, worrying, remembering, and analyzing. Importantly, paying attention is *not* the same thing as thinking, although we often equate the two. Because awareness doesn't belong to the thinking brain, we can train ourselves to pay attention to thinking brain habits just as we might pay attention to sounds or physical sensations. And, just like sounds and physical sensations, thoughts are fluctuating observable events that come and go.

These default modes such as autopilot, mind-wandering, and planning have powerful inertia. The more time we spend in these default modes, the more we reinforce their conditioning—and the more effort it takes to interrupt them and redirect our attention to the present moment. It's ridiculous to expect that we could rewire a mindful default mode spontaneously, without intentional training.

Fortunately, with training and repetition, it's possible to rewire the brain to make present-moment awareness our new default mode. Each time we "wake up" from daydreaming, worrying, planning, or thinking, we reinforce a new mindful default mode. As you practice, you'll discover it's possible to pay attention to *thinking as an activity* the mind habitually does, without getting caught up in the *content of the thoughts*.

Precisely because our existing mental canyons are so deep, it's important to set aside time for practicing mind fitness exercises without distractions. To start seeing our existing default modes and mental filters, as much as possible we need to create a training environment without them. In the beginning, formal practice of mind fitness exercises cultivates the new mindful default mode most efficiently. After we rewire a new mindful default mode, we can access this mode more easily during any life activity.

When we begin practicing mind fitness exercises, we need conscious intention, self-discipline, and consistent practice. Understanding how we got here helps us appreciate the powerful inertia of a Grand Canyon—so that we can be kind to ourselves when we're stuck in one. We can also remember that we're more likely to fall into one of these canyons when we're stressed and dysregulated.

After a few weeks of consistent practice, the new mindful default mode begins to develop momentum of its own. Indeed, after a few weeks many people tell me that they notice their mind spontaneously moving into the mindful default mode at random times—even when they're not actively trying to be mindful.

Conversely, when we stop practicing mind fitness exercises consistently, we usually notice some cognitive, emotional, or self-regulatory declines. After taking an extended break from practice, people usually report more distractibility, irritability, mind-racing, and memory problems. They also experience greater physical symptoms and sleep problems. They notice themselves getting "stressed" at minor things—as well as getting pulled toward unhealthy coping habits. Seeing these changes provides "proof" that mind fitness training had been working, which usually motivates them to begin practicing again.

Mind fitness exercises are not complicated—but they can be surprisingly challenging. Our attitude and motivation determine how well we can navigate these challenges. Just as the attitude we bring to physical exercise makes a tremendous difference to how much we can demand from our body during a workout, the attitude we bring to mind fitness training is critical to how effectively we can widen our window.

The attitude we're aiming for is *nonjudgmental curiosity about our present-moment experience.* Nonjudgmental curiosity means bringing a quality of interested, impartial observation to each moment—without criticizing, comparing, categorizing, or expecting this moment to be like a previous one. When we're willing to experience the present moment, just as it is, we can fully connect with what's happening.

Skepticism is a natural and healthy response to claims about the benefits of anything. It can also be helpful to bring a questioning attitude—as long as we have a genuine openness to the answers. If you've already decided this is a waste of time, you likely won't get anything out of it.

On the other hand, thinking that this is definitely the answer to all your problems won't help, either. Instead, it's best if you can suspend all judgments and assess the training benefits *based on your own direct experience.* Really check it out for yourself. Treat the whole process like an experiment in your own personal mind-body laboratory.

Just as physically overtraining can lead to physical injury and exhaustion, overefforting during mind fitness training can be detrimental, too. This usually happens when we sit down for mind fitness practice with a particular outcome in mind—such as "now I'm going to get relaxed, or control my pain, or become a mind-reading ninja." As Chapter 11 suggested, practicing with such a goal can be counterproductive, because the discipline we exert during the exercises gets hijacked and subverted by this goal-oriented agenda of trying "get somewhere."

When we bring this kind of energy into mind fitness practice, we introduce *striving* into the equation. *We've also told our survival brain that things are not okay as they are right now.* That's when we end up fighting with reality as it actually is.

In response, we usually get very tight and constricted; we may experience a headache or muscle tension. We exacerbate an adversarial relationship between the thinking brain and survival brain—impeding our ability to recover and often intensifying our symptoms of dysregulation. Overefforting also reduces our ability to see clearly what's actually here and then respond with flexibility and balance.

Thus, the paradox: Although consistent and disciplined mind fitness practice does provide benefits, the best way to achieve benefits is actually to back off from striving for them. Of course, given the cultural narrative of the Mindfulness Revolution, that "benefits come quickly," this may be difficult to do. Nonetheless, see if it's possible to simply engage in the exercises fully—without trying to "accomplish" anything. Focus instead on nonjudgmentally observing things as they are right now, moment by moment. Trust that you're cultivating the fungible qualities of a fit mind along the way.

This paradox also points to a major difference between physical exercise and mind fitness training: *"Progress" in mind fitness training is not linear.* In fact, it's common to perceive that we're getting "worse" before we get "better."

A typical trajectory is for someone to engage in mind fitness exercises consistently for a few weeks and see some "early wins"—such as better ability to pay attention while your boss or spouse is talking, a sense of feeling calmer, or an easier time falling and staying asleep at night. However, after a few weeks, most participants report experiencing *more* symptoms of stress activation. Especially when we have narrow(ed) windows, as we develop awareness with mind fitness practice, we are likely to experience more stress activation and increased symptoms of dysregulation—including nightmares, flashbacks, intrusive thoughts, panic attacks, hyperarousal, intensified chronic pain and gastrointestinal symptoms, or heightened restlessness, anxiety, or irritation. This increase in symptoms usually occurs sometime during the third through seventh weeks of practice—and it is often accompanied by doubt, discouragement, and introspection.

In most mindfulness programs, this "setback period" is often explained as someone becoming (more) aware of symptoms, emotions, or other aspects of themselves *that were already there* but had been outside their conscious awareness. Although this is frequently true, it's not the whole story.

Instead, this dynamic is actually related to survival brain bottom-up processing—especially if we've experienced chronic stress or trauma without recovery. As Chapter 13 will explore, our Paleolithic wiring knows innately how to bring itself back into regulation, *when we aren't overriding this wiring* habitually. Whenever the survival brain neurocepts safety, it knows how to direct the mind-body system how to recover back to a healthy baseline—and in the process, widen our window. Therefore, after practicing for a few weeks, you may notice that you're experiencing an increase in symptoms of dysregulation or intensifying symptoms. This doesn't mean you are doing anything "wrong"! Trust that this increase in stress activation is completely normal. *It's simply your mind-body system taking the necessary preparatory steps to discharge activation from prior dysregulation, so that you can eventually widen your window.*

The setback period is a crucial part of the reregulation and recovery process. During this period, many people are likely to stop practicing because they don't understand why they are now feeling "worse." In fact, this period can be uncomfortable in the mind-body system, especially if someone is extremely dysregulated and/or has decades of override conditioning, as I and my Marines did. The most important thing during this period is to continue practicing consistently and working nonjudgmentally with the dysregulation.

In later chapters, I'll give you concrete guidance for how to work with these symptoms skillfully, so that you can support your mind-body system in recovering, without flooding or retraumatizing the survival brain. For now, I'll remind you that *when* and *how* this recovery process unfolds is not under your thinking brain's control. That's because, as I've said before, neuroception and recovery are survival brain jobs.

For now, if you notice an increase in stress arousal when you practice, keep redirecting your attention to your contact points. It's most important to disengage your attention from your physical symptoms, emotions, and distressing thoughts, because paying attention to these things will likely increase your stress arousal and exacerbate your symptoms further.

HOW TO START MIND FITNESS TRAINING RIGHT NOW

The first MMFT exercise develops awareness of the contact between our body and our immediate surroundings.

Find a comfortable place to sit, preferably in a chair with your back toward a solid wall, rather than toward a door, window, or open space. Sit with your feet shoulder-width apart and flat on the ground. If it feels comfortable to you, close your eyes; if not, direct your gaze at the ground in front of you. Sit so that your spine is upright yet relaxed. Sometimes the easiest way to achieve this posture is to push your butt back in the chair, crunch your shoulders up by your ears for a moment, and then allow them to drop, with your arms and hands resting in your lap.

Allow yourself to notice the feeling of being supported by the chair and ground. You're aiming for the *felt sense* of this support, in your body, rather than *thinking about* or *analyzing* this support. Notice the contact between the backs of your legs and butt and the chair, and between the soles of your feet and the floor. If you have trouble feeling your feet, you can gently wiggle your toes or press your feet into the ground.

As you notice this support from your surroundings, briefly scan your body for any places holding tension or tightness. In particular, check your brow, jaw, neck, and shoulders. Without trying to make anything particular happen, see if by bringing attention to those places, the tension shifts. It may or may not, but either outcome is perfectly fine.

Now, bring your attention back to the *physical sensations of contact* between your body and your surroundings. You may notice things like pressure, hardness, softness, heat, coolness, tingling, numbness, sweatiness, or dampness. Pay attention to these sensations of contact at three different places: (1) between your legs, butt, and lower back and the chair; (2) between your feet and the ground; and (3) where your hands are touching your legs or touching each other.

After you investigate the sensations of contact at each point, select the place where you most strongly notice the sensations of contact. *This one contact point will now be your target object of attention.* If you find that you aren't noticing any sensations at all, try taking off your shoes and socks, sitting on a hard surface, or running your hands slowly along your thighs. These shifts may help you notice sensations more clearly.

Once you've selected your contact point, direct and sustain your attention there. Notice the sensations at this contact point in great detail. For instance, if you're focusing on contact between your butt and the chair, are the sensations similar or different underneath the left cheek and the right cheek? Without thinking about it, simply notice and investigate the physical sensations with nonjudgmental curiosity.

If you notice your attention wandering off, simply recognize that it's wandered and gently and nonjudgmentally redirect your attention back to the sensations of contact. Your mind may wander off a hundred times. That's perfectly fine. Simply choose to begin again. Each time you redirect your attention back to the sensations of contact, you're breaking up your old mental canyons and building attentional control. Think of it as a neuroplastic rep.

In the beginning, aim to practice for at least five minutes at least once each day. Over time, as you develop attentional control, you can build up to practicing the Contact Points Exercise for ten to twenty minutes daily. You can also use this exercise as a transition between different phases of the day, such as upon awakening, after getting home from work, or at bedtime.

To conclude the exercise, widen your attention to take in your whole body seated in the chair. Notice if anything has changed in your mind-body system from having done this exercise. With nonjudgmental curiosity, investigate: Is the body more relaxed or more agitated? Is there more or less muscle tension? Is your energy level higher or lower? Are you sleepier or more alert? Is the mind more focused or more distracted? Is it calmer or more anxious and irritated? You may notice shifts in your mind-body system after practice, or you may not. Either way is perfectly fine. The goal is simply noticing the state of your mind-body system right now.

For the first few weeks, I recommend you engage in this exercise while seated in a chair. However, you can also practice this exercise while standing or lying down. While standing, you'd notice the sensations of contact between your feet and the ground. While lying down, you'd notice the sensations of contact between the entire backside of your body and the surface you're lying on. You can also practice while lying on the floor with your legs up against the wall; this posture is especially calming for the nervous system and helps with venous and lymphatic drainage. However, if you have high blood pressure, it is *not* a good posture for practice.

The Contact Points Exercise is the first exercise in the MMFT sequence, for three reasons.

First, by developing the ability to notice the physical sensations of contact between your body and your surroundings—such as the chair you're sitting in or the floor beneath your feet—you'll have a portable target object of attention, available for all situations. Our body is always in contact with something!

Second, by noticing whenever your attention wanders and then nonjudgmentally choosing to redirect your attention back to the sensations of contact, you'll be strengthening your attentional control.

Finally, by directing your attention to the sensations of contact, you'll be focusing your attention on neutral grounding stimuli, which will show the survival brain and nervous system that you're grounded, stable, and safe. In turn, this creates the conditions most conducive for your survival brain to neurocept safety and then decrease stress arousal. If you'd exceeded your stress capacity threshold, this helps your mind-body system get back inside your window. It also sets the stage for a complete recovery.

For these reasons, the ability to bring steady awareness to the sensations at our contact points is a fundamental mind fitness skill, which we'll build on in later chapters.

As your mind-body system becomes conditioned to paying attention to your contact points, you'll find that you can also use your contact points as a touchstone—a means by which to arrive fully in the present moment, in an embodied way. Your contact points can be especially helpful when you're experiencing stress activation, caught in a loop of thinking, or engaged in interpersonal conflict. They can also help you interrupt impulsive or reactive behavior.

Most importantly, after consistent practice for a few weeks, your contact points can become a refuge for your survival brain. You'll need this refuge for the next mind fitness exercise, in Chapter 13.

CHAPTER THIRTEEN

Building Resilience through Recovery—Micro-Level Agency, Part I

In 2013, I rescued an eighteen-month-old traumatized dog, a Shiba Inu named Chloe. Even after two abusive homes, Chloe was willful, regal, and shockingly lethal. In our first three weeks together, she easily caught and efficiently executed seventeen squirrels in our small backyard—one quick shake and snap of their necks—without even breaking a pant.

Although she's significantly less traumatized and less aggressive now than when she first arrived, among my friends and family she's still known as the Deadly Diva. In fact, Chloe continues to express her assassin aptitude whenever animals wander into her yard—and I've become rather adept at disposing of bodies.

One night in August 2016, Chloe jumped on the bed and started clawing at my legs—her signal for needing to relieve herself at night. Since letting her outside alone in the night is a recipe for dead animals, I usually get up and walk her on the leash. That night, however, I was feeling especially tired, so I simply opened her gate and listened as she bounded downstairs and out her dog door. Shortly thereafter, Chloe started barking furiously. Not wanting to wake my neighbors, I jumped up, slid into my sandals, turned on the backyard light, and ran outside.

I found Chloe barking, nosing, and pawing at an eight-inch adolescent possum "playing dead" on the patio. The possum must have flipped involuntarily into freeze against her speedy onslaught.

Finding this near-fatal tableau, I felt a surge of mixed emotions. On the one hand, I was elated, because after a decade teaching about freeze—using possum as an example—here I was, finally getting an opportunity to observe a real live possum, up close, in freeze. At the same time, however, I was also quite alarmed, since this poor possum didn't stand a chance against Chloe's slaying skills once it came out of freeze and tried to get away.

In my most imposing alpha voice, I ordered her to "Come!" Of course, willful Shiba on the hunt, Chloe was having none of it and ducked instead under the nearby hydrangeas. Thus, I went to stand protectively over the possum while Chloe furiously dug a trench between them, frustrated at being thwarted.

For the next several minutes, we repeated a cycle while I protected the possum: First, Chloe would venture forward. In response, I'd spread my arms over the possum and tell her "No!" Retreating reluctantly back under the bushes, Chloe would grudgingly obey and whine while expressing her frustration through feverish trench-digging. Then, she'd venture forward again.

From my protective position, I watched the possum at my feet. I could smell the residual musky signature scent that possums emit to mimic the smell of dead animals and confuse their predators. After several minutes, its ears started twitching and rotating—the first sign it was coming out of freeze. A few minutes later, the possum opened its eyes, blinked, and looked right at me. Next, I watched its breathing deepen, as its hind legs jerked and twitched a few times. Finally, it lifted its head and started rotating its neck to look around.

I was captivated watching these dynamics unfold—especially that the possum was recovering from freeze even with Chloe barking less than three feet away. But the more the possum twitched, the more frenzied Chloe became, and the more anxious I got. I was sure that as soon as it started to scamper away, Chloe would outrun and dispatch it with ease, and I'd be powerless to intervene.

By now, Chloe was completely furious with me, literally frothing at the mouth and getting bolder in her forays out toward the possum. In response, I was now yelling my commands while smacking at the hydrangea branches above her head. Finally, in an effort to escape this cycle of torment, Chloe exploded down along the fence to the other side of the yard, where she sprinted speedy laps to expend her pent-up frustration. In turn, I assumed a defensive blocking position on the patio, hoping to intercept her inevitable breakout dash.

As I laughed at the sheer ridiculousness of this early-morning scene—Chloe barking and sprinting along the fence, me helplessly bobbing and weaving in my nightgown to block her—Chloe launched herself off the retaining wall and onto the patio past me. I spun around, horrified at the carnage to come.

The possum was gone.

Our diversion on the other side of the patio had given the possum the space it needed to get away safely. While Chloe clawed at the musky fluid left in the possum's wake, I breathed a sigh of relief, knowing we could both finally stand down and return to bed.

BUILDING RESILIENCE IS NOT BEING PAMPERED

By the end of this chapter, you'll learn MMFT's primary exercise for recovering from stress activation; over time, with consistent use, this exercise can also help you not just recover but actually widen your window. To teach this exercise well, however, first I must explain some basic neurobiological principles of resilience and trauma extinction, so that your thinking brain can understand *how* and *why* this exercise works.

When Chloe went outside that night, the possum's survival brain clearly neurocepted danger and started to mobilize energy to respond to the threat. Its survival brain and nervous system also perceived that the two active SNS defenses, fight or flight, were not going to be successful. Thus, it involuntarily "fell back" to a dorsal PSNS freeze.

After I stood over and guarded the possum, however, the possum's survival brain must have neurocepted enough safety to cue its *ventral* PSNS to turn on, in the process taking the *dorsal* PSNS out of defensive mode. This allowed the possum to begin recovering from freeze. Remember from Chapter 4 that the ventral PSNS, which controls social engagement and recovery functions, cannot be turned on at the same time as the dorsal PSNS in defensive mode.

We know the possum's ventral PSNS got turned on for two reasons.

First, it started to exhibit *orienting behaviors*—rotating its ears, opening its eyes, and then moving its head and neck to look around. *Paying attention to and engaging with the external environment through the senses, such as seeing and hearing, is a function of ventral PSNS.*

Second, the possum showed physical symptoms of discharging stress activation. Its leg muscles twitched, its hind legs jerked, and its breathing deepened. These physical movements helped the possum discharge stress arousal—in the process, moving its mind-body system from the *extremely* high arousal associated with freeze and back down to the relatively less high arousal associated with fight-or-flight.

As the possum rallied, it prepared to flee at the right moment. Thus, once Chloe was distracted on the other side of the patio, the possum successfully ran away to safety.

Had we continued tracking the possum, we'd likely have seen it run into my neighbor's yard and hunker down under his shed. There, in that protected place, the possum's survival brain would have neurocepted even greater safety and completed its recovery by discharging any remaining stress activation. Its heart rate and breathing rate would have slowed. It might have experienced more twitches and jerks. After a complete recovery, the possum might have rested for a while before foraging again for food.

In other words, while our early-morning encounter no doubt frightened the possum, there weren't any lingering detrimental effects on its mind-body system. Allostasis was functioning properly: It mobilized energy to deal with the threat and selected the most adaptive defenses to survive *this particular situation*—initially falling back to freeze, since neither fight nor flight were possible, and then, after I stood guard, discharging enough stress activation to come out of freeze and get safely away. Afterward, it fully recovered back to its neurobiological baseline. Thus, we can be relatively certain that this possum didn't develop PTSD or another stress-related disease since allostasis worked as it was supposed to.

In fact, rather than having negative repercussions, *our encounter actually conferred survival benefits for the possum.* Why? This possum had the opportunity to practice confronting a threat, successfully defending itself, and surviving. Such experiences allow the possum's survival brain to accumulate implicit learning of successful defensive strategies. And this kind of implicit learning conveys increased resilience for threatening experiences in the future.

Indeed, unless wild animals succumb to a predator, they undergo this process automatically all the time—encountering a threat, successfully defending themselves, and then completely recovering afterward. Each time they run through this cycle, they add to their toolbox of survival skills— *and widen their window.*

To be clear, mammals don't have to experience a freeze response and recovery to have this window-widening effect. All that's required is *(1) experiencing something stressful that pushes us outside our comfort zone, (2) moving through that stress activation, and then (3) recovering completely afterward.* In the process, the mind-body system learns to tolerate and function effectively amid more stress activation than before.

You may remember the stress-hardy rat pups from Chapter 3, who received licking and grooming from their attentive rat moms after a short separation. Rat pups who received such attentive licking upon reunion showed lifelong epigenetic changes for genes regulating their stress response. Later, as adults, these rats showed less stress reactivity and fear and lower stress hormone levels during stressful experiences. Not surprisingly, they were also better learners and showed delayed aging in the hippocampus.

Although I mentioned this study in Chapter 3 as an example of epigenetics, it's also relevant here. Each time the rat pups were separated from their moms, *they experienced stress arousal outside their comfort zones.* Upon reunion, they received maternal soothing and nurturing—licking and grooming—which *helped stimulate a full recovery.* Over time, with repeated experiences of separation and recovery, these rat pups widened their windows.

Likewise, the "safe home base" of secure attachment allows humans to wire a wide window. Well-attuned and well-regulated parents modulate their baby's arousal—calming him when his arousal is too high, or stimulating him when his arousal is too low. Over time, the child's survival brain and nervous system become conditioned to associate distress outside the comfort zone with the subsequent recovery experience of being soothed afterward. Then, as the child grows, he can try new things that take him outside his comfort zone, but his survival brain and nervous system will trust that he can always come back to his parents for soothing and support.

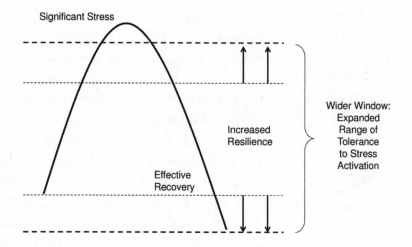

Figure 13.1: Widening the Window through Full Recovery

This figure depicts how we can widen our neurobiological window of tolerance to stress activation. When we experience stress or trauma that take us outside our comfort zone, followed by a complete recovery, we create a wider window.

Each experience outside the comfort zone, followed by a complete recovery, widens this securely attached child's window (see Figure 13.1).

In contrast, when we experience significant stress outside our comfort zone *without* a complete recovery, over time our resilience gets undermined. This can happen through a shock trauma that overwhelms the mind-body system all at once. Or it can happen through chronic stress or relational trauma that slowly depletes our resources over time. In either situation, if the PSNS recovery functions don't occur, our mind-body system gets conditioned to leave stress arousal turned on without ever turning it off. As a result, allostasis stops functioning properly.

That's when our window narrows—and we're more likely to exceed our stress capacity threshold, with four common consequences, as Chapter 10 explained. Each of these serves as an "alert" to warn us that complete recovery has not occurred. First, we're more likely to behave in ways that work against our values and goals, such as by procrastinating, acting out, or engaging in unethical behavior. Our lifestyle choices start to reflect the narrowed attentional focus of chronic stress arousal and we prioritize the immediate over the truly important.

Second, even in the face of "minor" stressors, we're more likely to rely on our default programming or automatically fall back to fight, flight, or freeze—rather than choosing the defensive strategy that's most appropriate for the current situation. For instance, that's when we might numbly acquiesce to a situation that's harmful to us or others, when standing our ground or enlisting help from others would actually be a more appropriate response.

Third, our thinking brain and survival brain end up in an adversarial relationship, with faulty neuroception, degraded thinking brain functions, and impaired social engagement—which decreases our ability to interact cooperatively with others. With impaired social engagement, for example, we're more likely to argue, interrupt others, become sarcastic or coercive, stop communicating, or withdraw. With an adversarial relationship, we're also more likely to experience *survival brain hijacking*, in which emotions and stress arousal drive our decision making and behavior. Conversely, we're also more likely to experience *thinking brain override* in its various forms, including suppression, denial, compartmentalization, and suck it up and drive on.

Finally, we tend to fall into stress reaction cycle habits, which may help us feel better in the short term but actually add to our allostatic load. For instance, we might rely on caffeine, sugar, nicotine, alcohol, or illegal, over-the-counter, or prescription drugs to increase or decrease our arousal and mask stress symptoms. Or we might overeat, skip meals, or choose unhealthy or fast food. We might use television, mobile devices, the Internet, or video games to numb or distract ourselves. Or we might compulsively engage in self-harming, high-risk, or adrenaline-seeking behaviors, such as cutting, extreme sports, gambling, aggressive driving, or infidelity. We might also rely on thinking brain habits such as chronic worrying, planning, or ruminating.

Until a complete and effective recovery occurs, these patterns will continue to narrow our window further (see Figure 13.2). Over time, we eventually develop symptoms of dysregulation.

There were horrible experiments with chicks, conducted many decades ago, which would never pass research ethics review today. Nevertheless, I mention them here since they illustrate these dynamics quite dramatically. In these experiments, chicks were divided into three groups. Group A chicks were held and immobilized so that they could neither fight nor flee, which

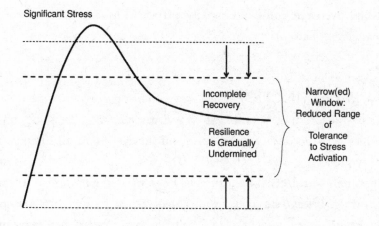

Figure 13.2: Narrowing the Window through Incomplete Recovery

This figure depicts how we can narrow our neurobiological window of tolerance to stress activation. When we experience significant stress or trauma that takes us outside our comfort zone, but we don't experience adequate recovery afterward—often because we override the recovery process—over time we narrow our window.

induced a freeze response. Afterward, they were allowed to go through the process of discharging their stress activation to recover from freeze without interference. Group B chicks were also immobilized to induce a freeze response, but before they could recover completely, the human researchers prodded them on the breast until they aroused. Finally, Group C chicks were a control group, experiencing neither the freeze nor recovery.

All three groups were then thrown in a tank of water and left to swim until they drowned. (This is the horrible part.) Which group drowned first?

Not surprisingly, Group B chicks—those that had been stressed but not allowed to recover—drowned first, because their window had narrowed. Group C chicks—those that had not experienced freeze—drowned next, because they had not had a window-widening experience. Group A chicks swam longest, because the prior stressful experience of freeze, followed by an effective recovery afterward, had widened their windows.

In short, experiencing challenge and recovering completely afterward conveys survival advantages. To achieve resilience, we have to experience hardship, challenge, and sometimes even failure. Resilience is not about being pampered.

TRAUMA EXTINCTION

Just like the possum, we can also experience recovery. Yet some unique dynamics between our thinking brain and survival brain may impede the recovery process.

Although we can't make our survival brain recover, we *can* create the conditions where it feels safe enough to start recovery on its own. By choosing to direct our attention consciously in ways that support recovery, we can facilitate trauma extinction. As Chapter 5 explained, rather than erasing an existing memory, trauma extinction actually involves forming a new implicit memory of a situation when we were no longer helpless or lacking control.

This "learning by doing"—taking active steps to direct our attention in ways that regulate our physiological and emotional arousal—teaches the survival brain that we are no longer helpless or lacking control. By overcoming the helplessness connected to our unresolved memory capsules, the survival brain can finally understand that traumatic events are truly in the past. This understanding is what corrects the survival brain's corrupted implicit memory.

In Julio's story from Chapter 5, his survival brain had generalized from the drive-by shooting that he couldn't successfully defend himself whenever he felt "pinned down." Thus, when Julio watched an animal overtaken and "pinned down" by a predator in the video in an MMFT class, it evoked the same sensations in his body that he'd experienced when he was "pinned down" by his cousin's grip during his fall. That similarity triggered his unresolved memory capsule from the drive-by shooting and sent him into a freeze response in the classroom, even though, on the surface, the drive-by shooting and animal video seem to be unrelated.

When Julio and I met after class, he recounted what had happened during the animal video, which triggered another flashback. However, I was able to interrupt his narrative and help guide his mind-body system to stay inside his window. To do this, I began by directing Julio to pay attention to *present-moment sensory input.* Specifically, I asked him to move his head and neck, look around the room, and report what he was seeing.

Just like the possum on my patio, as Julio engaged in these orienting behaviors, it helped his mind-body system turn on the ventral PSNS—and turn off the dorsal PSNS in defensive mode, preempting the possibility of another freeze. With the ventral PSNS back online, I knew that Julio's mind-body system was back inside his window. Now Julio's mind-body system was in the proper stance for recovery.

All we needed to do next was help Julio's survival brain feel safe enough so that it would naturally start recovery on its own.

With this in mind, I directed Julio to pay attention to the sensations at his contact points. As Chapter 12 explained, this is a terrific way to cue the survival brain to neurocept stability and safety. As Julio did this, soon he started experiencing some symptoms associated with the discharge of stress activation. For instance, he started yawning and his arms started twitching. He also reported waves of heat through his torso and head. His face flushed and his breathing deepened. As these sensations of release were happening, Julio simply paid attention to them and to his contact points.

Table 13.1 gives a complete list of the signs of discharge; you'll recognize some of these symptoms of release from the possum's story.

Table 13.1: Signs of Nervous System Discharge and Recovery

• Shaking/trembling	• Crying
• Twitching	• Laughing/giggling
• Slower, deeper breathing	• Yawning
• Slower heart rate	• Sighing
• Relaxation in the chest or belly	• Stomach gurgling
• Tingling/buzzing	• Burping
• Waves of warmth/heat	• Farting
• Chills	• Coughing (and phlegm)
• Flushed skin/sweating	• Itching

After completing one round of discharge, Julio reported feeling grounded and stable. He immediately launched back into his story of the drive-by shooting.

Once again, I asked Julio to slow things down. I asked him to share just one small component of the flashback while he paid attention to what was happening inside his mind-body system. For instance, as he recounted seeing his cousin fall forward, still holding on to Julio's arm, he also reported noticing his heart rate and breathing rate increase and butterflies in his stomach. In other words, this one small component of the flashback stimulated the unresolved memory capsule again—triggering new stress arousal of enough intensity to take him outside his comfort zone once again.

As his stress arousal intensified, I directed Julio to direct his attention back to his contact points, to help his survival brain neurocept safety and stability once again. Shortly thereafter, Julio experienced a second round of stress activation discharge—this time, yawning, crying, and more waves of heat.

As we worked together, what was most important was for Julio's mind-body system *to experience stress activation only slightly beyond his comfort zone, so that he could then discharge that activation and recover completely.* This process of working with only a small amount of stress activation at a time is called *titration.*

By titrating his stress activation, Julio was able to hold *dual awareness*—simultaneously being aware of the images in his flashback *and* the sensations in his mind-body system in the present moment. In other words, metaphorically, Julio had one foot in the past and the other foot in the present—the stance where trauma extinction is possible.

To be clear, we were not changing Julio's memory of the drive-by shooting. Instead, we were changing how Julio's survival brain *reacted* to the memory and, in the process, creating a new implicit memory, which is what bottom-up processing is all about.

In effect, by experiencing stress activation outside his comfort zone, followed by cycles of complete recovery, Julio showed his survival brain that he was no longer helpless, being "pinned down" by his cousin. Instead, his survival brain learned that Julio has the inner resources—micro-level agency—to work with the memory skillfully. In turn, Julio's memory capsule of the drive-by shooting could finally lose its "charge," because his survival brain could finally understand that this event was truly in the past and no longer threatening to Julio's survival.

GRAND CANYONS AND RECOVERY

One of the biggest misconceptions I had while experiencing PTSD was the expectation that I could recover from my traumatic history in one big cathartic moment. However, recovering from chronic stress and trauma doesn't happen once and for all. *Just as we need repeated experiences with chronic stress and/or trauma to build allostatic load and narrow our window, we also need repeated experiences with complete recovery to widen it.*

What this means is that *there's no quick-fix way to achieve these transformations.* The very nature of allostatic load, accumulating gradually through chronic stress and trauma without recovery, means that there cannot be any quick fixes.

As Julio's story shows, there is no "once and for all" from the perspective of our neurobiology. Rather, *all we have is the stress activation we're currently experiencing and the choices we make, in this moment, for working with it.* That's why we don't need to reexperience traumatic events to widen our windows. We only need to work skillfully with and recover completely from stress activation *each time* we find that our mind-body system is experiencing it. It doesn't matter if we're currently experiencing a flashback, irritation about being cut off in traffic, or anxiety about an upcoming test. *Any time* we experience stress activation for any reason, we always have the choice to work with it skillfully and recover completely, to widen our window over time.

The thinking brain cannot force this recovery process, because it is a survival brain job. Instead, the best way our thinking brain can facilitate recovery is by directing our attention in ways that increase the likelihood that our survival brain will neurocept safety. As soon as it neurocepts safety, the survival brain and the nervous system will naturally move to discharge stress activation.

In other words, through where, when, and how we intentionally direct our attention, we can train our thinking brain to structure a safe and supportive environment for our survival brain. This is how we teach our thinking brain to become our survival brain's ally.

As Julio's story suggests, recovery is possible only when we're inside our window, with interoceptive awareness online. Being inside our window allows us to integrate top-down (thinking brain) processing and bottom-up (survival brain) processing simultaneously. When they work together as allies, we can mindfully pay attention to and integrate information from both the internal and external environments. Internal information includes physical sensations, emotions, body postures, and motor impulses. External information includes sensory input from our surroundings.

Julio's story also shows the importance of titration, stimulating only enough stress activation to take us *slightly* outside our comfort zone and then discharging that activation for a complete recovery. Because of the adversarial relationship between the thinking brain and survival brain that we develop when we're dysregulated, however, we can often override our capacity for titration.

When this happens, we may flood the survival brain and body with too much stress activation—moving so far outside our window that the ventral PSNS is no longer online and recovery is no longer possible. That's why it can be helpful to work with a clinician skilled in body-based trauma techniques, such as sensorimotor psychotherapy or Somatic Experiencing, to help you titrate how much stress activation you try to discharge during each recovery cycle. These forms of therapy were explicitly designed to help clients with bottom-up processing in a safe, efficient, and effective manner. This can be especially helpful if you're experiencing extreme symptoms of dysregulation. In effect, the clinician provides surrogate interoceptive awareness to facilitate their client's recovery.

Although working with a clinician can help us recover efficiently, it's important to remember that we actually have the inborn capacity for recovery. All we need is attentional control and interoceptive awareness; nonjudgmental curiosity about our symptoms of stress activation; and understanding about where, when, and how to direct our attention to stay inside our window.

However, we *don't* need to "figure out" or "analyze" what's causing our stress arousal. This thinking brain habit could actually exacerbate our stress, because these thoughts could lead our survival brain to trigger more stress arousal. It also pulls our attention away from noticing physical sensations and emotions in our body and sensory input from our surroundings. Finally, any thoughts or beliefs we have about what's causing our stress will likely be biased by the stress arousal.

So if you find your thinking brain ruminating or worrying about your stress, see if you can disengage from this habit. In the moment that we're stressed, it doesn't really matter what's causing our current stress activation. The only thing that matters is that the thinking brain and survival brain work together as allies to modulate our current stress activation, using interoceptive awareness.

Moreover, should an unresolved memory capsule get triggered, this alliance lets the survival brain "update its files" and experience micro-level agency in the face of stress activation. As Julio's story shows, we can metaphorically stand with one foot in the traumatic past—where the survival brain and the body believe that the traumatic event is still ongoing—and one foot in the present, where interoceptive awareness can notice the stress activation arising and then being discharged.

Whenever we're activated, *for any reason*, and we choose to work skillfully with our current stress activation, over time we can widen our window, in two ways.

First, by directing our attention to facilitate the survival brain in neurocepting safety and initiating recovery functions, we can complete one cycle, as in Figure 13.1. We can be just like those Group A chicks that experienced freeze and then experienced discharge of that stress activation. With each cycle of Figure 13.1 we experience, we widen our window incrementally.

With enough cycles of recovery, we can bring our everyday activation level down—from where it might have habitually hovered during chronic stress right below our threshold, back down to our "true" neurobiological baseline (see Figure 13.3).

Second, by working skillfully with our stress activation *each time* we notice it, we also increase how much we can tolerate with interoceptive awareness online. As we do this, our survival brain learns that the symptoms of stress activation are not as frightening as they once were—and that we can experience stress activation in our mind-body system, without collapsing into overwhelm or freeze.

In other words, *the survival brain learns that it's no longer helpless.* It begins to trust that stress activation will not escalate beyond our control, as it had before—because now we have agency in modulating our stress activation.

This is the key ingredient for the survival brain finally recognizing that the traumatic events are in the past. It's what finally lets the survival brain

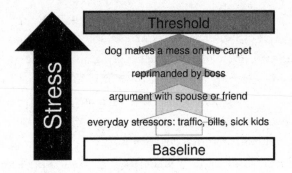

Figure 13.3: The Stress Capacity Threshold

The Stress Capacity Threshold depicts how we can move outside our window when we experience chronic stress and trauma without adequate recovery. The width of our window is the space between our baseline and our threshold; someone with a wider window can tolerate more stress arousal before they exceed their stress capacity threshold. If we've been living with a narrow(ed) window—hovering just below or even exceeding our threshold—repeated cycles of recovery can help us return to our neurobiological baseline. Furthermore, with enough cycles of significant stress or trauma followed by a complete recovery, we can also increase our stress capacity threshold. Both of these dynamics can help us to widen our window.

"update its files" and dissolve unresolved memory capsules, rewire default traumatic programming, and release kindling patterns.

Thus, as we grow our capacity to pay attention to and tolerate ever more stress activation in our mind-body system, *we also raise our stress capacity threshold*, as seen in Figure 13.3. With a higher stress capacity threshold, our mind-body system can tolerate more stress activation before it moves outside our window—and experiences all those consequences of exceeding our threshold. In other words, we set the conditions for even better performance during progressively greater stress in the future.

Do you remember Chapter 10's discussion about how to influence the Stress Equation? (See Figure 13.4.) In the short term, we gain the most leverage with the equation's third component, from *working with stress activation once it's arisen*—by disengaging our attention from any thinking brain habits that inadvertently make stress worse, and instead redirecting our attention to facilitate the survival brain neurocepting safety.

$$\text{Stressor} \quad + \quad \begin{matrix}\text{Perception}\\\text{of threat}\end{matrix} \quad \longrightarrow \quad \text{Stress}$$

$$\left(\begin{matrix}\text{internal or external}\\\text{event}\end{matrix}\right) \quad \left(\begin{matrix}\text{survival brain's}\\\text{neuroception}\end{matrix}\right) \quad \left(\begin{matrix}\text{activation in the}\\\text{mind-body system}\end{matrix}\right)$$

Figure 13.4: The Stress Equation

Whenever we experience a (1) *stressor*, which is an internal or external event that (2) the survival brain perceives as threatening or challenging, then our mind-body system turns on (3) stress arousal, which is physiological activation in the body and mind. In the short term, we can gain the most leverage with the third component, by working with stress activation once it's arisen. However, by skillfully working with our activation each time we're stressed, over time we also change the second component, our perception of the stressor.

Interestingly, however, by skillfully working with our stress activation, over time *we also change our perception of the stressor*, the equation's second component. In effect, our survival brain learns that we can experience a particular stressor and any resulting stress activation, yet still have the self-mastery to navigate through it and recover. Over time, after repeatedly encountering a stressor, experiencing stress activation, and recovering completely afterward, the survival brain stops perceiving that stressor as quite so threatening. Ultimately, this gives us our greatest influence over the Stress Equation, as we teach our survival brain to shift its relationship with particular stressors.

To build these new Grand Canyons for recovery, we need to rely on the two warrior qualities, wisdom and courage. That's because in the face of intense stress activation, it's extremely tempting to give in to our old patterns of ignoring, denying, self-medicating, masking, or avoiding the stress activation with stress reaction cycle coping habits.

With wisdom and courage, we can find the motivation to forgo our default habits that exacerbate our dysregulation. Instead, we can notice the stress activation in our mind-body system—and then intentionally choose where and how to direct our attention to support complete recovery. *Although we can't control when stress activation arises, we can always choose what we do with it.*

Resilience is an active process—an embodied domain-general skill that we can practice and learn—not some buzzword or silver bullet.

DISCHARGING STRESS ACTIVATION

Using the Ground and Release Exercise (G&R) each time we experience stress activation can help us develop new Grand Canyons that help us recover. The following steps describe how to release stress activation safely. You'll recognize aspects of these instructions in how I guided Julio during his flashback. It's the same process, only you'll be doing it for yourself.

Although this process may seem unfamiliar or uncomfortable at first, it's an essential tool for resetting the nervous system and calming yourself when stressed. Over time, when we use G&R repeatedly and intentionally, the mind-body system begins to restore allostasis. Once allostasis gets restored, the discharge of stress activation and complete recovery can happen naturally, just as it did with the possum.

It's helpful to use G&R *after mild to moderate stressors on a regular basis*, for two reasons. First, it allows us to discharge our current stress activation so that we can recover completely to baseline. Second, when we use G&R after mild or moderate stressors, we train our mind-body system to access this deliberate recovery skill even after extremely stressful events. In that way, we can proactively build this domain-general skill.

With G&R, you'll need to use the two core skills of mind fitness training: attentional control and tolerance for challenging experience. Attentional control, you may remember, is the ability to sustain our attention on a chosen target. Tolerance for challenging experience, whether internal or external, is the ability to pay attention to, track, and stay with such experience without needing it to be different. If you've been practicing the Contact Points Exercise consistently, you've been developing these two core skills. Now, you'll use them as you track stress activation and discharge in the mind-body system.

As this chapter has explained, recovery is possible only when we're inside our window. *You'll know that you're inside your window if you can keep these two core skills active and engaged throughout the exercise.* If you can't, you'll need to redirect your attention to other target objects until you can bring yourself back inside your window. To say that differently, as soon as you find that you *cannot* notice, track, and stay with challenging internal experiences—physical pain, symptoms of stress activation, intense emotions, distressing thoughts, flashbacks—you have moved outside your window. That's when you need to redirect your attention somewhere else—preferably to sensations at your contact points or to neutral sights, sounds, or smells in your surroundings. You don't want to inadvertently flood your mind-body system.

Remember that the thinking brain doesn't control the recovery process, and *the survival brain won't discharge stress activation unless it feels safe and grounded.* Therefore, we need to direct our attention to increase the likelihood that the survival brain will neurocept safety. This also implies that whatever the survival brain appraised as threatening or challenging in our external environment must already be over. Our external environment must be safe for recovery to happen.

For instance, I once had a student who tried to use G&R while he was flying during severe turbulence. Afterward, he asked me why he didn't experience any discharge during G&R. I replied that as long as the turbulence was still under way, his survival brain was still neurocepting danger!

In any situation where the external stressor is still present, the best that the thinking brain can do is to redirect attention to a target object that is safe, stable, or neutral—such as sensations at the contact points. Alternatively, you can redirect your attention outward to *neutral* or *pleasant* sights, sounds, or smells in your surroundings. Although discharge won't be possible until the external stressor passes, at least you won't be adding any additional stress activation beyond what you're already experiencing. In other words, you can stop your mind-body system from moving any closer to your stress capacity threshold.

THE GROUND AND RELEASE EXERCISE

To use G&R, say that you've just had an argument with a loved one, or you've just awoken from a nightmare, and now you're experiencing stress activation.

Move yourself to a quiet place where you can be alone. Think about structuring the external environment to facilitate your survival brain neurocepting safety. For example, you should find a comfortable place to sit, preferably with your back toward a solid wall, rather than toward a door, window, or open space. However, G&R can also work in bathroom stalls, parked cars, or sitting against a big tree trunk. If you're in a chair, sit with your feet shoulder-width apart and flat on the ground. Alternatively, you might sit with your back against the wall and your legs stretched out on the floor in front of you. Either way, sit so that your spine is both upright yet relaxed.

With interoceptive awareness and nonjudgmental curiosity online, bring your attention to your symptoms of stress activation (see Table 13.2). For instance, you might notice a clenched jaw, a pounding heart, shallow breathing, a wave of nausea, dizziness, or tension in your neck and shoulders. Or you might notice that your body posture is hunched and collapsed. You might notice feeling sad, angry, irritated, impatient, anxious, frazzled, overwhelmed, or ashamed. You might notice racing thoughts. Even with anxious, angry, or depressed thoughts, remember that *all stress activation has a physiological component*, so pay attention to the physical sensations. The goal is to notice the stress activation without judgment—*and without trying to suppress, deny, ignore, distract from, reframe, compartmentalize, or otherwise override it.*

Table 13.2: Signs of Stress Activation

• Faster breathing	• Reduced visual field/tunnel vision
• Difficulty breathing	• Hair standing on end
• Tightness in chest or belly	• Losing bladder/bowel control
• Faster heart rate/pounding heart	• Racing thoughts
• Nausea	• Anxious thoughts
• Butterflies in stomach	• Rumination/looping thoughts
• Dry mouth	• Anxiety or panic
• Clenched jaw	• Impatience, irritation, or rage
• Pale and cold skin	• Sadness
• Sweaty palms	• Shame
• Sweating	• Overwhelm
• Hunched or collapsed body posture	• Restlessness/fidgeting
• Dizziness	

You might also notice that your mind-body system is restless and distracted, such as when your legs are bouncing, your hands are fidgeting, or you can't sit still. In fact, restlessness and fidgeting are excellent ways to distract ourselves from our stress activation—which is why they are so common in our society. However, restless energy cannot be discharged through G&R. Therefore, if you notice yourself experiencing restlessness, see if you can sit perfectly still. When you do, the stress activation that had been coursing through you as fidgeting energy will almost certainly shift to become autonomic arousal. For instance, as you sit still, you may notice increased heart rate, shallow breathing, nausea, muscle tension, anxiety, irritation, or even waves of panic. Although this may not feel comfortable, trust that this is actually an excellent development, because in order to discharge our stress activation, we first need it to manifest as autonomic arousal.

Once you've noticed the physiological component of your stress activation, consciously recognize that you're activated. It can be helpful to note this to yourself, as in "Oh, I'm activated right now." However, you *don't* need to make an exhaustive list of all of your symptoms of activation. You also *don't* need to focus on the activation in great detail, as you might with other mindfulness practices. If you do, that will only amplify your stress activation. Instead, simply notice from a more global perspective what's happening in the body.

After consciously recognizing that you're activated, you can redirect your attention. Locate a place in the body where you feel the most solid, stable, grounded, and strong. This is usually at a contact point—with your butt, your lower back, or the backs of your legs touching the chair, with your feet touching the floor, or perhaps with your hands touching each other or your legs. If you cannot feel contact with the chair or floor, consciously push your butt into the chair and your feet into the floor until you feel that support.

Notice the sense of being grounded and supported by the chair and floor. Just as in the Contact Points Exercise, you're aiming for the *felt sense* of this support in your body, rather than your thinking brain trying to *think about* or *analyze* this support. For instance, you may notice pressure, hardness, softness, dampness, heat, tingling, or coolness at these places of contact. Keep your awareness on your contact point or another place in the body that feels most stable, comfortable, safe, and grounded.

Of course, your attention may get pulled back to the physical sensations of stress activation, or back to the storyline, image, flashback, emotion, or thoughts that are fueling the activation. When this happens, keep redirecting your attention back to the contact point or other solid place in the body. *The goal here is to let the thinking brain be the survival brain's ally, by disengaging attention from the stress activation and redirecting attention toward stimuli that will facilitate the survival brain neurocepting safety.*

Keep redirecting your attention in this way until you feel more relaxed or more settled, or until you notice one of the signs of release (see Table 13.3). As you notice symptoms of release, there is no need to try to control or stop the release. Recognize that this is a beneficial part of the process.

Table 13.3: Signs of Nervous System Discharge and Recovery

• Shaking/trembling	• Crying
• Twitching	• Laughing/giggling
• Slower, deeper breathing	• Yawning
• Slower heart rate	• Sighing
• Relaxation in the chest or belly	• Stomach gurgling
• Tingling/buzzing	• Burping
• Waves of warmth/heat	• Farting
• Chills	• Coughing (and phlegm)
• Flushed skin/sweating	• Itching

As you pay attention to symptoms of release, you may notice that they amplify. If you can tolerate it, this is perfectly fine, because it's discharging more stress activation. For instance, if you bring your attention to yawning, you're likely to experience several more yawns in quick succession. Or if you focus your attention on trembling in your hands, the trembling will likely intensify.

Once you notice some sensations of release, you can shuttle your attention between the sensations of release and your chosen contact point.

Shuttling in this way may help you feel more stable while the release is going on. If the release is quite intense, or if you find shuttling your attention to be uncomfortable, then simply keep your attention on your contact points.

If you've been paying attention to sensations at your contact points but you don't experience any sensations of release, open your eyes. Move your head and neck to look around the room. Observe your surroundings and name to yourself what you're seeing, just as Julio did with me. You can also pay attention to any smells or sounds in the environment—or shuttle your attention between noticing external stimuli and the sensations at your contact points. The goal here is to redirect your interoceptive awareness to *neutral sensory input in the external environment*, while also using orienting behaviors, just as the possum did to come out of freeze. *When you do this, you'll also be stimulating your ventral PSNS circuit and encouraging the survival brain to neurocept safety.*

Anytime we stimulate one aspect of the ventral PSNS circuit, we stimulate all of it—which can help the survival brain initiate recovery. Therefore, you should keep your attention focused on these target objects, not on any sensations of stress activation, until you feel more relaxed or notice one of the symptoms of release.

It is possible for release to feel strange, unfamiliar, or even frightening. Remind yourself that discharging stress activation is a natural part of the recovery process. If you find that you're getting caught up in the narrative, images, or thoughts that triggered the activation, redirect your attention to the physical sensations at your contact points.

You can stay with the release process as long as you are noticing signs of release. Sometimes a cycle of release can complete in less than a minute, but sometimes it can continue for twenty minutes or even longer. As long as you have the time, space, and privacy to allow the discharge to happen, let it happen. Once the signs of release stop, you have completed one cycle of Figure 13.1. Only use G&R for one cycle of recovery.

It's common for our mind-body system to move from stress activation into release, and then back into activation. This is actually a sign that the nervous system is cycling between the sympathetic and parasympathetic branches, as it's wired to do. The narrower your window, however, the quicker this cycling back into activation will occur. *When we use G&R, it's important to complete only one cycle of recovery and then stop the exercise.*

Therefore, if you find your mind-body system cycling back into activation again, or if you never experience any signs of release, redirect your attention away from the activation. Instead, keep bringing your attention back to sensations at your contact points. Sustain your attention there. You can also open your eyes; move your head and neck; notice sights, sounds, and smells around you; and name to yourself what you're noticing. As you orient to your surroundings through one of these senses, also notice the felt sense of being grounded and supported by the chair and floor. Keep directing your attention in these ways that support the survival brain from cycling back into more stress activation.

If you've tried all of these steps and still not experienced any signs of release, it's likely that you're experiencing stress arousal outside your window. In this situation, the best way to expend some of the excess stress activation is to engage in cardiovascular exercise, such as running, walking briskly, biking, rowing, swimming, jumping rope, doing high-intensity interval training, climbing stairs, or dancing. Aim for at least fifteen to twenty minutes during which you are slightly out of breath, to expend some stress hormones and move back inside your window. Since we each have different cardiovascular capacity, we'll differ in terms of how much and what kind of exercise we need to attain this state. Then, during the cooldown after your workout, you can try G&R again. Many people, including me, have experienced the most amazing discharges during cooldown after their workouts.

The discharge process differs for everyone, depending on our cardiovascular fitness level, how much stress activation we're currently experiencing, and the current situation. You will need to get to know your own mind-body system. For instance, if you're extremely activated, you may need a long and vigorous workout before trying G&R again. Alternatively, during a wave of anxiety before an important meeting at the office, sometimes walking briskly up and down flights of stairs will be enough before trying G&R again. Chapter 14 will explore other ways to adapt G&R for your current situation.

Finally, it's important to remember that we cannot reregulate our mind-body system in just one G&R session. There's no "once and for all" from the perspective of our neurobiology! In the beginning, therefore, *slower is faster.* Only work with a small amount of activation at once, so that your survival brain can successfully discharge it. When you do, your survival brain will implicitly learn that it's no longer helpless in the face of stress activation. This is micro-level agency in action. Over time, as you complete more cycles of recovery and widen your window, you'll be able to guide yourself through this recovery process, even after extremely stressful and traumatic events.

CHAPTER FOURTEEN

Accessing Choice with Stress, Emotions, and Chronic Pain—Micro-Level Agency, Part II

"Michael," a Georgetown student in my semester-long course that includes MMFT, arrived at my office hours a frazzled mess. Since middle school, Michael had been medicated for generalized anxiety disorder. Now in his twenties, he was a thoughtful man with a delightful sense of humor. Michael pushed himself very hard: In addition to full-time graduate studies, he also held a full-time job and maintained proficiency in two foreign languages.

Sinking into the chair across from me while juggling a Starbucks cup, backpack, and bike helmet, he inadvertently spilled coffee on my desk. Although I assured him it was no big deal, he was horrified and burst into tears.

Since the beginning of the semester, Michael had enthusiastically practiced MMFT exercises at least thirty minutes daily. He'd rebooted his exercise regimen and adjusted priorities to get more sleep. He'd also noticed a lot of stress activation discharge using G&R; as a result, for the first time in his life, he was experiencing relief from his anxiety and working with his doctor to taper down his medications. In our previous meeting, he'd told me that he felt more spacious, empowered, and regulated than ever before.

Now, two months into the semester, Michael was just past the perfect storm of a romantic breakup, a major work deadline, and his Georgetown midterms. He'd also caught a cold. During our meeting, he acknowledged that taking time off to rest and heal was a new choice. Yet Michael was quickly moving into another intense period—upcoming work travel and approaching deadlines for the semester's final papers. He knew the breakup was for the best, but he really missed his ex-girlfriend. To make matters worse, his boss was micromanaging him after expressing disappointment in Michael's recent performance. The last straw: a massive fight the day before, after his roommate was extremely inconsiderate.

"I was really pissed. I'm not sure I can take living with him anymore, but I don't think I can afford my own place," Michael reported. "The worst part was that I tried Ground and Release, but it didn't work. That really scared me, because G&R has always worked in the past. I distracted myself with TV during dinner, and I then tried G&R again. But it *still* didn't work. I just felt so powerless, like back when I'd have panic attacks in high school.

"My mind started racing about everything, the breakup, my boss, how I'm ever going to finish all the work for this semester. I couldn't stop. None of the old CBT [cognitive-behavioral therapy] tricks worked, either. It was weird, because I could *see* how the stress activation was creating the anxiety, and then how the anxiety was fueling the racing thoughts. That's progress, I guess.

"But then, I started beating myself up for not being able to get to sleep. I had a big meeting this morning, and I was really worried that without enough sleep, I'd screw something up and then my boss would ride me even more. I know I was being self-critical, but I just couldn't help it. I finally fell asleep about an hour before the alarm went off. Just as I expected, I forgot to bring something important to the meeting, and then my boss chewed me out."

Michael's experience is not unusual when someone's in the process of widening their window, especially if they have dysregulation from chronic stress and trauma. For two months, Michael had been using mind fitness techniques to widen his window, honing his attentional control and tolerance for challenging internal experiences—and then applying these skills to discharge activation. In addition, by tapering down his meds, Michael was masking his symptoms less, creating more opportunities for discharge and recovery.

By willingly confronting the deeper layers of dysregulation that were ready for discharge and healing, paradoxically, Michael was primed to experience *more* stress activation now than when he'd been suppressing and masking it.

The fact that Michael developed a cold during this period—when he'd been consciously choosing to eat well, exercise, and get adequate sleep—told me that his body was in a cycle of cleansing, eliminating stored emotional and physical toxins to move itself into a place of greater regulation. Breaking up with his long-term girlfriend, who he knew was not the right partner, was another sign that Michael was facing his truth and acting on it. To me, each of these things was *an important sign of healing and growth*—only to Michael it looked and felt like he was getting worse.

Clearly, this particular week, he'd been adding to his stress load without enough recovery. Thus, by the time he fought with his roommate, their argument pushed him over his stress capacity threshold—and outside his window.

In this hyperaroused state, it's not surprising that he didn't experience any discharge when he tried G&R—his stress arousal levels were simply too far outside his window for the exercise to be effective at that time. Trying G&R again after watching TV, unsuccessfully, only compounded his survival brain's neuroception of helplessness, adding fuel to his hyperaroused fire.

"So what should I have done?" Michael asked.

To be effective, our reregulation and recovery activities *must be aligned with our current arousal levels.*

When G&R failed to help, Michael was hyperaroused and his survival brain perceived helplessness. Thus, what he most needed was to expend his excess stress activation.

I asked if he considered exercise.

"Actually, that thought did occur to me, but then I remembered you saying that we shouldn't do vigorous exercise within three hours of bedtime, so I didn't," he responded.

I apologized. It's true that exercising within three hours of bedtime is counterproductive to getting restful sleep, since it releases a surge of stress hormones. However, in Michael's case, his stress hormones were already surging! Moreover, I knew that Michael didn't have a habit of *mis*managing his stress by compulsively overexercising to mask underlying depletion.

Thus, he actually needed to *discharge* those stress hormones. After at least twenty minutes of running, climbing stairs, swimming, or cycling, he would have expended the stress hormones, exhausted his body, and quieted his racing mind. At that point, during his cooldown, he could have tried G&R again—continuing the discharge from his workout.

Then, after a shower, he could have engaged in some other *present-moment-focused* activities to keep his attention aimed at something other than his stress-inducing roommate and boss—cooking, soaking in the tub, or listening to relaxing music. Any of these activities would have protected him from adding new stress activation to his mind-body system, further helping him unwind and move toward restful sleep.

Importantly, given how hyperaroused he'd been, venting to a friend or journaling that evening would likely have spun him right back up, reinforcing his sense of powerlessness and exacerbating his insomnia. Especially with the important morning meeting, getting restful sleep was the priority that night—not figuring out a long-term solution. The time to journal would be the next day, *after* the meeting. At that point, he'd naturally see things differently. Back inside his window, he'd be able to access creative problem solving and develop a workable plan for dealing with his boss and his roommate.

SKILLFUL MEANS VERSUS SURRENDER

This chapter focuses on finding agency in several situations in which people often feel helpless, powerless, or lacking control, including when we experience strong emotions, distressing thoughts, or chronic pain. The next two chapters focus on working skillfully with limits and boundaries, resistance, uncertainty, and change.

In these chapters, I'll be moving fluidly between concrete strategies and the conceptual frameworks and context for those strategies. The tools in these three chapters are advanced moves for when you find yourself facing specific "problems." *They do not replace MMFT's core skills for widening the window using attentional control and tolerance for challenging experience.* In fact, many of these tools will be available to you *only* after you've been consistently practicing the Contact Points Exercise and G&R for several weeks.

Thus, if you choose to make only one change after reading this book, let it be to develop a consistent, daily practice of those two MMFT exercises.

I also don't want you to think these tools "fix" the "problem." For example, by sharing ideas for skillfully aligning his reregulation activities with his stress arousal levels, I wasn't suggesting that Michael could control his anxiety and racing thoughts, or willfully overcome his insomnia. Michael didn't take it that way during our meeting, and I don't want to give you that impression, either.

We *can't* control our stress arousal, emotions, distressing thoughts, or physical pain. At the same time, however, I don't want you to assume that what we choose to do or not do is irrelevant. *We can't control what arises, but we can always choose what we do with it.*

Whenever we struggle against reality, reality always wins. We can't always have things how we want them. Eventually, we must allow them to be exactly as they are. Nonetheless, it's much easier to put down our fight if we feel like we've chosen to. Skillful means, like those I offered Michael, allow us to access agency. In turn, from this empowered stance, it's easier to surrender to our present-moment experience.

Whether it's our health, job, relationships, thought patterns, or chronic pain, we usually know, somewhere deep in our gut, if there's something else we could do to shift our situation. If there's something you could do and you know it—even if you want to ignore or deny it—then it's your responsibility to do it: Have the uncomfortable confrontation. Get the exercise your body needs. Spend the time investigating, uprooting, and releasing the distressing thought patterns that are causing your suffering. Make the effort. Attend to the things that are up to you.

However, once you've implemented the skillful means at your disposal, you'll likely arrive in some moment when you realize, honestly and sincerely, that there's nothing else you can do. When you've reached this point, your mind-body system needs to absorb this truth: *There is nothing else you can do.* It's not under your control. Trying anything else only delays the inevitable. That's when you know it's time to surrender—or, as some people say, let go and let God. Jesuit priest Anthony de Mello put it this way: "Enlightenment is absolute cooperation with the inevitable."

To assume that we're in control, or that we can use skillful means and effort to manipulate experience to always match our preferences, is delusional. Often we take this pathway when we're dysregulated toward *hyper*-arousal. We rouse *too much* energy and effort to cope, expecting that somehow, through our striving, we'll be able to create the exact circumstances that we want. At some point, we need to let go of all the techniques and just submit to reality.

To surrender *without* attending to things that actually *are* under our control—this is counterproductive, too. Often we take this pathway when we're dysregulated toward *hypo*arousal, when life's beaten us down and we've given up rousing *any* effort to cope, abdicating responsibility and feeding a victim mentality. Alternatively, we may engage in spiritual bypassing, using spiritual insights or beliefs as an excuse to avoid healing psychological or emotional patterns and taking responsibility for our behavior. Both variants of this second pathway refuse to acknowledge that we always have choice.

In truth, we need both: skillful means *and* surrender, effort *and* grace. We need to take responsibility for our choices—*and* we need to surrender to reality as it is, not as we want or expect it to be. This balance is at the core of how the warrior traditions define courage. When we use both skillful means and surrender, Providence moves to meet us halfway. It may not reward us with the *outcome* we want, but it will certainly gift us in other ways—such as through the skills and strengths we acquire, the relationships we deepen, or the insights we realize.

Keep this balance in mind as you read this and the next two chapters.

STRESS AROUSAL AND UNSKILLFUL VERSUS SKILLFUL COPING

Since the survival brain isn't verbal, it can send messages only through physical sensations and emotions in the body and racing thoughts in the mind. By extension, we can communicate back to the survival brain only though mindful awareness and where, when, and how we direct our attention.

Which choices are available to us, and which tools will best help us recover back into the optimal performance zone, will depend on our current situation. In Michael's case, he actually had time to attend to his hyperarousal with vigorous exercise, followed by G&R to get back inside his window, followed by sleep.

Finding a long-term solution for his roommate situation needed to wait for another time. On a different day, different conditions might have allowed other options. Without a morning meeting, for example, he might have chosen to journal—creating space for the different waves of emotion to wash through and completing more cycles of recovery—before turning in for the night.

Or let's say their fight happened the next morning, right before Michael needed to leave for work. In this case, getting vigorous exercise followed by G&R was not an available option. Instead, Michael might have walked to work by way of a park, letting nature support some downregulation. Or he might have called his boss and asked if they could delay the meeting, to buy himself a small window for recovery. Then he might have called a friend for emotional support.

Or he might have put in his earbuds and danced to a fifteen-minute playlist devised for expending activation and coaxing his nervous system down from angry/anxious hyperarousal (for instance: Linkin Park's "By Myself" or Eminem's "Rabbit Run") through well-being hyperarousal (Heavy D & the Boyz' "Now that We Found Love" or Ricki Lee's "Can't Touch It") down to empowered moderate arousal (Booker T's "Sound the Alarm" or Caro Emerald's "Back It Up"). You can experiment yourself—as you play a song, notice the different flavors of activation and different energy levels in your mind-body system that result. (If you play these six songs, you'll likely see what I mean!) You can also develop playlists that guide the nervous system in the opposite direction, *up*regulating from apathy, numbness, or depression to empowered moderate arousal. In either direction, it's important to play only one or two songs in your current arousal zone (either angry/anxious hyperarousal or depressed/numb hypoarousal) before playing songs that help you move incrementally toward empowered moderate arousal.

Or let's say Michael had absolutely no wiggle room in his schedule that day, with back-to-back time-sensitive meetings. In this worst-case scenario, after fighting with his roommate, he could have gone into his room, jumped up and down vigorously for sixty seconds to increase his heart rate, and then used G&R to discharge at least *some* of the *physiological* stress activation. Then his thinking brain could have made a conscious pact with his survival brain. He could have said to himself: *Yes, survival brain, I know you're distressed right now because I am noticing . . . {anxious thoughts, pounding heart, churning stomach, etc.}. I wish it were otherwise, but I must compartmentalize this distress until I get through this challenging day. This evening, I promise to focus on getting back into regulation. In the meantime, as much as possible, I promise not to add any more activation on top of what's here now.*

With such a pact, Michael would reaffirm his thinking brain and survival brain's alliance. Then, to make good on this promise, he could have postponed any tasks or meetings that were not *truly* time-sensitive—and there always are *some*—to buy some breathing room. He also could have directed his attention throughout the day to the sensations at his contact points. If thoughts about his roommate arose, he could have disengaged from them and redirected his attention back to the task at hand.

In other words, his intention would be to *keep his attentional focus in the present moment as much as possible.* These small choices would have kept his stress level relatively steady without piling more on, while signaling to his survival brain that its message had been received and the situation was under control—helping his survival brain stand down. Then, after work, he could exercise, have dinner with a supportive friend, sit quietly in nature, and use G&R.

As these examples suggest—all of which I personally have used at different times—as long as we hold the intention to stay inside our window, Providence will guide our thinking brain to devise a workable path toward our intentioned goal.

Devising a workable path requires acknowledging the reality of where the survival brain and nervous system are in this moment. As we contemplate our options, we must take our current stress arousal and emotions into account. *Our body doesn't lie. It always points us to our unvarnished reality right now.*

If we reach for self-regulation tools that diverge too much from our current physical and emotional reality, we'll encounter the mismatch that Michael experienced. Furthermore, our survival brain will perceive the large disparity between the selected tool and our nervous system's truth as threatening. Believing that its message isn't getting through, the survival brain will then increase our arousal even further! Therefore, *the reregulating tool or activity we select must align with the survival brain's and the nervous system's truth in this moment.* Only then can we up- or downregulate our arousal level back inside our window, one step at a time.

As you get to know your mind-body system, you'll start to learn which self-regulation tools work for you depending on how much arousal you're experiencing.

In general, when we're *hyper*aroused with anxiety/fear or irritation/anger, we need to use tools that raise our heart rate to expend the excess energy and stress hormones, so that we can then *down*regulate into the moderately aroused zone of optimal performance. You might run, dance, or use high-intensity interval training, followed by G&R. Remember that the more often we use G&R to downregulate after moderate stressors or physical exercise, the more available and effective this tool will be when we're experiencing higher arousal levels.

Conversely, when we're *hypo*aroused with apathy, shame, or depression, we need to build energy and *up*regulate into the moderately aroused zone. Starting from this place, you might try yoga, walking in nature, singing, or cooking, followed by some activity to raise your heart rate somewhat, such as dancing to a cheerful song. You get the idea.

In effect, we become a securely attached friend to our distressed self. Just like the securely attached mother who consistently and accurately attunes to her toddler's needs—using soothing techniques for downregulating the toddler's distress or playful tools for upregulating his curiosity and joy—we can train ourselves to accurately attune to our own survival brain and our stress arousal level.

As Chapter 9 explained, the more dysregulated we are, the greater the temptation to choose an *unskillful habit,* something that will provide short-term relief by masking, distracting from, or self-medicating the arousal. These habits aren't "bad" or "wrong"—but they are unskillful because they ultimately build allostatic load.

Before choosing to engage in an unskillful habit, see if you can tolerate your discomfort—as well as the craving for your unskillful habit—for thirty seconds while paying attention to sensations at your contact points. After those thirty seconds, it's likely something will have shifted, so that you can regroup and reach for a *slightly more skillful* choice.

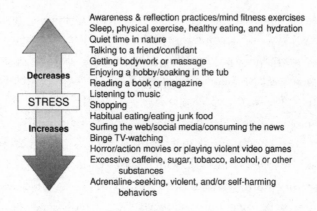

Awareness & reflection practices/mind fitness exercises
Sleep, physical exercise, healthy eating, and hydration
Quiet time in nature
Talking to a friend/confidant
Getting bodywork or massage
Enjoying a hobby/soaking in the tub
Reading a book or magazine
Listening to music
Shopping
Habitual eating/eating junk food
Surfing the web/social media/consuming the news
Binge TV-watching
Horror/action movies or playing violent video games
Excessive caffeine, sugar, tobacco, alcohol, or other
    substances
Adrenaline-seeking, violent, and/or self-harming
    behaviors

Figure 14.1: The Skillful Choices Hierarchy

The Skillful Choices Hierarchy depicts many of the common habits that we rely on to cope with stress. The habits at the bottom of the spectrum tend to suppress or mask stress activation, adding to our allostatic load and contributing to dysregulation. The ones at the top of the spectrum tend to provide true recovery and help us move back inside our window. The ones in the middle are more neutral—they aren't really helping with reregulation, but they aren't adding too much additional activation, either.

It can be self-defeating to reach for a significantly more skillful choice because of the arousal mismatch. You'd likely just be setting yourself up for failure, another round of shame and self-recrimination, and *more* stress activation.

The Skillful Choices Hierarchy is a nominal representation of the many coping habits that we typically rely on when we're stressed (see Figure 14.1). The habits at the bottom of the spectrum tend to suppress or mask the activation, contributing to further dysregulation. The ones at the top are more skillful—they help us move back inside our window. The ones in the middle aren't really helping us reregulate, but they aren't adding too much additional activation, either. As you get to know your mind-body system, you can devise your own hierarchy populated with your own skillful and less skillful habits.

Notice how interacting with our electronic gadgets is at the more dysregulated end of this spectrum. This is not an accident. Our nervous systems get revved by traffic, noise, lights, and contact with the electromagnetic spectrum—including computers, mobile devices, televisions, movie theaters, video-game arcades, and plasma screens/blaring music in restaurants. More-over, they get *especially* activated by violent video games, horror or action movies, and even reading or watching the news. In contrast, our nervous systems are soothed by being in nature, listening to peaceful music, and spending time with well-regulated people.

In other words, our nervous systems are constantly resonating with and being influenced by social and environmental stimuli. Therefore, one effec-tive approach for diagnosing your current arousal level is to notice which level of intensity you're being drawn to in this moment. This can provide some pretty reliable clues about the survival brain and the nervous system.

Each time we intentionally choose to up- or downregulate arousal in this way, we become conscious and intentional participants in widening our win-dow and rewiring our brains and nervous systems. As Chapter 13 made clear, significant stress followed by a complete and effective recovery leads to a wider window.

Ultimately, the goal is to apply this range of tools to keep ourselves in the moderately aroused zone of optimal performance as much as possible. Not only is this where we make our most effective decisions, it's also where we can best access clarity, curiosity, creativity, connection with others, and *choice*.

EMOTIONS AND DISTRESSING THOUGHTS

Just like stress arousal, emotions are essential and valuable parts of being human, since they point us to our truth in this moment. We can't transcend this aspect of our humanity simply by choosing to limit our experience of our emotions—this only exacerbates an adversarial relationship between the thinking brain and the survival brain. Indeed, when we disregard how we feel, we often make choices we later regret or set ourselves up for repeating an emotional pattern again. For this reason, one of the MMFT modules fo-cuses on working skillfully with emotions.

Emotions create tendencies toward action. Their evolutionary purpose was to propel our Paleolithic ancestors toward opportunities/safety/pleasure and away from threats/danger/pain.

The survival brain generates emotions and stress arousal together. Stress arousal points to a *primary emotion* of either approach or avoidance/withdrawal. Beyond this impulse, we usually also experience one of the *categorical (or secondary) emotions*—happiness, sadness, anger, contempt/disgust with others, shame/disgust with self, surprise, and fear. Love is not a categorical emotion, since, like awareness, it's available to us in any moment. In contrast, a *mood* is the general tone of our emotions over time. With a depressed mood, for instance, we experience a range of negative emotions, such as anxiety, irritation, and sadness.

Which categorical emotion gets triggered in any situation results from each person's neurobiological conditioning from earlier life experiences. That's why different people may experience different emotional responses to the same stimuli. Nonetheless, whichever emotion gets triggered will influence which strategies that person pursues.

Emotions are also a form of social communication, allowing us to reveal our internal states to others—and for us to perceive and resonate with others' internal states, too. We reveal our emotions externally via our *affect*, that is, our nonverbal behavior, facial expressions, and voice tone. We also communicate emotionally through *mirror neurons* and *nervous system resonance*, both of which are experience-based, prereflective, and automatic forms of understanding others. Our brains have a system of mirror neurons, which help us understand the intentions, behaviors, and emotions of others. We also experience stress contagion through nervous system resonance, as Chapter 6 explained. These mechanisms undergird how we're wired to connect, which I'll explore further in Chapter 18.

Each categorical emotion has its own "tells," which we can learn to identify in ourselves and others. When we investigate carefully, we see that each emotion comprises three components: (1) particular physical sensations, body postures, and arousal patterns; (2) a particular flavor in the mind; and (3) a particular "voice" and belief structure, which usually manifests as storylines, narratives, and thought patterns.

When we disaggregate our emotions into these components, we obtain three different doorways into recognizing an emotion's presence. With anger, for instance, we may notice heat in our neck, tension in our chest, glaring eyes, or the fact that our body is leaning forward with fists. Other times, we may notice that we're "seeing red." And other times, we may notice our mind racing with self-justifying thoughts, blaming thoughts, or an impassioned narrative about how we're right and someone else is wrong. With interoceptive awareness and nonjudgmental curiosity, we can become acquainted with the common tells of each emotion.

Each categorical emotion also exists along a spectrum of intensity, as Table 14.1 shows. *We have the most capacity to interrupt, change the trajectory, or release emotions at the mild end of the spectrum.* With fear, for instance, it's much easier to work skillfully with anxiety than it is with terror.

Table 14.1: Spectrum of Emotional Intensity

Categorical Emotion	Range of Emotion from Mild to Intense
Happiness	contented—satisfied—glad—happy—thrilled—ecstatic
Sadness	disappointed—disheartened—sad—depressed—devastated
Anger	impatient—irritated—annoyed—angry—furious—enraged
Surprise	baffled—startled—surprised—shocked—stunned
Fear	uneasy—anxious—afraid—panicked—terrified
Shame/disgust with self	self-conscious—embarrassed—ashamed—mortified—humiliated
Contempt/disgust with others	disapproving—disdainful—disgusted—antagonistic—hostile

Emotions can build just like stress activation. Impatience at breakfast can build to annoyance by lunch, setting the stage for explosive rage later in the day. Similarly, when we shove an emotion down and out of awareness, it often reappears later, usually with greater intensity, when it's harder to manage.

Therefore, the sooner we notice an emotion's presence at the mild end of the spectrum and support its release, the more likely that we'll safeguard our ability to work with it skillfully and choose the most effective response.

With practice, we can learn to ride a wave of emotion out—allowing it to pass through our mind-body system while neither suppressing nor indulging in it. In the process, we gain access to the information that our survival brain is trying to convey.

Antonio Damasio's research into *somatic markers*—the physical-sensation component of emotions—demonstrates how sensations focus our attention in ways that unconsciously eliminate certain courses of action from conscious consideration. As a result, the thinking brain focuses its deliberate reasoning process on a smaller set of options, decreasing the risk that we'll overtax our executive functioning capacity. Negative somatic markers function like alarm bells, whereas positive somatic markers are like beacons. We condition and accumulate somatic markers throughout our lives—from both actual events and "as-if" events, those anticipatory stressors that the thinking brain only imagines.

Inside our window, the thinking brain uses this survival brain input *and* conscious information informed by thought to understand the situation and then choose the best response.

In contrast, when we're outside our window, we tend to default to an adversarial relationship—either suppressing the emotion (thinking brain override) or indulging in the emotion (survival brain hijacking). Outside our window, emotions filter our perceptions, bias our information search, and influence our risk-taking tendencies and behavior. The emotion also affects how we make sense of the situation—influencing which analogies, metaphors, or past events we remember, apply, and consider. In turn, this emotionally biased understanding shapes which courses of action we deliberately consider and which get excluded from consideration.

Outside our window, any thoughts we have—or meaning we make—will be biased by the emotion. It can be especially hard to remember this when we're in thinking brain override, because we *feel* like we're being "rational" or "reasonable." Nevertheless, the suppressed emotion *is* biasing our thinking and sense making. Thus, whenever we're outside our window, we must guard against believing the *content* of thoughts, because they simply aren't true.

Thinking brain override, when we compartmentalize, avoid, and suppress our emotions, is especially common in stressful environments, because it helps us switch rapidly between widely different circumstances and behavioral codes. In fact, emotion suppression may actually be *adaptive in the short term*—allowing us to function effectively and survive during extreme events. Nevertheless, emotion suppression cuts us off from valuable information, including our intuition.

Habitual emotion suppression is linked with many allostatic load–building effects—including somatization, more stress hormones, more inflammation and chronic pain, and a greater risk of cardiovascular disease. It also increases our vulnerability to worry, depression, anxiety disorders, self-harming behavior, and suicide attempts.

Moreover, we often use addictive behaviors to avoid, mask, numb, or suppress uncomfortable emotions. That's why habitual emotion suppression is also linked with substance abuse, overeating and other eating disorders, adrenaline-seeking and risk-taking behaviors, inappropriate aggressive behavior, violent outbursts, and physical or sexual harassment and abuse of others.

With *survival brain hijacking*, on the other hand, emotions drive our behavior almost completely. Depending on the emotion, we may become overwhelmed, fueling avoidance, addictions, or paralysis. Or we may "act out" and become violent. When we indulge in an emotion, we make impulsive or reactive decisions—the destructive tweet we can't take back, the condescending comment that devastated a colleague, the punch we threw that lands us in jail. Emotional arousal degrades our impulse control and executive functioning. Thus, when we indulge in—and intensify—an emotion, it may completely inhibit thinking brain functions. One clue that we're indulging in an emotion is when we get caught up in racing and/or distressing thoughts, catastrophizing, and rumination.

In practice, thinking brain override can actually flip quickly to survival brain hijacking. For instance, many MMFT participants report a habitual pattern of suppressing emotions during the workday, followed by violent outbursts after work or lying awake at night with rumination and worry. *These are simply two sides of the same outside-the-window coin.*

Whenever we experience emotional intensity outside our window—as either thinking brain override or survival brain hijacking—we can always find agency by disentangling the emotion into its three components.

Many people tell me that they're afraid to let themselves feel their emotions—either because they're worried the emotion will never stop, or because it will cause them to lose control. Although this *fear* is genuine, the *beliefs that drive* this fear are false. With wisdom, we can remember that even intense emotions are energetic waves—they arise, eventually peak, and *always* dissipate. Perhaps not according to our preferred timeline, but they do.

Wisdom also shows us that whatever we try to push away, we're actually energizing—thereby ensuring its persistence. As the saying goes, what we resist will persist.

That's what happened with Michael, for example, when he got entangled in the *content* of his anxious thoughts. Given how hyperaroused he was, he also couldn't successfully employ CBT techniques. Cognitive-behavioral therapy, cognitive reappraisal, and positive psychology techniques tend to fall short when we're outside our window. That's because these top-down techniques require thinking brain effort, and when we're outside our window, our executive functioning capacity may be too depleted or degraded to support such cognitive effort.

We may experience three other common patterns when we're outside our window with emotional intensity. First, our thinking brain may try to figure out or "solve" the emotion, give it meaning, or assign its cause to an external event. Of course, emotions are never *caused* by an external event, but by our survival brain's neuroception of a complex convergence of factors. Furthermore, thoughts at this arousal level are emotionally biased, so letting the thinking brain spin this way is counterproductive.

Second, our situation may feel incredibly urgent, like we must figure it out and fix it *right now*. Usually, this *internal* urgency is fueled by our arousal, not by *true* urgency compelled by the situation at hand. Alternatively, if the urgency comes from outside us, it's often because of *other people's internal* urgency. Either way, internal urgency is emotionally biased. We must first discharge the emotional arousal to get back inside our window—*then* the thinking brain can focus effectively and plan.

Third, at high arousal levels, we tend to focus narrowly on information we perceive to be psychologically central, while screening out other contextual information. This narrowed attentional focus may actually exacerbate our sense of uncertainty and ambiguity. We also narrow our information search, relying more heavily on fewer inputs and stereotypes. We tend to draw consequential, and often sweeping, conclusions from small amounts of information.

In this place, *negative information is more likely to capture our attention.* A single negative event will produce stronger, more enduring reactions than a single positive event. In fact, empirical research across a diverse range of phenomena suggests that only "many good events can overcome the psychological effects of a single bad one." *This negativity bias is especially strong when we're emotionally aroused*—because we're predisposed to learn more rapidly and easily from negative rather than positive events. When we focus on threatening or negative information, however, we're likely to ignore, miss, or devalue positive information. We're also likely to misperceive neutral stimuli as threatening. Thus, any decisions we make during high emotional arousal will be disproportionately pulled toward negativity.

For all these reasons, when you're outside your window, try to disengage from the content of your thoughts and postpone any decision making, problem solving, and planning until you're back inside your window.

Focus instead on physical sensations, grounded in the present moment. Allow the emotional wave to wash through your awareness while tracking physical sensations. You might use G&R to discharge and release the emotional arousal. Alternatively, you could shuttle your attention between sensations at your contact points and the sensations associated with the emotion moving through your body. As much as possible, give the emotion permission to fully blossom in your field of awareness and then wash through.

Just as you'd sit quietly with a close friend during her emotional turmoil, you can be a true friend to your own emotions. Held in awareness with love and compassion, we can experience and release the emotion, without trying to change or add anything to it.

Sometimes an emotional wave *is* too intense to bear—in which case, skillful avoidance may be the right choice for a while. That's when it could be helpful to call a friend for some support, walk in nature, do some gentle stretching, or soak in the tub. This is *not* the time to vent or journal about the emotion—such activities will only feed its intensity. If you find yourself experiencing intolerable emotional intensity frequently, it can also help to work with a therapist. Therapists trained in body-based techniques are especially well equipped to help you uncover and release old emotional patterns.

Experiencing the wide range of emotions available to us is part of being human. We don't have to avoid, deny, or make excuses for them. However, our emotions don't need to define us, either. I've trained many people who referred to themselves as "an angry person" or "an anxious person"—allowing one emotional pattern to comprise their self-identity and thereby painfully limit who they actually are. Similarly, many people expect they should be capable of an "evolved" or "optimistic" emotional response in every circumstance—in the process denying their capacity to feel "childish" or "irrational" or "negative" emotions.

When we have such underlying beliefs about emotions, we're not actually shielding ourselves from those emotions arising. They will. *We're only making it harder to allow these emotions into our awareness*, where they can be nonjudgmentally observed and released.

These denial patterns are also socially and culturally conditioned. When I've trained people working in high-stress professions, for instance, it's fascinating to see which categorical emotions are normatively appropriate and therefore easily experienced—and which ones are not. Not surprisingly, emotions on the anger spectrum—from impatience to rage—are easily accessible for most men, since anger is a culturally sanctioned emotion for them.

When I ask them to recall when they felt sad or anxious, however, many men initially notice sensations associated with these emotions—such as muscle weakness, collapsed posture, butterflies in the stomach, or a dry mouth—but almost immediately, these sensations get replaced by sensations and thoughts related to anger. Through their cultural conditioning to "stay strong" or "keep a stiff upper lip," anger is permissible—while sadness and anxiety are not.

I've observed the opposite pattern in many women, who may easily recognize the components of anxiety or shame but have difficulty recognizing and accessing anger. When I ask them to recall when they felt angry, many women initially notice a clenched jaw, muscle tension, or a hot neck—but almost immediately, these sensations get replaced by sensations and thoughts related to anxiety, sadness, or shame.

Thus, you may have to peel back many layers of the onion to actually notice, identify, willingly allow, and release *all* the emotional waves you experience.

Especially if you have difficulty identifying emotions, tend to get caught in the content of distressing thoughts, rely on emotion suppression, or have beliefs that deny certain emotional experiences, it can be helpful to journal. Ideally, you might journal once you're back inside your window, after the emotional intensity has subsided. However, if you choose to journal when you're outside your window, make sure you have the space and time alone to experience the emotional wave fully. Many empirical studies have shown the benefits of "expressive writing" for working with emotions.

For instance, when you notice that you are unsettled or upset, you could write about an unresolved incident. As you notice the thoughts, beliefs, and physical sensations associated with emotions, you might put down your pen and seamlessly transition into a cycle of G&R, to support the movement of the emotional wave through your body. Continue to alternate between journaling and G&R, until you feel like the wave has completely passed. Eventually, by uncovering the unconscious beliefs that drive the emotion, we can extinguish the conditioning associated with that emotional pattern.

When we encounter emotional intensity that seems disproportionate to the current situation, it's often linked to emotional patterns from earlier in our life. Triggers in our current life circumstances can help us access, heal, and extinguish older patterns. It may be quite painful to allow this stored emotional debris to come to the surface, but as much as possible, willingly stay present with and *feel* the emotion. To be clear, things come up for release from the age at which they were conditioned. Thus, emotional patterns from when we were very young—and didn't have many internal or external resources—can feel especially overwhelming. This is another reason it can help to work with a therapist.

When old emotional patterns surface for release, we may also notice smells, sounds, images, and memories. These are simply aspects of an unresolved memory capsule expressing themselves in awareness for discharge. The thinking brain may try to assign meaning or figure out why this pattern is coming up for release now. As much as possible, disengage your attention from such thoughts. Instead, just willingly allow the emotional wave to wash through awareness.

The past can be healed only by release through present-moment awareness. As we do, we naturally feel lighter and more spacious. The more our window widens, the more emotionally open and available we become. We can trust that we've developed the inner strength to fully experience and release even excruciatingly intense and painful emotions.

CHRONIC PAIN

Emotional and physical pain use the same neural circuits. So just as we can exacerbate physical pain through emotion suppression, we can also use emotional processing to reduce physical pain. Although pain is not just "in our heads," unresolved emotions *can* intensify pain.

Chronic pain affects more than 100 million Americans. Not only is chronic pain stressful, debilitating, and depleting, it also fuels anticipatory stress—such as when we worry that it will never get better, it may leave us disabled, or we may lose our job because of it.

For many people, chronic pain began with an injury or accident but then continued long after their injury healed. Without a clear diagnosis, it becomes easy to catastrophize. This creates a vicious cycle: Empirical research shows that catastrophizing is linked with greater pain intensity, more fatigue with pain, and a greater likelihood that pain becomes chronic. The fact that catastrophizing amplifies pain shows that pain can have both physical and emotional components.

In fact, we activate the same *pain distress network* in the brain whenever we experience physical, emotional, and/or social pain. This network includes parts of the ACC and the insula, two brain regions discussed in Chapter 12 that play a role in regulating stress arousal and emotions—and that MMFT has been shown to modulate.

As Chapter 2 explained, people working in high-stress environments may be prone to expressing emotional distress via somatization, because they may perceive less stigma for seeking help for physical rather than psychological or emotional concerns. Remember that study about the 82nd Airborne Division soldiers, before the 2003 Iraq invasion? Compared with combat-naïve soldiers, combat veterans tended to cope with the stress of the upcoming deployment by creating relatively more *physical* symptoms—including chronic pain—while simultaneously denying *emotional* distress. Chronic pain is among *the most* common symptoms of dysregulation I observe when teaching.

Culturally, we tend to believe that pain is caused by injury or physical damage, but recent research paints a more complex picture. Physicians have long understood that chronic pain can exist in the absence of physical harm. For instance, around 85 percent of people with lower back pain have nothing diagnosable wrong with them; conversely, many people who've herniated a spinal disc—a common explanation for lower back pain—experience no pain at all.

Because they share a common pain distress network and use the same neural circuits in the brain, emotional pain and physical pain interact. In people who have chronic pain, these circuits become hypersensitive, in what's often called *central sensitization* of their pain pathways. Recent research shows that central sensitization comes from systemic and chronic inflammation in the mind-body system.

Highlighting this complex interrelationship, about half of people with major depression also experience chronic pain, while 65 percent of people with chronic pain also experience major depressive disorders. Even worse, depression and chronic pain together increase the risk of disability, substance abuse, and suicidality. Among U.S. veterans of the post-9/11 wars, for example, chronic pain and drug misuse significantly correlate with suicidal ideation and violent impulses.

Unfortunately, the way most Americans cope with chronic pain is often ineffective—fueling a sense of powerlessness and moving the survival brain toward traumatic stress.

Although Big Pharma markets many pills to "solve" the problem, these prescriptions are rarely effectual with chronic pain. Worse yet, chronic use of *almost any* analgesic—including Tylenol and nonsteroidal anti-inflammatory (NSAID) drugs, like aspirin and ibuprofen—increases hypersensitization of the pain receptors. Chronic use of pain pills also fosters gut permeability ("leaky gut"), undermining the microbiome's health and increasing inflammation, something I'll discuss further in Chapter 17. Opi-oids (like Vicodin, Percocet, and morphine) and benzodiazepines (like Va-lium, Ativan, Xanax, and Klonopin) share these downsides, but they're also addictive, contributing to substance abuse and our nation's exploding opioid epidemic.

Alternatively, many people turn to surgery, which can be expensive and risky. Pain often subsides temporarily after surgery—the well-known pla-cebo effect—but then it often returns worse than before, feeding a sense of helplessness. For instance, about 20 percent of back surgery patients still have chronic pain after their expensive procedures. Likewise, laparoscopic surgery for knee pain related to osteoarthritis has been shown, in several well-controlled studies, to be no more valuable than a placebo—while often accelerating the need for a knee replacement.

So how can we find agency with chronic pain? I'll offer concrete sugges-tions for decreasing systemic inflammation and rebalancing our gut flora in Chapter 17. Here, I want to offer some suggestions for working skillfully with intense physical sensations.

First, recognize that *pain will usually be worse when you're outside your win-dow*, such as when you're sleep-deprived, overwhelmed, or experiencing in-tense emotions. *Chronic pain is a symptom of dysregulation.* Thus, by using tools for discharging stress activation and recovering back inside your window—such as getting more sleep and exercise, eating well, and using G&R—you'll also expedite pain reduction.

Second, you can modify G&R for working with places of tension and pain. Say, for instance, you're experiencing neck pain. You could practice rolling your neck *very slowly*, coming right up to the place where you feel it "catch." It's important to stretch slowly enough that you don't inadvertently override this place. As you stretch right up to the edge of the painful place, hold still and focus your attention there. As you hold, the pain may intensify momentarily, but then it will release.

You'll know this has happened when you notice the symptoms of release from Chapter 13. You can also use these same principles while pressing a

tennis or golf ball against a particularly sore or tender spot. If you can tolerate the discomfort for about ninety seconds, it usually releases.

You can use this G&R stretching variant to supplement body-based
therapies, such as sensorimotor psychotherapy and Somatic Experiencing.
You may also benefit from bodywork that brings your body back into biomechanical alignment and targets deeper myofascial, energetic, and emotional releases.

Third, many of us cope with chronic pain by blocking it out; however,
suppressing and fighting with our pain consumes a tremendous amount of
energy. By slowly training ourselves to allow physical pain into our awareness, we can free up that energy previously used to block it out.

To work with pain mindfully, you first need to develop attentional
control, such as by using the Contact Points Exercise. While you initially
cultivate attentional control, it's important to practice in a posture that is
relatively pain-free. For instance, many people with back pain find that
practicing the Contact Points Exercise while lying down, with the knees
bent and together, minimizes pain during practice. Returning your attention again and again to sensations at the contact points helps cue the survival brain that you're safe, stable, and settled. As the survival brain
neurocepts safety, it can help the pain subside.

Once you've developed attentional control, you can begin to work directly with the pain, building your capacity to tolerate painful sensations.
To be clear, *this exercise is best done when you're feeling relatively rested, regulated,
and balanced*, such as first thing in the morning.

Begin by paying attention to a neutral target object, such as sounds or sensations at your contact points. Then you can direct your attention to the *edge* of the
place where you feel pain. It's not helpful to take your attention right smack into
the middle of the most painful place—this will only cause the survival brain to
neurocept danger and trigger more stress arousal and pain!

As you pay attention to sensations at the *edge* of the pain, it's important
to distinguish between the *actual physical sensations*—such as throbbing,
burning, stabbing, or heat—and *your thinking brain's narrative about* the pain.
As much as possible, disengage your attention from any thoughts about the
pain and keep your attention focused on the sensations.

When paying attention to the edge of the pain becomes too much, simply redirect your attention back to the neutral target object. While you practice, you can shuttle your attention back and forth between the neutral target object and the edge of the pain many times.

As you pay attention at the edges of the pain, you may notice signs of stress activation, such as increased heart rate, shallow breathing, nausea, a dry mouth, sweaty palms, or chest constriction. If this happens, *redirect your attention away from the pain and back to the sensations at your contact points—followed by one cycle of G&R.*

Indeed, chronic pain often relates to unresolved memory capsules and incomplete defensive strategies that we've not yet discharged. Sometimes when we pay attention at the edge of the pain, we tap into one of these. No problem. Just use G&R to discharge and release it.

If this happens frequently, however, I *strongly* recommend that you find a therapist trained in body-based techniques, such as sensorimotor psychotherapy or Somatic Experiencing. They can help you move through and discharge these memory capsules safely—and more efficiently—than you could on your own.

Regardless, as you practice paying attention to the edges of the pain, over time you'll increase your capacity to tolerate discomfort. In turn, greater tolerance for challenging experiences will shift how you relate to pain. You'll experience less distress and fear about it. This is one reason why mindfulness-based interventions have been shown to reduce pain intensity and decrease the distress associated with pain. *When we can attend to a physical sensation without fearing it, the pain naturally decreases.*

Since emotional and physical pain use the same neural circuits, you could also work with the underlying emotions directly, using ideas from the emotions section of this chapter. You could work with a therapist. You could also journal about your pain, asking the pain to share its message: Write down a question, and then let "the pain" write the answer using your nondominant hand, which tends to sideline your inner critic. While this sounds strange, I know from personal experience it really works.

Finally, you might journal about any limiting beliefs you hold about the pain—such as assumptions about what you can and can't do physically—and then test these beliefs with small experiments. Are these beliefs really true? In the process, you may move things from the "can't" to the "can" category, minimizing the pain's negative repercussions in your daily life. I'll have further tips about working with limits, which also apply to chronic pain, in the next chapter.

CHAPTER FIFTEEN

Working Skillfully with Limits and Resistance— Macro-Level Agency, Part I

I once trained a military leader who beat himself up for taking a nap.

As we talked, "Tom" was nursing a cold. He explained that he and his troops had just returned from several weeks in the field, where they'd conducted intensive training exercises and worked extremely long hours. When they got back, after securing weapons and ammunition, Tom released his troops early to take hot showers, find a decent meal, and get some overdue rest. In contrast, he'd planned to stay at the office to knock out some paperwork and make a dent in his long email queue.

When Tom started in on those tasks, however, he found he had no motivation. He berated himself for his lack of self-discipline. "Other leaders certainly would've been able to rouse the effort to stay at their desks," he reported. "Instead, I gave in to my laziness," by moving over to his office couch, picking up a mystery novel and, after reading a few pages, promptly falling asleep. When he woke up that evening with the book on his face, he went home and slept deeply for several more hours that night. Tom slumped in his chair as he confessed this "lazy" behavior to me.

"Was any of that work so time-sensitive that it had to be finished last night?" I asked.

He shook his head. "No."

"Did you feel more rested when you woke up this morning?" I asked.

"Well, yes."

"Is your cold feeling better today than it did yesterday?"

"I guess so."

"Are you more alert this morning after giving in to your supposed laziness yesterday?"

"Absolutely."

"Then I don't see what the problem is," I declared. By listening to his body's true needs and napping, he was now more alert and motivated. Today, his tasks would likely take less time to complete and be completed more effectively. That wasn't laziness; it was wisdom.

Tom nodded. Yet he still seemed ill at ease.

Why?

Although *Tom* had paid attention to his survival brain's and his body's input—and honored that input by sleeping—his thinking brain remained caught up in *thoughts and beliefs* about what "laziness" is or what a "good leader" would do. These thoughts were at odds with the actual situation—a disconnect that was energizing his distress. Tom was still fighting with reality, adding a tremendous amount of self-flagellation, shame, and stress to a rather straightforward example of adaptive decision making.

THE STRUGGLE WITH "SHOULD"

Whenever there's misalignment between reality and the thinking brain's thoughts, opinions, beliefs, assumptions, expectations, and preferences, it's a setup for suffering. I'll refer to these thought-based phenomena collectively as *the thinking brain's agenda*. Indeed, while the thinking brain is an excellent tool—and we rely heavily on its talents as we move through the world—it's often our own worst enemy.

This chapter and the next explore situations in which the thinking brain can get us into trouble—in the process, leaving many people feeling helpless and lacking control. Unlike stress arousal, emotions, and chronic pain, however, the topics in these chapters are *mind-made stressors*, mostly created *by* the thinking brain. The distress we experience with them usually lies in the gap between reality *as it actually is* and the "reality" that our thinking brain expects, believes, or prefers.

This chapter focuses on finding agency with limits, boundaries, and resistance. Although MMFT includes some experiential practices for recognizing and working skillfully with limits, which are not included in this book, much of this chapter's discussion extends beyond the formal MMFT course.

As Chapter 14 explained, since the survival brain is *threatened* when our thoughts and behavior diverge too widely from this moment's reality, it always responds by intensifying stress arousal to ensure that its message is getting through. Thus, it's not surprising that Tom's survival brain amped up stress, self-criticism, and shame when his thinking brain expected that he could simply sit down at his desk and dig in.

One of the best clues that our thinking brain is struggling with reality is when we have a thought involving "I should . . ." or "I'm supposed to . . ." When a "should" thought arises in the mind, it's a clue that the thinking brain is ignoring or denying some aspect of our present-moment experience. This "should" thought is pointing us to the *gap* between present-moment reality and the thinking brain's agenda.

Perhaps we *don't really want to do* what the thinking brain thinks we "should" do, but rather we're unconsciously motivated by a desire to meet external expectations, please others, or not let others down. Unless we become aware of these mixed feelings—the part of us that wants to please other people *and* the part of us that doesn't—we're setting ourselves up for internal division. Eventually, such internal division will likely manifest as resentment. Furthermore, if we continue to *deny and suppress* that resentment—because as a "strong leader" (or "good parent" or "helpful colleague") we "shouldn't" be feeling this way—we're also setting the stage for habitual emotion suppression. In turn, as Chapter 14 explained, habitual emotion suppression has many allostatic load–building consequences, including more stress arousal, inflammation, chronic pain, and reliance on stress reaction cycle habits.

Alternatively, perhaps we physically, emotionally, or cognitively *can't do* what the thinking brain thinks we "should" do. This can happen when we come up against the mind-body system's *actual* limits—such as the physical limits of our range of motion; how long we can go without food, water, or sleep; or the number of facts we can hold in working memory at once.

In this *can't* category, I'm talking about *actual constraints that arise from the laws of nature*, not the thinking brain's limiting beliefs. We can't magically produce a week with eight days, or stop our mind-body system from aging and eventually dying. Again, unless we become aware of the actual constraint and the thinking brain's tendency to deny it with "should," we widen the gap between reality as it is and the thinking brain's agenda.

Of course, "should" thoughts can also be expressed in the negative, as in "I shouldn't . . ." Such statements approach the internal division from the opposite perspective. For instance, if the thinking brain thinks, "I shouldn't make waves and speak up," it may be trying to keep us quiet and accommodating to please others—even as some part of us *really wants* to speak up. Or, if the thinking brain thinks, "I shouldn't eat that cookie," it may be ignoring the underlying stress arousal and emotions that our survival brain wants to soothe with sugar. Either way, the gap between reality as it is and the thinking brain's agenda is still there.

To be clear, it's not a problem that the thinking brain may have an agenda that isn't aligned with present-moment reality. Just as when the survival brain generates stress arousal and emotions, *it becomes a problem only when we unconsciously let our decisions and choices be driven by these things.* That's when we start to struggle with reality.

For this reason, unless there is an ethical imperative to act, most "should" thoughts provide a choice point: They present us an opportunity to nonjudgmentally investigate what's driving the gap between our thinking brain's agenda and reality's *whole* picture in this moment. If there's part of us that *doesn't want* the "should" directive, then we can make a conscious choice: Sometimes we may choose to act on the "should"; other times, we may not. Either way, we've made the choice conscious. Similarly, if the gap results from the hard constraints of a *can't*, then we can notice this limit and make our peace with it for now.

With both kinds of gaps, once we take the whole picture into account, we can access the empowered stance of agency.

Interestingly, sometimes we externalize the "should" onto our external environment. Say, for instance, you're at a prolonged work meeting. Nearby, a colleague plays solitaire on their phone underneath the table, and you notice the thought that "they should pay attention instead of goofing off." Another colleague drones on, and you think, "they should come to the point already and stop talking!" Although these "should" thoughts are directed externally, both of them are actually clues that there's a gap between reality as it is and *your* thinking brain's agenda.

In fact, these thoughts point to the likely unconscious impatience and irritation in *your* mind-body system. Your thinking brain is struggling with present-moment reality: It doesn't want to sit in this long meeting, it wants to leave. While impatience creates these externalized "should" thoughts that disapprove of your colleagues' behavior—*something over which you have absolutely no control*—in the process it adds stress activation to *your* mind-body system.

Instead, if you were to observe these externalized "should" thoughts as the clues that they are, you could bring your attention to your underlying irritation, thereby allowing the whole picture of present-moment reality into awareness. Noticing the irritation, you might then choose to direct your attention to your contact points, allow the wave of impatience to wash through, and recover back to a regulated state. Although we never can control the people around us, we always have the choice to work skillfully with what's arising in our own mind-body system.

As you start to pay attention to "should" and "supposed to," you might notice how much these thinking brain opinions facilitate our playing the victim or the martyr. The thinking brain can be very sneaky this way! Even when we experience horrific things, reality never requires us to be a victim or a martyr—we do that to ourselves.

I coach for a faculty writing boot camp. In this role, I once coached a professor who hated his job. "Chris" hated writing, teaching, his colleagues, and his academic discipline. He didn't even want to be in our boot camp, but his dean had "strongly encouraged" him to participate. As the weeks went by, during our weekly conference calls, Chris whined about his job and the boot camp assignments, while I and the other professors in his small group encouraged him to investigate his unhappiness and/or consider leaving academia altogether. His response was always the same: "I'm supposed to be in this boot camp because my dean is making me," or "I should stay in this job, because my family depends on my salary."

After several weeks of this, I realized his constant negativity was creating a harmful effect on the group. With genuine curiosity, I finally asked Chris what benefit he got from being so unhappy, both as a participant in the boot camp and in life more generally. The question stunned him into silence. I suggested he consider my question, while the rest of us continued the conversation.

Later in the call, Chris courageously admitted that being unhappy helped him feel "powerful"—because "everyone in his life could see he was always making sacrifices." I next asked why he liked being seen this way. After a short pause, Chris replied that it let him avoid considering making any changes. This insight created space for a tremendous shift in Chris's life.

When we're caught up in "should," we're consciously or unconsciously fighting with some aspect of present-moment reality. Therefore, if we make a decision based on "should," we're frequently overriding important inputs from the whole picture. Indeed, someone who's caught up in "should" may even be overlaying her experience with thinking brain opinions about how *other people would choose* with current inputs. When she does this, these opinions not only disconnect her from seeing the current situation clearly. They also divide her from her own innate wisdom about *the best choice for her* to make right now.

When awareness and wisdom are present, "should" is not driving our choices. In this place, we can clearly apprehend both the internal and external environment and use *all* inputs—rather than "should"—to choose the best course of action.

Especially when we're feeling stressed, the thinking brain's agenda can cloud us from seeing what's actually happening. *That's why one of the most important inputs is our physical sensations*—because the body is our *best* doorway into recognizing our truth in this moment. Using interoceptive awareness to tune into this body-based knowing, we can access the mind-body system's organic intelligence, which extends way beyond the thinking brain's agenda.

LIMITS AND BOUNDARIES

A major reason why Tom, the napping military leader, got caught up in "should" is that he unconsciously overrode several limits. His thinking brain's agenda was not realistically aligned with his actual capacity at the time. His thinking brain assumed he was capable of peak performance possible under optimal conditions—let's call it Situation A performance—such as when he's rested, relaxed, enjoying his work, and inside his window. In reality, however, Tom found himself in Situation B—exhausted, sleep-deprived, cranky, nursing a cold, and outside his window.

This struggle with limits is one of the *most common* issues I encounter when teaching both professionals in high-stress occupations and high-achieving Georgetown students. It's endemic to our "no pain, no gain" culture. It also fuels our individual and collective overreliance on powering through and suck it up and drive on.

Part of this struggle comes from the fact that when we're dysregulated, we're more vulnerable to getting caught up in the thinking brain's agenda. But part of our confusion comes from an incomplete understanding of how limits work.

Before I continue, let me define some terms. *External limits* include the space-time continuum, gravity, the speed of light, and other environmental phenomena governed by the laws of nature. *Internal limits*—also governed by laws of nature—relate to the brain's and body's biological constraints, along physical, cognitive, energetic, emotional, and psychological dimensions. Some examples of internal limits: How much pain we can bear before passing out. How fast our heart can beat. How much weight we can carry. How long we can sustain our attention on a task. How quickly we deplete our executive functioning. How much stress arousal in ourselves, or reactivity in others, we can tolerate before becoming overwhelmed, withdrawing, or acting out. Indeed, *the current width of our window is one of our most important limits.*

In contrast, *boundaries* are relational. They include how physically close we prefer being with another person, and whether we're comfortable with touching and being touched by them. Boundaries can also be cognitive, energetic, emotional, and psychological. With healthy boundaries, we can separate our thoughts, emotions, and nervous system activation from those of other people. For instance, we can be aware of another person's thoughts, emotions, and stress arousal without being overly swayed by—or taking responsibility for—their internal states. We can also allow and accept differences between us, while still remaining connected. We neither let their perspective drive our decisions nor try to convince them of ours.

As with all conditioning, we instinctively internalized our beliefs and habits about limits and boundaries through our early social environment. If our parents and care providers respected our limits and boundaries—*and their own*—then we likely developed a healthy relationship with limits and boundaries. In contrast, if we grew up in an environment where we observed people around us habitually overriding their own limits—or we felt pushed to override our own, either to avoid punishment or to receive attention and/or love—then we likely developed unhealthy patterns about limits. Likewise, those of us who experienced relational trauma as children, or were required to do what others wanted at the expense of taking care of ourselves, likely developed unhealthy patterns about boundaries.

Limits and boundaries have been a particularly important growth edge for me—in large part because I experienced repeated traumatic violations starting when I was a toddler. As older, stronger, more powerful people repeatedly denied my limits and transgressed my boundaries, I unconsciously learned that my boundaries and limits *must not be real*.

Several neurobiological adaptations flowed from this. Through these events, my survival brain conditioned its faulty neuroception, its inability to accurately perceive safety and danger. Moreover, denying, suppressing, and compartmentalizing my pain—and overriding my own limits—were some of the only defensive and relational strategies available to me at that age.

My thinking brain also played a role in protecting me in these situations. It created several unconscious beliefs—most of them revolving around a core narrative that something must be wrong with me to have caused these events to happen. Put differently, my thinking brain created *beliefs, thoughts, and assumptions* that allowed me to convince myself that I was safe in environments that *actually were unsafe*. At that age, it was so much easier to believe that something was wrong with *me* than to believe that something was wrong with *the situation*. This was how my thinking brain could square the circle between an unsafe reality and the fact that I was too young and powerless to get out of it.

Of course, this whole conditioned package—my survival brain's faulty neuroception, my unconscious strategies to deny my pain and override my limits, and the credence that I gave my thinking brain's agenda—*was also a setup for trauma reenactment*. Which is exactly what transpired, many times over, each time further narrowing my window while also reinforcing my circle-squaring pattern.

All of this conditioning, working in tandem, is what allowed me to survive. Yet this deeply ingrained and unconscious pattern—always trying to square the circle between reality as it actually is and the "reality" of my thinking brain's agenda—was extremely costly.

Returning to my examples from Chapter 1, for instance, this was how I could impale a hammer's claw one inch into my heel and then, seven days later, choose to complete a marathon, running the entire way—even though the wound started bleeding again at mile ten. This was how I could have seven chapters of a Harvard Ph.D. dissertation left to write with only ten weeks to go and then choose to deliver the 461-page manuscript on deadline—even though I pulled the equivalent of a multimonth all-nighter, vomiting on my keyboard along the way.

Just as with Tom, the gaps between reality and my thinking brain's agendas were very large, indeed.

Was I successful in squaring those circles? At the time, my thinking brain sure thought so. Even more, the world around me rewarded me for these choices.

Yet behind this circle-squaring pattern—and its corresponding drive to achieve—was always that *very young* unconscious thinking brain belief that something must be wrong with me. And the cost of this circle-squaring pattern? Well, that was borne by my body, my relationships, my inner turmoil, and my ever-growing allostatic load.

This is how suck it up and drive on *works*.

As my story suggests, many of our conditioned gaps between reality and our thinking brain's agenda about "reality" originated in childhood. Over time, as we allow more and more of these unconscious patterns into conscious awareness, we can nonjudgmentally examine and heal them—and then finally choose something different. What a relief!

In the process, we come to recognize that all we're ever responsible for is seeing the current situation clearly and then working with it *to the best of our ability right now*. Of course, "to the best of our ability right now" is going to change moment to moment, depending on current conditions, *including the current status of our limits and boundaries*. We're not aiming for perfection here. We're aiming for wholeness.

When we start to investigate our limits and boundaries with nonjudgmental curiosity, we quickly realize three things. First, they *vary*. With some limits, for example, we may have astonishing capacity to bear tremendous demands, while other limits are minuscule, getting depleted almost immediately.

Moreover, as Tom's story illustrates, they're *context dependent*: When we're inside our window, our limits are a certain size. When we're sick, exhausted, rushed, or stressed outside our window, our limits are much smaller. Likewise, we naturally move physically closer to and share more with friends than we do with work colleagues or strangers.

Finally, they're *dynamic*. Our bodies and minds are considerably more fluid than we usually presume. In the morning rush before a big client pitch, for instance, your boundaries may be extremely restricted: Caught up in your own anxiety, you have virtually no capacity to connect meaningfully with your spouse. After a successful presentation wins the big contract, however, your boundaries expand dramatically. Now, you're perfectly happy to listen and offer your spouse emotional support after their stressful day.

It's partly because of this dynamic quality that mindfulness-based practices can be so helpful. By engaging in these practices daily, we can observe how much our internal limits are always in flux. Some days we can get into the challenging yoga pose; other days we can't. Some days the mind is calm and focused; other days it's a firestorm of racing thoughts and distractions.

Observing these shifts helps us not take our limits and boundaries so personally. More importantly, *it helps the thinking brain relax its agenda about achieving Situation A peak performance when we actually find ourselves in a Situation B reality.*

Although we need to accept our limits and boundaries as they are right now, we can also intentionally extend them over time. We see an incomplete echo of this truth in the motto "no pain, no gain."

Why is "no pain, no gain" a partial truth? Because it focuses only on the first half of the process—taking ourselves outside our comfort zone—which is to say, deliberately choosing to override a current limit. However, "no pain, no gain" misses the second half of the process—following that deliberate override with a complete recovery afterward. Importantly, *this second half of the process is what actually extends the limit.* You've probably caught on here:

Intentionally extending our limits over time is *the same process by which we widen our window.*

If you've ever worked with a physical trainer or coach, you've probably heard of "first stage" and "second stage" muscle failure. At the first stage, your muscles scream at you while the thinking brain spins distressed thoughts to convince you to stop *right now.* You're in *agony,* but the coach insists you keep going, so you willfully override your pain and fatigue.

As you continue pushing, your muscles eventually reach second-stage failure. At this point, you may notice waves of heat as the working muscles begin to quiver and shake uncontrollably. These "shakes" are the reason you're working out—tearing down tiny muscle fibers so that, over the next few days of recovery and soreness, the muscles can repair themselves. By choosing to work these muscles beyond your comfort zone, you set the stage for recovery back to a *stronger* baseline.

Importantly, however, *to build the muscle and extend our physical limits, we must have the recovery time.* In contrast, when we train too hard without enough recovery, we set the stage for repetitive strain and injury.

Empirical research has explored how humans can push beyond physical exhaustion. This research posits that the fatigue and discomfort we feel at first-stage muscle failure don't come from the working muscles—they actually come from the brain. In effect, the brain acts as the "central governor" of our exertion, protecting the mind-body system from damage and preventing us from using any internal system to its absolute maximal capacity.

Thus, when the brain senses that we're spending our energy stores too quickly, it creates distressed thoughts and physical pain to tell us to stop. This signaling is *conservative*—the brain creates fatigue and discomfort long before the muscles reach their actual limit. Just as the out-of-gas light comes on in our car when we still have a few gallons to make it to the gas station, these thinking brain *thoughts about* our limits do not align with our *actual* limits.

In fact, the brain paces our performance based on its expectations and beliefs about the duration of our exertion. That's why we can feel like we're running on fumes but then still find energy to sprint at the race's end: *The brain expects that the race is almost over.*

There's related research about the placebo effect, which shows how thinking brain *expectations* can produce *physiological* outcomes—including improvements in the immune system, more endorphins that reduce sensations of physical pain, and decreased levels of anxiety and fatigue. In this research, the doctor knowingly uses a bogus drug or treatment but *creates expectations in the patient* that the intervention will be successful. Although the patient doesn't know it's a sham, the doctor's enthusiastic introduction of the intervention helps the patient mobilize internal resources and thereby improve physically and psychologically. In other words, holding a particular thinking brain expectation can lead to material outcomes consistent with it.

Taken together, this research highlights the gap that exists between *actual* limits and the thinking brain's *agenda about* limits. This gap manifests in two different ways.

First, as already discussed, someone may habitually, often unconsciously, *override the actual limit.* This is the "no pain, no gain," powering-through approach. Just as athletes may seriously injure themselves when they "play hurt," habitually overriding actual limits without adequate recovery builds allostatic load and narrows our window. Moreover, it sets the stage for that awful pendulation between overwork and exhaustion, followed by procrastination, resistance, and self-defeating behaviors. Then, feeling "behind," we scurry to catch up. Rinse and repeat.

To be clear, there will certainly be times when we *must* override our limits, such as to respond effectively during a crisis or emergency, meet a deadline that gets changed unexpectedly, or care for a sick family member during the night. In each of these examples, life happened—and *acute* circumstances require flexibility. Not a problem. We can easily choose to override our limits in these situations, *as long as we do it consciously.*

If we do it consciously, we'll be more aware of how the overridden limits may be biasing our perceptions, narrowing our attention, and depleting our executive functioning capacity. We'll also be more aware of how this may fuel resistance, procrastination, anxiety, irritability, and other emotional effects. With this understanding, we can more easily access self-compassion and not beat ourselves up. Most importantly, *we consciously put ourselves on notice that we need recovery soon.* In turn, we can deliberately schedule time for sleep and other recovery activities, to bring ourselves back into a regulated state inside our window.

The second way that the gap manifests between actual limits and the thinking brain's agenda about limits is when we habitually, often unconsciously, *stop short of an actual limit*. Here, the thinking brain's agenda includes *limiting beliefs* that keep us from taking risks. Common limiting beliefs I encounter when teaching include the following:

- I'm just an anxious (stubborn, angry, impatient, sad, self-reliant, introverted, restless, hyperactive, disorganized, spacey, perfectionistic) person.
- The depression (anxiety, diabetes, heart disease, addiction) genes run in my family, so I can't help but have it.
- I don't have enough time (money, skills, connections, experience, friends, self-discipline, willpower, support).
- I don't have the right resources.
- If you can't do it right, it's not worth doing at all.
- I'm too busy/I have too many competing demands/I don't own my calendar.
- I've never been a people person.
- I'm not good with feelings (children, animals, new situations, public speaking, sports, relaxation, taking time off).
- I'm not strong (thin, coordinated, wealthy, smart, charismatic, creative, resilient, emotionally intelligent, self-disciplined, worthy) enough.

Limiting beliefs are persuasive aspects of the thinking brain's agenda that keep us from doing what we set out to do. And as we begin to recognize and release certain limiting beliefs, we often uncover deeper beliefs that are even more global.

Limiting beliefs get in the way more often than we might realize, impeding us from taking risks and extending our limits. So what can we do? First, bring limiting beliefs into conscious awareness and *nonjudgmentally* own them. Although we may feel embarrassed by their content, we need to recognize how they served us in the past. As my story suggests, many limiting beliefs were adaptive earlier in our lives, when they helped protect us from something painful.

Next, we need to investigate: Is this limiting belief actually true? Asking this question—and remaining genuinely open to whatever we learn—we usually discover that the belief *is not actually supported by empirical evidence.* Alternatively, *one aspect* may be supported by the facts, but *the entire generalized statement* is not. As soon as we choose to examine the belief deeply, carefully comparing it to today's actual reality, the limiting belief usually dissolves like a puff of smoke.

At this point, we could replace it with something that's actually true today. For instance, if we notice the limiting belief that "I never have enough time," we might replace it with "When I have clear intentions, I have more than enough time for what really matters." The clearer we are about the *actual* situation and our *actual* limits, the less we'll buy into limiting beliefs. In the process, we create possibilities for change and growth.

Limiting beliefs come in many different flavors of *insufficiency.* At its core, the thinking brain's agenda is dissatisfied with some aspect of reality—and thus believes it must add, do, or have something to feel complete. Limiting beliefs also fuel impostor syndrome, shame, and self-hatred.

With this in mind, often the most efficient path to uprooting limiting beliefs is to work directly with the underlying feelings of insufficiency, inadequacy, and unworthiness. To begin this inquiry, you might journal about the limiting belief and any memories you have about when it first appeared in your life. You might discover, for instance, that it's actually something you absorbed from someone else. As you nonjudgmentally write the narrative of its first appearance, you'll soon uncover any underlying emotions linked to the belief. Then you can work directly with the emotions, using the tools from Chapter 14.

I want to make one last point about how we habitually relate to limits. In effect, we often combine two things: (1) overriding *some* limits, while simultaneously (2) buying into limiting beliefs about *other* limits, which lets us avoid the core issues. In my own life, and in observing others' experience, I've come to understand how much these opposing energies can keep us trapped in situations that are, ultimately, an artificial status quo. *These situations need to fall apart, but with our thinking brain's agenda, we won't let them.* For instance, frequently we buy into those limiting beliefs about emotions from Chapter 14—we're afraid to feel our pain for fear that the emotions will never stop or that we'll lose control. In turn, because we trust these limiting beliefs, we refuse to take risks with our emotions, psychological patterns, or relationships—*and we square this circle by overriding other limits and boundaries.*

For example, these situations include the job you hate while continuing to "make it work." The relationship built on patterns of abuse or addiction. The marriage or business that needs to be either radically redirected or allowed to fail. The organization or creative endeavor in which the people involved are willing to do "whatever it takes" to accomplish the mission, even as they damage their health and obliterate their personal lives.

We often choose to perpetuate these situations out of a sense of responsibility, service, dedication to the mission, or obligation to others. Nonetheless, when our perseverance and commitment to "making it work" gets abused or taken for granted, we set ourselves up for burning ourselves out, generating resentment, relying on stress reaction cycle coping habits, and building allostatic load. We also create relationships of (co)dependency, in which everyone involved colludes—consciously or unconsciously—in maintaining the increasingly shaky status quo. Most importantly, we blind ourselves and everyone else involved to the lessons that reality is patiently trying to show us.

The longer we artificially prop up these situations through our thinking brain's agenda, *the greater the tension that builds from the ever-widening gap between reality and our thinking brain's agenda—and the greater the course correction eventually needed to release that tension and bring the situation back into alignment with reality.*

Trust me, I learned this the hard way: In my case, it took losing my eyesight and leaving a marriage. The gaps between reality and my thinking brain's agenda were very large indeed.

The underlying issues in situations that need to fall apart won't magically disappear through willful effort or doubling down on our thinking brain's agenda. We can't just strive our way out of it. Until we allow the underlying issues to surface—and willingly *feel and release* whatever pain we've been avoiding—it's only going to continue getting worse. The longer we struggle, the greater we suffer, while the tension keeps accumulating until we're finally ready to face reality. In the meantime, all that accumulating tension must get expressed *somewhere*—through "acting out," accidents, health crises, or major ethical lapses.

In the struggle with reality, reality always wins.

So how do we allow these situations to fall apart gracefully? As thoughtfully and compassionately as possible.

First, we must *equally* consider and honor everyone's needs, *including our own*. We must be willing to be truthful with ourselves—because until we can do that, we can't be truthful with anyone else. We must recognize that this situation has continued because we remain attached to an outcome at odds with reality. For this reason, the thinking brain agenda that maintains the pretense is likely extremely subtle.

Next, we must uncover and test our thinking brain's agenda—every belief, assumption, and expectation that is holding this whole unsustainable facade together. Belief by belief, we must methodically bring each one into awareness and ask: Is this actually true? Until we can answer for sure, one way or the other, we can't really surrender to reality.

Most of all, we must willingly allow the answers to be whatever they are.

If we don't have enough information to answer, then we need to gather data. By way of example, as part of my journey with MMFT, I founded the nonprofit Mind Fitness Training Institute. We kept the institute operational for eight years—mostly by overriding limits, especially mine. For eight years, I juggled two jobs on one salary. I consciously knew I was overriding my limits, but I chose to persist out of a deep commitment to our mission. Like any start-up, we grew through the "sweat equity" of our people.

In large part, our choice to persevere without enough resources was driven by one *explicit* collective assumption: that generating evidence of MMFT's efficacy, through rigorous scientific research in high-stress environments, would eventually lead to MMFT's wider implementation in those environments.

Eventually, MMFT became the best-validated resilience training program empirically tested with the U.S. military. Yet for many reasons outside our control, a contract for wider implementation had still not materialized. In the meantime, the tension and financial instability we'd been accumulating reached a breaking point.

In response, the board and I spent time carefully unearthing and examining *all* our assumptions, *especially the semiconscious ones*, through a deliberate path of experimentation that lasted more than two years. During this process, there were failures and messes, but we gained valuable information that helped us find clarity. Although closing the institute was heartbreaking, it was ultimately the right choice.

Did we make some clumsy choices while testing our assumptions? Of course we did—we're human. But did we learn from those choices? Absolutely.

Letting things fall apart can be a painful process, especially if we've been avoiding reality for quite some time. Nonetheless, we must trust that we have the wisdom and courage to face and move through it. We must trust that when we sincerely consider everyone's needs, reality will reveal the best course of action for everyone involved, even if our thinking brain can't see that right now. We must trust that any loss always brings with it new possibilities.

Reality works in mysterious ways, way beyond our thinking brain's narrow agenda. Paradoxically, allowing such situations to fall apart usually provides us the opportunity to grow into a new structure, better aligned with who we really are.

RESISTANCE

Resistance is anything that interferes with our manifesting our intentions and goals, including the goal of healing, growing, and evolving. Thus, resistance may show up any time we direct ourselves to do something that may involve physical, intellectual, emotional, spiritual, or relational discomfort—such as beginning an exercise regimen, changing our diet, learning a new skill, relinquishing an unskillful habit, showing up more authentically in our workplace, or committing to deeper intimacy in a relationship. Resistance is also common with creative work, such as writing or composing music. Even something as basic as forsaking short-term pleasure or comfort to reach a long-term goal can create resistance. To be clear, I'm not talking here about "good" resistance, like opposing oppression or preventing unethical or harmful behavior.

Although it comes in many flavors, resistance impedes our growth, creativity, and highest purpose. At least some part of us is fighting with the reality of the task at hand. It's often a sign that we've narrowed our window. The following are some common forms of resistance:

- Procrastination, overwhelm, and avoidance behaviors—including engaging in excessive busyness and low-priority activities, while putting off work for important long-term goals
- Anxiety, irritability, restlessness, and other emotions that drive us to seek "comfort-based solutions" and/or quick and easy gratification through entertainment, shopping, food, sex, alcohol, or other substances
- Self-sabotage and self-defeating behaviors—including addictions, affairs, accident-prone behavior, and dramas that we manufacture at work or in our personal life, that allow us to play the victim and/or distract us from our work
- Perfectionism, self-criticism, limiting beliefs, negative attitudes, rationalizations, and other thinking brain narratives for why we shouldn't be doing our work
- Self-doubt, fear of failure, and—paradoxically—fear of success

High-stress environments are especially likely to induce resistance, because they often involve dangerous, demanding work that forsakes short-term comfort to achieve individual and collective goals. These environments are rife with sleep deprivation, long work hours, cranky colleagues, impatient customers and clients, sore feet, burdensome equipment, bad weather, harsh and toxic environments, crappy food, and intense training regimens that push the mind-body system to its limits. And that's *before* we add Murphy's Law into the equation—computer meltdowns, canceled flights, flat tires, power outages—because if it can go wrong, at some point it will.

Resistance thrives in high-stress, high-demand environments. For this reason, the warrior traditions uniquely organized their teachings around outwitting resistance because *warrior environments are characterized by resistance-evoking situations.* It's not an accident that the important ingredients for overcoming resistance—conscious intentions, consistent practice, and self-discipline—are central to any training regimen associated with the Warrior's path. Furthermore, the warrior qualities of wisdom and courage help us choose to persevere toward our goals despite the presence of resistance.

Many observers have commented that resistance is often fear-based—trying to protect us from failure, social exclusion, reprisal, or unwelcome attention. Here, resistance is trying to keep us safe. Especially for those of us who developed early childhood survival strategies that rely on people-pleasing, hiding, or making ourselves invisible, we were conditioned to block ourselves from expressing our creativity and authenticity. Interestingly, with fear-based resistance the thinking brain's agenda—the insidious narratives it relies on to keep us quiet and small—usually contains *an element of truth*, which is what gives it its power.

It's *true* that rocking the boat with new ideas and innovations may not be appreciated by everyone—the status quo doesn't like it when we offer suggestions that would require change. And although many people will appreciate our creative efforts, there will always be at least some critics and haters. We can't please all of the people all of the time.

These statements may be true, but we don't have to let them stop us in our tracks. Rather than engaging with the *content* of the thinking brain's fear-based agenda, it can be helpful instead to work with the fear in the body, using tools from Chapter 14.

Here, too, the gap between the thinking brain's agenda and reality can be an important clue. In fact, when the inner critic usually appears, we can be sure that we're engaged in something that's challenging us to grow and manifest our best self. Fear and self-doubt often arise *exactly when* we're doing our most creative work—engaged in projects that employ our unique gifts and talents, follow our heart's true calling, and reveal our authenticity.

Paradoxically, therefore, fear and self-doubt often indicate we're doing exactly what we're meant to be doing. Writer Steven Pressfield argues that resistance is proportional to love—the more passion we feel for a project or a calling, the more it terrifies us. Yet these projects and callings are often critical for our growth. That's why the warrior traditions cultivate wisdom, courage, and self-discipline—so that we can understand what's happening and still choose to move forward.

When resistance is fear-based, it's important to understand why it's been triggered, acknowledge the fear with self-compassion, and choose to turn toward it. *Resistance naturally builds as the day wears on*, so facing the fear and starting the task early in the day can help. Conversely, if we give in to avoidance behaviors, we usually squander our energy, build tension in the mind-body system, and feed shame and self-judgment. Furthermore, as with any habit, if we give in to avoidance today, we're strengthening it—making it that much harder to face the resistance and do the task tomorrow.

Therefore, when resistance is present, identify and name it; notice its physical sensations, emotions, distressing thoughts, and behavioral impulses with nonjudgmental curiosity; and then—get to work! *Find the easiest, most effortless way to begin*, with the least struggle possible. When I experience fear-based resistance with writing, for instance, I've learned that I struggle less if I let myself free-write in longhand without editing myself, at some point transitioning effortlessly to the computer.

We can also set a timer for fifteen to thirty minutes to face the work directly. Steady incremental progress can feel like such a relief. Even better, it accumulates momentum and confidence—showing us that progress is possible despite the fear. Then when resistance arises again, we'll trust ourselves more deeply not to be jerked around by it.

Over time, we can train ourselves to show up every day, no matter what—fully committed to the *practice* and the *path*, without overidentifying with the *product* or the *outcome*. This is another place where balancing skillful means and surrender comes into play: Choosing to sit down and do the work naturally creates the space for inspiration, insight, guidance, creativity, passion, and momentum to arise. I often light a candle when I sit down to write, as a small reminder that my only job is to show up with awareness and invite the Muse in.

Importantly, however, not all resistance stems from fear or self-doubt. *Resistance may also result from having overridden our limits.* You've probably noticed that you're more likely to procrastinate, self-sabotage, and engage in avoidance behaviors when you're sleep-deprived, stressed, burned out, sick, experiencing chronic pain, or dysregulated. As Chapter 2 explored, when we collectively equate "being stressed" with being busy, successful, and important, we tend to disconnect "being stressed" from its eventual consequences. We also tend to override many limits, creating the perfect setup for this kind of resistance.

A major clue that resistance is related to overridden limits is when we waste time, spin our wheels, and dissipate our energy in too many directions—unable to focus our attention or effectively harness our energy. Having observed this dynamic in myself and many others, I believe resistance from overridden limits is one ingenious way that our mind-body system tries to restore a sense of equilibrium: Having pushed too hard or too far, we now find ourselves brought to a standstill in compensation. We may even catch a cold to force ourselves to take a break.

For example, during my fellowship year after pushing to complete my Ph.D. dissertation on deadline, I often found myself "wasting time"—knitting sweaters and watching endless reruns of *Law & Order* with my housemates. Today, I recognize that my mind-body system was creatively enforcing a long-overdue recovery period after my workaholic multimonth all-nighter. At the time, however, I beat myself up mercilessly for this "lazy" behavior, just like Tom.

This kind of resistance also tends to fuel that nasty pendulation between procrastinating and wasting time, followed by burning the midnight oil or working through the weekend because we feel "behind." Such pendulation is almost always a sign of overridden limits!

Therefore, when you notice resistance, don't assume that it's only related to fear. Start by investigating whether you unconsciously overrode a limit. You may have said yes to something, when your mind-body system really wanted you to say no. Or you may have been slowly accumulating sleep deprivation. You may have even *consciously* overridden a limit but still haven't made the time to balance it out with the necessary recovery.

If you discover overridden limit(s), now's the time to face this kind of resistance directly. If you can't attend to recovery and replenishment immediately, then at least use Chapter 14's Skillful Choices Hierarchy to choose a slightly more regulating activity in the short term. At the same time, be sure to block off time very soon in your calendar for some deeper recovery, sleep, and time off.

By taking a break to give yourself needed replenishment, rest, recovery, and rejuvenation, you can come back stronger with renewed vigor, transformed perspective, and focused energy.

CHAPTER SIXTEEN

Thriving during Uncertainty and Change— Macro-Level Agency, Part II

"Jenna," a woman I trained in MMFT who works in the national security community, was a major worrier. Her thinking brain was always caught up in planning—usually anticipating a range of things that could go wrong and what she would do in response.

When Jenna raised her hand to answer a question, by the time I called on her, she'd often forgot what she'd planned to say. When we met privately, I asked her whether she experienced these memory lapses in other parts of her life, and she nodded.

Jenna said that it was difficult to remember sometimes, because her mind kept returning to its endless contingency planning. She also admitted that her planning made her anxious and kept her awake at night. Most nights, she slept only three or four hours, and she'd recently gained almost twenty pounds from the stress.

Jenna told me that she wanted the anxiety, insomnia, memory lapses, and weight gain to go away. Yet she also shared many examples of having imagined improbable scenarios that eventually came to pass. Jenna adamantly believed these examples were evidence that her worry kept her and her family safe.

"Have you ever considered the possibility that your thinking brain's planning could be anxiously creating self-fulfilling prophecies?" I asked.

"Absolutely not." Jenna vigorously shook her head. "I'm keeping my family safe by imagining these scenarios before they happen."

"Okay. But doesn't this strategy seem costly?" I asked. "Given all you now know about the links between the thinking brain's planning habit and the anticipatory stress arousal that planning creates, don't you think there might be an easier way? It's possible that the anticipatory stress from your planning is feeding your memory lapses, anxiety, and insomnia. And you were just saying that you wished these things could be different."

"Then I guess those things are just the cost of doing business," she replied.

"But what if I told you that you could let go of this constant planning and still be all right?" I asked. "It would only require trusting yourself to respond effectively in the moment, with the range of amazing talents that I know you have."

Jenna frowned and shook her head again. "No, the only way I can respond to things is if I have planned out all of the different options ahead of time."

"But what about when something happens that doesn't match any of the options you'd planned out ahead of time?" I asked. "To put a fine point on it: Say, for instance, terrorists learned to fly planes so they could drive them into the sides of buildings."

"Well, then, I guess I'd just be screwed," she shrugged and relaxed back in her chair. With that, we both laughed, the moment of tension dissolving.

Before long, Jenna's laughter morphed into tears. I handed her the tissue box, and we sat together quietly as she cried out a spontaneous discharge of stress activation.

After a moment, Jenna wiped her tears. "But the United States was still all right after 9/11, wasn't it?" she asked quietly. "It wasn't what we wanted, but we were okay."

"Yes, we were," I nodded. "While what happened didn't match any of our contingency plans, we were still able to draw on our collective strength. I don't want to dismiss the pain and deaths of that time. But there were some incredible acts of courage and community that emerged spontaneously afterward. We had adaptive capacity we could rely on."

UNCERTAINTY

This chapter focuses on finding agency with uncertainty, the sense of time scarcity, and change. As with Chapter 15, the topics in this chapter are *mind-made stressors*, mostly created *by* the thinking brain. Although much of this chapter's discussion extends beyond the MMFT course, its topics and strategies are included in my semester-long, MMFT-related course at Georgetown. This chapter is especially important for the book's second goal, a wider reflection about how we, individually and collectively, approach stress and trauma.

As Jenna's story highlights, especially during times of uncertainty and change, we humans tend to rely on anticipation—thinking about, rehearsing, or planning for events that *may* occur in the future.

Anticipation is a *future-oriented anxiety-management system*. We usually anticipate future events that we expect to be unpleasant—by catastrophizing, doing worst-case-scenario planning, and trying to strategize and manipulate our experience to avoid discomfort and pain. It's also possible to anticipate future events we expect to be pleasant, such as fantasizing about our upcoming vacation. Either way, I call this *Planning 1.0*, our thinking brain's habit of imagining and preparing for future contingencies.

In our techno-centric culture, we've taken this anticipation approach to the extreme, trying to predict unwanted events and prevent them from occurring. We want science and technology to explain the uncertainty and unpredictability—and give us tools for controlling it.

In fact, as Chapter 2 explained, embedded in modern scientific practice is the cultural belief that uncertainty endures because of incomplete human knowledge—not because of random nature or the inherent unknowability of phenomena. With enough measurement, data, calculation, and/or analysis, we assume certainty *is* possible, because phenomena *can be known*.

Consider, for instance, the massive rush today toward "big data" and "data-mining" methodologies. These tools rely on turning information into something indexable, searchable, and amenable to algorithmic, automated analysis. Part of big data's allure is *its presumed ability for prediction*. Correlations may not tell us precisely why something is happening, but we like that they identify patterns that may help predict and prevent future unwanted events.

At its core, our collective overreliance on technology is driven by our collective intolerance of uncertainty. Our collective consciousness is aligned with Jenna's: We assume that we can predict and plan our way to certainty, with technology's help.

As Jenna insightfully acknowledged in our conversation, thinking that we *can know* what will happen—and having a plan for how to deal with it—gives our *thinking brain* a sense of safety. And since many of us identify with our thinking brain's understanding of the world, it gives *us* a sense of safety, too.

Planning 1.0 is a habit perpetuated by negative reinforcement. We feel stressed by the sense that life is uncertain and time is scarce. The thinking brain defaults into spinning up mental to-do lists and contingency plans. As a result, we *perceive* ourselves to be more certain and in control—the reward that soothes the anxiety.

Importantly, however, *this reward is illusory*. The next time you notice your thinking brain engaged in Planning 1.0, observe it with nonjudgmental curiosity. What you'll likely discover is that Planning 1.0 is usually distracted and frazzled, fueled by an internal sense of urgency and focused on handling the immediate "fires," not our long-term priorities. The thinking brain also tends to loop endlessly through the same things. Moreover, *because it's biased by our current arousal levels*, it's rarely seeing the whole picture, taking all available information into account. Just as Jenna experienced, when the thinking brain engages in Planning 1.0, it revs the survival brain—increasing our stress and exacerbating our anxiety.

For these reasons, while many people have active Planning 1.0 habits, the "planning" actually accomplished is ineffectual, at best.

So why do we do it? Most of us don't like to face the truth that we aren't in control. Externally, for instance, we can't control whether there will be another terrorist attack or school shooting or, more mundane, what other people think about us. Internally, we can't control which thoughts or emotions arise in our mind-body systems, or when we will get sick or die.

Inevitably, however, there will come a time when reality inconsiderately refuses to correspond with our thinking brain's expectations. Anticipation does not reduce uncertainty as much as it denies, ignores, or obscures it. Instead, it increases the likelihood we'll be caught off-guard when events don't align with our expectations. The result is that when an unexpected or unwanted event inevitably occurs, the world can seem even more uncertain, hostile, and threatening than before.

The narrower our window, the more intolerant of uncertainty we'll be—and the more likely we'll rely on robust, well-organized, well-developed, and rather inflexible plans, response sequences, templates, and strategies. Furthermore, *the more physiological and emotional arousal we'll experience at having these plans interrupted.*

In other words, anticipation comes with a certain amount of rigidity, both individually and collectively. The more invested we are in the thinking brain's agenda about what "should" happen, the more difficult it is to pivot when our preferred course of action gets thwarted. This gap between reality and our thinking brain's agenda impedes our ability to perceive other potential courses of action and flexibly adapt.

Conversely, the wider our window, the more we can access adaptability when our plans and expectations get interrupted. Here, it's much easier to see alternative courses of action, improvise, and flexibly adjust in response. As a result, the survival brain doesn't perceive the interruption of our plans as threatening—*so we don't experience as much stress or emotional arousal.* We can align more quickly with reality as it is and simply move forward.

Here's the rub: The future is unknown. The thinking brain only creates its *sense of knowing* through its concept of "the future" and its plans, expectations, scenarios, strategies, and agenda for that future. Yet the only situation we can actually be certain about is this present moment.

For this reason, we need to balance our overreliance on anticipation with resilience—*functioning effectively before and during unwanted events, and recovering completely, learning, and adapting afterward.*

In contrast to anticipation, resilience is a *present-oriented anxiety-management system.* Even when the plan goes off the rails, we can still rely on our adaptive capacity—the relationships we've nurtured, the skills we've learned, and a wide window that naturally brings awareness, wisdom, courage, creativity, and confidence to the situation. With this adaptive capacity, we trust that we can encounter life as it really is—uncontrollable, ambiguous, uncertain, unpredictable, volatile, surprising, messy—and flexibly respond to whatever happens. As Jenna put it, things may not be how we want them to be, but we'll still be okay.

To be clear, I'm not suggesting we jettison anticipation. Rather, I'm advocating for *a balance between anticipation and resilience,* which requires several things.

First, we need to rein in our hyperactive *unconscious* planning habit—Planning 1.0—and replace it with *mindful, resilient* planning, what I call Planning 2.0.

Second, we need to build our adaptive capacity, so that we can meet life with resilience. We can develop adaptive capacity through *the process* of making plans and rehearsing scenarios. In fact, we can structure stress inoculation training (SIT) in order to promote domain-general learning, so that we pair stressful training scenarios with window-widening recovery. When we deliberately devise SIT for this resilience-building purpose, we can practice the skills and self-confidence—and deepen the relationships—that we'll need to rely on when shit hits the fan in the future. *These* are the adaptive capacities that we can employ in any future situation, not the plans or scenarios themselves.

Third, we can widen our windows using the skillful means throughout the chapters in Part III, but especially with the next chapter's window-widening habits. Collectively, we can also work together to widen our collective window, something I'll address in Chapter 18.

PLANNING 2.0

Reining in our hyperactive overreliance on Planning 1.0 requires noticing when our thinking brain engages in this habit and then choosing to disengage. By repeatedly noticing when Planning 1.0 arises and directing your attention to another target object—such as your contact points—over time the habit will atrophy. If the thinking brain fights you on this, tell it to relax, because Planning 2.0 provides a new structured outlet for all those planning impulses.

Planning 2.0 relies on six principles. The first is that we naturally make better choices—most aligned with manifesting our *long-term* success and well-being—when we make them from inside our window. Just as making tomorrow's lunch the night before helps us get out the door calmly in the morning, setting aside time when we're in a regulated, rested, and unhurried place to plan for the upcoming week sets the stage for more peace and productivity.

These deliberate preparations increase the likelihood that we'll manifest our *important* intentions, while minimizing the chance of being sidetracked by the *immediate*. Therefore, it's helpful to find time when you're unhurried and rested to create your written plan for the upcoming week. Some people like Friday afternoons or Saturday mornings before the weekend; others prefer Sunday afternoon after some weekend recovery.

Conversely, when we let the week unfold without Planning 2.0, we're more likely to make choices from a rushed, exhausted, frazzled, or irritated place. In this place, we're more likely to choose "low-hanging fruit"—smaller tasks that may be easier to accomplish but less important—while putting off what really matters. We're also more likely to procrastinate, succumb to resistance, and engage in self-defeating behaviors, preparing the way for shame and self-judgment. Then we feel "behind" and scurry to catch up—while eating poorly on the run, skipping our workout, canceling plans with our friends, and depriving ourselves of sleep. In other words, we climb back on the Gerbil Wheel.

The second principle is intentionality—getting clear about what we want to create. With intentionality, our path becomes clear and our energy becomes focused and effective. Planning 2.0 helps us nonjudgmentally observe whether our time expenditures align with our deepest intentions and goals.

Planning 2.0 is about creating a healthy, happy, balanced life, not just succeeding at work. Thus, I like to group intentions into four buckets: *physical intentions* relate to our body, health habits, finances, and surroundings; *emotional intentions* relate to our personal and professional relationships, hobbies, vacations, and leisure activities, and the way we want to *feel* throughout the week; *spiritual intentions* relate to our spiritual life and practices; and *intellectual intentions* relate to our educational and professional goals.

Ideally, we can hold our intentions loosely, as a heartfelt commitment toward a particular goal or aspiration—while leaving open the possibility that there may be several different pathways to reaching that destination, some of them entirely beyond what the thinking brain could imagine. In other words, we remain committed to the intention, without getting too attached to manifesting it in any one particular way.

We can create plans that span different time intervals. As you play with Planning 2.0, I suggest you try weekly planning and some form of strategic planning, from a month to a year at a time. If your workplace operates on fiscal quarters, for example, you might create a quarterly plan that aligns your personal and professional intentions with what will be expected of you at work during that quarter. Personally, I rely on three plans of differing lengths—making sure that I have alignment among them. For instance, I create an annual plan on my birthday for the upcoming year. It's very simple—only two or three major intentions in each of the four categories. I write them on an index card, so that I can easily refer to them throughout the year. My second plan, which I create on each new moon for the upcoming lunar cycle, sets the basic structure for the upcoming month. Finally, I use a weekly plan.

Critically, the weekly plan *needs to match our list of intentions and goals to the actual space-time continuum.* Thus, the most important aspect of weekly planning is aligning our intentions with the calendar. I know many people who have running to-do lists, which they add to all the time but would actually take *years* to complete. Using something like this as your "weekly plan" is a setup for ending each week feeling discouraged, depleted, and constantly behind. At the same time, matching our intentions/goals to the actual space-time continuum can help reveal gaps between our thinking brain's agenda and reality—such as when we're being perfectionistic or compulsively trying to control outcomes that are actually beyond our control.

The third Planning 2.0 principle is that not everything is going to fit on the calendar, so you must make choices. Planning when you're inside your window increases the likelihood that you'll make choices aligned with your intentions. Better to make the necessary trade-offs when you're grounded and unhurried. I recommend scheduling *first* the tasks for meeting long-term personal and professional goals—including activities for keeping your window wide. *This includes scheduling time off.*

Aim for *at least* one day each week without any work, errands, or household tasks—focused exclusively on recovery, leisure activities, relationships, and your spiritual life. As you consistently use Planning 2.0, you'll naturally notice yourself harnessing your energy and effort more effectively throughout the week, with less procrastination, resistance, and other self-defeating behaviors. Over time, this leaves greater space to prioritize periods for recovery, rest, relaxation, and replenishment.

Ideally, you can schedule some time for window-widening activities and long-term goals *each day*, even if only for a short period of time. You might save the time of day when you're feeling most energized, focused, creative, and joyful for making progress on important goals. For instance, since I'm a morning person, I always schedule at least one writing block during this time.

With goals that trigger procrastination or resistance, try scheduling them first thing in the morning, since external distractions and resistance always grow as the day wears on! After a few weeks, you'll be amazed at how much progress you can make on long-term goals by intentionally and patiently engaging them each day in small, consistent chunks. Paying yourself first by attending to your long-term goals every day—even for fifteen minutes—can also dissipate resentment you may feel about other periods when you feel less in control of your schedule.

After populating the calendar with time periods for manifesting your intentions—including window-widening blocks for exercise, mind fitness training, relationships, nutritious food, sleep, and time off—you can add the non-negotiable items that must happen during the upcoming week.

Then it's time to make the hard choices about whatever's not yet on the calendar: You might push things off to a future week, delegate, ask or hire someone to help, renegotiate deadlines, or decline to do something. In fact, clearly observing the realities of the space-time continuum on the calendar is a terrific way to strengthen skills for saying no.

Remember, we live on planet Earth, where each day is twenty-four hours, and seven days make a week. Since we can't change this space-time continuum, surrender to it. Relinquish the magical thinking that you'll somehow accomplish everything on the running to-do list and still have time for sleep, health, and happiness. In addition to your weekly plan, you might also create a "parking lot"—a running list of tasks for future weeks that you have *consciously acknowledged will not happen this week*. Having a parking lot can really support your thinking brain in disengaging from its old Planning 1.0 habit completely.

The fourth Planning 2.0 principle is *deliberately building plenty of white space into the schedule*, so that you'll have ample time for unhurried transitions, unexpected opportunities, and, of course, the inevitable interruptions, flat tires, and other manifestations of Murphy's Law. With white space already available, you'll feel less pressured when something unexpected happens—*increasing the likelihood that you'll stay inside your window* and allowing you to approach the unexpected with more resilience, flexibility, and balance. Building in white space also leaves us more available for synchronicities, unexpected moments of joy and wonder, and heartfelt connections with colleagues, friends, and even strangers.

Deliberately scheduling in *and protecting* white space sets the stage for flowing more easily with whatever unfolds. It also lets us spend at least part of each day completely liberated from the tyranny of the to-do list and its corresponding sense of time scarcity.

Even better, if an unexpected curve ball—or terrific opportunity—comes your way that takes more time than the scheduled white space allows, you'll easily be able to make bigger shifts to accommodate. Why? Because with Planning 2.0, you'll already have identified the lowest-priority tasks most easily deferred to another day (or week).

The fifth Planning 2.0 principle is building a few hours into each week to attend to the "squeaky wheels" in your physical environment. A cluttered desk covered with unopened mail, an overcrowded closet, a messy car, accumulating piles of laundry, unread stacks of newspapers and magazines—these things can leave us feeling anxious and out of control. Being in chaotic surroundings, where we can't find what we need when we need it, can be draining and irritating. Likewise, having our financial records, important household files, and estate planning documents in disarray can contribute to an underlying sense of angst. We also feel depleted when surrounded by things that need repairs—a dripping faucet, a broken stove, a stuck car window, or a watch that needs a new battery.

Facing and tackling these tasks directly can give us an outsized boost of energy and motivation, enhancing our productivity and sense of self-efficacy. Organized, orderly external surroundings can also promote an internal sense of calm and focus. There are so many aspects of the external world *not* under our control—but these are some that truly are.

Therefore, one extremely straightforward way to feel more empowered is to schedule a few hours into each weekly plan to purge and donate items you no longer use, repair or toss things that don't work, and put everything else into its place. Whichever tasks like these you've procrastinated about, now's the time to make that "parking lot" list, and then set aside a few hours each week to attend to them patiently and methodically.

Most people find that this process of decluttering and repairing broken items leads to a surge of energy, which they can then harness for other things. Best of all, *these tasks tend to populate and fuel Planning 1.0*, so attending to them directly can naturally shrink your Planning 1.0 habit. If simplifying your surroundings feels challenging, you could hire someone to help, which may cost less in terms of your sanity than waiting to create time to do it yourself.

The final Planning 2.0 principle is grouping tasks of a similar energy level together, to minimize distraction and keep your window wide throughout the day. It's also helpful to match tasks of a similar energy level to the energy that you typically have at different times throughout the day. I already suggested using the time when you're most energized, focused, and creative for working on your long-term goals or projects that require focused attention. Likewise, you might schedule meetings with other people during periods when your energy levels are naturally lower, such as right after lunch, because talking will naturally raise your energy.

It can also help to group administrative tasks requiring less mental effort together at day's end, when your executive functioning capacity tends to be more depleted.

Finally, I recommend creating two time periods for responding to emails and phone messages, perhaps right before lunch and again at day's end. Once you become aware of your energy levels, you'll probably notice how exhausting it can be to begin your day with the email queue. Inevitably, this path leads toward starting the day in a reactive mode, since you'll gravitate to putting out fires, handling low-hanging fruit, and responding to *other* people's priorities rather than your own. It can also be a recipe for resentment.

If you must open email first thing, hold yourself to no more than thirty minutes—use a timer—and triage your queue. Only respond to things that are genuinely urgent and need immediate attention. After several days, you may notice that very little meets this threshold of *true* urgency. Most things could likely wait until right before lunch, leaving you time to focus on your own priorities first.

Planning 2.0 allows our intentionality to shine through—giving us more capacity to flow resiliently, productively, and peacefully through each week. It blends skillful means and surrender: Although we can never know what will happen, it's much easier to surrender to this reality when we've deliberately prioritized what's most important and scheduled plenty of white space. By making the hard choices about the week we want to create from a rested, grounded place *before the week actually starts*, we prioritize our intentions and align our expectations with what's actually feasible within the space-time continuum. This principle applies individually *and* collectively, in our schools and workplaces.

This alignment reduces busyness, so we spend more time connecting with others, accessing stillness, keeping our window wide, and doing meaningful work that brings us joy. We *choose* to do less, while focusing on what really matters. With less striving, racing around, and scurrying to catch up, we create a life with less anxiety, fewer distractions, and fewer self-defeating behaviors.

This is a recipe for agency.

CROSSROADS AND CHANGE

Just as mindful, resilient planning can help us relate in a different way to uncertainty and time scarcity, so, too, can we train ourselves to approach change more skillfully.

While the thinking brain understands, and usually welcomes, self-initiated change, it is less comfortable for the survival brain. That's because the survival brain neurocepts *all* changes—even "positive" ones, like getting a promotion, buying a home, starting a new relationship, or having a baby—as challenging or threatening, leading it to trigger stress arousal. If we aren't aware of the thinking brain's and the survival brain's differing reactions to self-initiated change, we may inadvertently exacerbate an adversarial relationship between them.

Whenever we plan to make a major change, therefore, we need to create space to experience our survival brain's reaction—without judging, devaluing, or overriding it. The stress activation and emotions aren't necessarily telling us to second-guess our decision. We just need to understand that *this is how the survival brain processes change.*

During the transition, we might benefit from additional space and time for awareness and reflection practices, such as journaling, walking in nature, meditating, or using G&R to discharge stress activation. I recognize that many self-initiated changes, like a new job or new baby, may likely shrink how much time we have available for these practices. Nonetheless, by giving ourselves space and time to notice and release our survival brain reactions, we prepare the way for our planned changes to occur with ease.

When change occurs that we did *not* initiate—especially unexpected change—the survival brain's reaction will likely be even more intense. For instance, we might be laid off, experience a flood that destroys our home, discover that our partner had an affair, or receive a terminal medical diagnosis. With such large, unexpected shocks, we understandably need more time to find our grounded footing again.

This is a time to be extra kind to ourselves and reach out for help from our supportive relationships. The thinking brain will likely want to figure out how to fix and solve things *right now*, but instead of reacting from that internal urgency, it can be helpful instead to focus first on using skillful means to get back inside our window. Before we can effectively plan and act, first we must just *be.* Moreover, until we willingly allow and release our stress arousal and emotions, all of the thinking brain's planning and decision making will be biased by the arousal, anyway.

The thinking brain may also vehemently tell you that you *must* have your plan *all mapped out,* as if you can know exactly what you'll need to do. This is just not true. You don't know—you *can't* know—not after a large shock has just rocked your world.

Instead, as much as possible, disengage from the thinking brain's agenda entirely, and *just be still.* That's when the internal wisdom and guidance will flow in. Trust that you will know *exactly* the next step to take and when it's time to take it. And then, the next one. And then, the one after that. By choosing skillful means to disengage from counterproductive thinking brain machinations, Providence will meet you halfway and help you through this difficult time.

Many of us survived our upbringing, excelled in school, and succeed in our workplace by overriding our instincts in order to keep striving, meeting external expectations, or pleasing people. As a result, we've not only habituated ourselves to ignore our inner wisdom, but we've also conditioned ourselves to look for answers externally—seeking and listening to others' advice and opinions.

While external guidance can provide *supplemental* information when we need to make a decision, *only we know the best choice for us.* Our inborn intelligence, which reaches beyond the thinking brain, is always available. Accessing it only requires that we disengage from the thinking brain's agenda and arrive fully in our body, in the present moment.

Say, for instance, you're debating between two different courses of action: Accept the new job offer, or stay where you are? Buy this house, or that one? End the relationship, or not? I've had dozens of students ask for my advice when they encounter forks in the road like this.

Whenever they do, I always tell them that only they can know the right choice for them, but I'd be happy to help them access their internal wisdom and discover it. Then I guide them through the following exercise, which you can also use—anytime you find yourself facing a crossroads.

Begin by noticing the physical sensations of stability and support at your contact points. As you do, if you're feeling stressed or depleted, you might spontaneously shift into a cycle of G&R. After completing some discharge of stress activation, return your attention for a few minutes to the sensations at your contact points.

Then, in this grounded place, visualize the first course of action you're considering. For instance, using the example of a job offer, imagine yourself accepting and starting the new job. Allow your thinking brain to embellish the visualization—really put yourself in that new job. As you hold the visualization in your mind's eye, scan throughout the body with nonjudgmental curiosity. Notice your posture, physical sensations, and/or emotions that might be present. Also notice your body temperature and energy level (e.g., are you energized or drained?). With nonjudgmental curiosity, allow yourself to catalog everything you notice in the body.

After taking stock of your body's reaction to the first course of action, return your attention to the sensations at your contact points. Keep your attention there until the sensations associated with that course of action have subsided. Then you can repeat this process with the second course of action, and any others you might be considering.

By the time you're done, you'll have your internal guidance. It may not be what you wanted or expected to learn, but you'll know your inner truth.

In general, courses of action that your inner wisdom is nudging you toward are ones where the body feels relaxed, relieved, calm, spacious, energized, empowered, joyful, or at ease.

Conversely, courses of action that it's warning you away from are ones where the body feels drained, exhausted, hesitant, stressed, activated, anxious, depleted, collapsed, irritated, depressed, resigned, powerless, tense, trapped, or in turmoil.

I've never had a student regret following the guidance they learned from this exercise. However, several students have second-guessed their internal guidance and instead pursued a different course of action—usually one that their families endorsed or that made more "logical" sense, as if it were the outcome of a balance sheet weighing pros and cons. Not surprisingly, every student in this second group who stayed in touch eventually told me that they'd made a "mistake" and regretted their decision.

In fact, they didn't make a mistake: Having chosen to override their internal guidance in that situation—and then experienced the consequences—they've since learned to trust their inner wisdom.

The thinking brain is quite proficient at making *simple, rule-based* decisions. But most of the crossroads in our lives are not straightforward, rule-based choices. Rather, they're usually quite complex, when we must consider multiple factors and weigh their relative importance.

Complex decisions always benefit from unconscious processing. With complex decisions, we can enlist the thinking brain as the excellent tool that it is: Send it out to do research, compile the necessary information, and answer any *factual* questions that you have. Don't try to make the decision until you've had a chance to gather this necessary information. Then kindly thank the thinking brain for its hard work and *tell it to stand down*. Now that you've used skillful means to assemble the relevant information, it's time to wait for your organic intelligence, "intuition," to metabolize the input and present you with its answer.

Indeed, empirical research has demonstrated that with complex decisions, people are generally happier when they choose the intuitive answer that arrives from unconscious processing, rather than an answer derived from a conscious, thinking-brain decision process. Experts who rely on their intuition use this same principle—their unconscious is simply working from a larger and longer-standing base of professional expertise and experience.

In sum, particularly with major life decisions, we can't successfully decide only with our heads. We must listen to our entire mind-body system, especially our hearts.

CHAPTER SEVENTEEN

Choosing Habits That Widen the Window— Setting Structural Conditions for Agency

Although there's no silver bullet for decreasing our allostatic load, the habits in this chapter can help us reduce that load over time. Using these habits, we also build adaptive capacity, so that we'll have more internal and external resources available during stress—even when we're facing uncertain, unpredictable, and uncontrollable stressors.

When we first learn a new behavior or routine, our thinking brain is very involved in guiding our behavior. Over time, however, we stop consciously thinking about the behavior, and it becomes automatic. To be clear, some habits actually begin in the survival brain, outside conscious awareness, through neuroception and the implicit learning system. Regardless, with repetition, habits get stored in the *basal ganglia*—a survival brain region responsible for recalling and acting on patterns—where it gets encoded as a habit loop: *trigger, behavior, reward.*

By pushing habits into evolutionarily older brain structures, we free up our thinking brain for new information and new tasks. Thus, the thinking brain stops participating fully in habitual decision making, which is why habitual behavior is non-intentional, automatic, and impulsive—the very definition of life on autopilot.

Habit loops rely on positive or negative reinforcement. In a study of individuals who worked out at least three times a week, for example, 92 percent said they exercised frequently because it made them "feel good," while two thirds said that working out gave them a sense of "accomplishment." In other words, the reward—an endorphin rush and positive self-conceptualization—led them to crave their exercise habit. Conversely, with a negative reinforcement habit loop, we might see danger/pain/discomfort (trigger), avoid (behavior), and feel better (reward). Most stress reaction cycle habits from Chapter 10 follow this trajectory.

For instance, "Tim," a firefighter I trained, had developed a habit of drinking alcohol with his buddies to unwind after work. Tim's trigger included both the discomfort of working for a micromanaging boss *and* unpleasant physical sensations of stress activation, including chest tightness and a pounding headache that would build each afternoon. Drinking alcohol and venting with his buddies was how Tim "let off steam"—and, not coincidentally, suppressed and masked his stress activation. Drinking helped him feel better in the short term but was actually building his allostatic load.

As these two examples suggest, triggers can be almost anything—a sight, sound, taste, touch, smell; physical sensation; emotion, a memory, a flashback, a sequence of thoughts; the company of particular people; a particular location; or even a particular time of day. The reward is usually a hit of the "feel-good" neurotransmitters—oxytocin, dopamine, and/or endorphins—which feeds the craving that keeps us addicted to our habit.

Habits are pervasive. In several studies in which people kept behavioral diaries, they classified up to 47 percent of their behaviors as habitual. Other meta-analyses suggest that intentions account for only about 30 percent of our behavior.

Since there's no way to avoid having habits, why not choose habits that widen our window?

CHANGING HABITS

Once the neurobiological structures in our brain and nervous system get established by our twenties, we become less influenced by our external environment. Instead, as adults, we act to *preserve* these internal structures—seeking out information and experiences that confirm what we already believe, and avoiding information and experiences that don't. We also tend to ignore, forget, reinterpret, or try to discredit information that isn't consistent with our internal structures. That's why we pick like-minded people to affiliate with and like-minded information sources to read, watch, and listen to. Avoiding experiences that don't match our internal structures—often unconsciously—is the first line of defense against dissonance between our internal wiring and our external reality. For this reason, we tend to experience familiar things as pleasant and, conversely, unfamiliar things—or the loss of familiar things—as unpleasant.

In cross-cultural studies of adults, the three most stressful life events are death of a significant other, divorce, and marriage. These three situations are times when an adult's external reality has simply changed too much to match their internal structures. That marriage—something most adults willingly choose—ranks just below divorce highlights how stressful for adults the incongruence between internal and external reality is. It takes time for the new reality to become familiar—and hence, more pleasant—as we restructure our internal world to match the altered external world.

These dynamics explain why most adults find external structural change to be painful and difficult. And, of course, one of the most pervasive external structures in our daily lives are behavioral habits and routines! The principles of neuroplasticity explain not only why it can be difficult to cultivate a new habit but also why it's so challenging to relinquish a habit we've outgrown or want to abolish.

Nevertheless, experimentation with new habits can produce outsized rewards. Experimentation itself strengthens important window-widening qualities—because it takes us outside our comfort zone, which can build our resilience when it's paired with recovery.

Be kind to yourself. Recognize that change is difficult; there may be days when you can't follow through on your intentions. No problem. People who show themselves more self-compassion are better at simply recommitting and beginning again.

In contrast, being hard on yourself is the best trajectory for backsliding and eventually giving up. There are downsides to pushing yourself with grit and willpower alone. Relying too much on willpower depletes executive functioning capacity—thereby making it easier to succumb to temptations, especially when you're busy, tired, or stressed. In this depleted place, once we "give in" to an unhealthy habit—or skip one day of a healthy habit—the inner critic is ready and waiting, fueling shame, resistance, self-sabotage, and procrastination.

You *know* this cycle—you start New Year's Day with a long list of resolutions, but by January 8 you've already conceded defeat on a quarter of them.

The best way out of this cycle is to establish some structural conditions to keep yourself from starting it in the first place. First, spend some time getting clear about *what* you want to create and *why*. You might write your intentions down. Approach your chosen shift with heartfelt commitment. Mixed feelings usually create mixed results—undermining your success, setting the stage for resistance, and creating obstacles. Better to take the time to resolve ambivalence before starting any changes, so that you can settle on a clear intention for action.

Second, beginnings are powerful moments when it comes to habits. The first few times we do anything shape the baseline—so that it becomes familiar and, therefore, pleasant. Afterward, it takes considerable effort to deviate from this status quo. You can use this understanding to your advantage, by deliberately creating circumstances that facilitate a fresh start. If you're starting a new diet, for example, begin by purging your home of foods you won't be eating. Or if you want to change your study or work habits, set a new routine during the first week of a new school year or new job.

Third, develop some accountability and support mechanisms. You might find an exercise partner, register for a class, hire a personal trainer or life coach, or agree to text a friend after meeting some daily goal. Accountability and support don't need to cost money, although many people find that spending money reinforces their commitment to change. It's a lot harder to justify magical thinking—like "I can start tomorrow" or "this doesn't count"—when we have to explain it out loud to someone else!

Conversely, when we've slipped on a self-imposed standard, external support can help us interrupt shame and self-judgment—and access self-compassion instead. That's why Alcoholics Anonymous and other addiction support groups can be so powerful, because they provide *both* accountability and support. External accountability is especially important for people who find it difficult to honor their own needs, self-motivate, and persist in working toward long-term goals.

Fourth, you can protect your planned changes. When you're in a rested and centered place, you might brainstorm strategies to work with eventual resistance. This way, when the going gets tough later on, you've already developed a range of options to draw on. It's a lot easier to implement a planned workaround when you're tired or frazzled than it is to create one in that moment. For instance, you might not keep sweets or soda in the house so that it's easier to ride out a wave of craving sugar.

Chapter 14's Skillful Choices Hierarchy is another safeguard. If you've developed your own personalized version, then you can easily choose a slightly more skillful activity when you're depleted and feeling drawn to stress reaction cycle habits.

Even with these structural conditions in place, there still may be times when we choose to "cheat" on our goals. That's perfectly fine, *as long as we choose to do it consciously.* We're not victims of our cravings, addictions, stressful circumstances, or exhaustion. We're always responsible for our choices. Own it.

So the next time you choose to eat that candy bar you told yourself was off-limits, fully embrace this choice. Don't call it "cheating" and make yourself wrong, because that will only fuel self-judgment and shame. Eat it with awareness and *joy.* Then, after you've savored every last bite, it will be easier to recommit to your intention and simply begin again.

After you read about the window-widening habits in this chapter, check them out for yourself. You might keep a log of the changes you implement—and any effects that you notice afterward—such as the following:

- How do you feel physically? Are your symptoms getting better? Do you notice any shifts in your energy level throughout the day? Are you more grounded ("in your body") than before? Is it easier to use tools from previous chapters to up- or downregulate your stress and bring yourself back inside your window?
- How do you feel emotionally? Have you noticed any shifts in your mood? Do you feel more connected with the people around you? Can you regulate your emotions with more ease? Do you feel more

capable of responding flexibly to interruptions, reactive people, or unexpected curve balls during the day?

- How do you feel cognitively? Are you more focused? Is your memory and "brain fog" getting better? Are you noticing less rumination, worry, planning, and other looping thought patterns? Do you feel more capable of creating, implementing, and following through on plans?

- How do you feel spiritually? Are you more in touch with your highest self's guidance and your life purpose? Are you noticing more stillness, contentment, equanimity, clarity, compassion, creativity, love, joy, and gratitude in your daily life? Are you feeling more connected and generous with the world around you?

Try to give any change you make at least two to three weeks of consistent implementation and monitoring. That's how long it takes for most people to notice some effects. As any scientist knows, the only way to be sure of an experiment's results is to go "all in." Therefore, whichever shifts you undertake, commit to giving them your all. After gathering data for a few weeks, you can analyze them and, if necessary, make some tweaks. Keep repeating this process of evaluating your progress and adjusting until you're happy with the results.

Especially with changes that are physically or emotionally uncomfortable, let them really show you their potential benefits before jettisoning them. If you're implementing a healthier diet or healing an addiction, for instance, your body will naturally go through a period of detoxification—as years of stored toxins get released from your tissues. This is an uncomfortable and tiring process! Likewise, if you're starting a new exercise regimen, your body is likely to feel achy. Most importantly, if you've narrowed your window through chronic stress and trauma without recovery, starting mind fitness practice is going to bring stress activation up to the surface for discharge.

In each case, you're likely to feel worse for a while, but actually be getting better, just like Michael's experience in Chapter 14. Be patient and kind with yourself. Give these kinds of mind-body shifts *at least a month*. If you're noticing a lot of resistance, use the tools for working skillfully with resistance from Chapter 15. Or you can try the habit diagnostic.

THE HABIT DIAGNOSTIC

Stress reaction cycle habits are *pseudo-regulators*—soothing in the short term, while actually making our stress load worse. We're usually drawn to pseudo-regulators, rather than true regulating activities, when we're depleted or dysregulated. Chapter 14 explored how to use the Skillful Choices Hierarchy to choose a slightly more regulating activity or substance in the moment you're drawn to a pseudo-regulator. Here I'll explore how to make deeper, structural shifts to replace the pseudo-regulators you rely on most with healthier choices, more conducive to keeping your window wide.

When you're in a rested and regulated place, take some time to investigate, with nonjudgmental curiosity, a major habit you'd like to shift. You may find it helpful to investigate with reflective writing.

First, describe the habit. How much time do you spend engaging in this habit? If you can't answer this question with precision, track your time engaging in this habit over the next week and keep a log. What triggers you to engage in this habit? Are there particular people, places, times of day, activities, environmental circumstances, emotions, thought patterns, beliefs, or physical sensations that cue you to do it? What do you notice in the mind-body system, both before and after you engage in the habit? You can also query your habit using the four categories of questions from the previous section.

Second, inquiring at a deeper level, what benefits and functions does this habit provide for you? *Often pseudo-regulators are a symptom of an unacknowledged truth or an unmet need.* For instance, they may mask loneliness, boredom, depletion, fear of failure, or resentment. Spend some time figuring out which functions this habit performs for you, so that you can choose to address the underlying issue directly. This will let you take charge of the situation and access agency.

Armed with this diagnostic information, you can then tune in to your inner wisdom and brainstorm: What are some alternative ways that you could fulfill these functions and get your needs met, with fewer detrimental consequences for your window? Develop a list of new options to replace this habit. While it's almost impossible to think of new options when you're in a depleted, overwhelmed, and dysregulated state, it's pretty easy to implement something from a list created beforehand. Then, the next time you're tempted by your pseudo-regulating habit, you can experiment with these alternatives and monitor the consequences. Be willing to let go of what no longer serves you energetically and replace it with something that will serve you better.

When I was trying to kick my head-banging habit, for instance, initially I replaced it with television binging—which *was* an improvement from the perspective of my window. After about a year of this, however, I knew I needed another change.

One day, I decided to count how many episodes I'd watched in the last month and calculate how much time I'd given over to this habit. I was stunned. Seeing this large number, my inner critic was off to the races, but I quickly asked it to pipe down so I could get genuinely curious: What on earth could TV binging be providing for me that I was choosing to give it so much of my life energy? I pulled out my journal and started writing.

After a long reflective writing session, I realized that TV watching was meeting needs in four categories. First, it was a way to check out when I was depleted, such as after a long day of writing or back-to-back meetings.

Second, TV watching helped me suck it up and drive on through depleted states. It was, in effect, a bargaining tool. Facing a task that I didn't like, I'd push myself to complete a chunk of work, reward myself with some TV, rinse and repeat.

Third, sometimes TV watching was an avoidance mechanism, when I was feeling overwhelmed and anxious. Finally, sometimes I watched TV when I wanted some TLC, like when I was sick or feeling lonely.

As I soon as I grouped my triggers and motivations for watching TV into these four categories, I could easily understand how television was filling *many* different needs at different times. Now it made sense why I'd chosen to give so much of my time and energy to this habit.

Then it was straightforward to brainstorm new alternatives to meet each of these different needs. For checking out during depleted states, for instance, I could replace TV with reading a novel, walking in nature, or doing some yoga. For bargaining with myself to keep pushing, I could reexamine my deadlines, lower my standards on less important tasks, or take something off the schedule. For TV as an avoidance mechanism, I could engage in some vigorous exercise, use G&R afterward to discharge my anxiety, and then set a timer for thirty minutes to tackle directly whatever I was avoiding. You get the idea.

With the habit diagnostic, you'll quickly see that your reasons for engaging in a pseudo-regulating habit are *particular to you*. Thus, only by investigating your habit with nonjudgmental curiosity and then opening to your own inner wisdom will you be able to discover your best alternatives.

For instance, I've worked with many people who believed they drank too much alcohol but felt stymied for how to change. Until they examined *what* was driving their alcohol use, however, they couldn't figure out exactly *how* to shift their habit.

Remember Tim, the firefighter who'd drink with his buddies and vent about his boss after work? When we talked through a habit diagnostic, he realized two things. First, drinking with his buddies was his major outlet for emotional connection; he often ordered another round simply to avoid going home. This insight helped him recognize that he could just as easily meet a friend at the gym and talk while they lifted weights. Second, he realized that alcohol helped him avoid acknowledging how toxic his workplace truly was. Avoiding this truth, instead of proactively doing something about it, was wearing him down. By the time we finished our conversation, he was coming to terms with needing to find another job without such a constant energy drain.

"Isabel" spoke with me about her overreliance on wine while fixing dinner each evening. She admitted to consuming two bottles each week, way more than she wanted to be drinking. In addition to a demanding full-time job, Isabel had two busy teenagers. Her husband was building a new business, which often kept him focused on work. Thus, Isabel usually cooked dinner alone, and many family members ate via "drive by" later.

The problem wasn't cooking, which she actually enjoys. Instead, she recognized how much she resented cooking *alone*. Initially, she devalued these emotions, explaining why everyone else was so busy and pointing out that they did the dishes. Nonetheless, as we sat with her resentment and loneliness, Isabel began to see how much the wine was helping her mask these feelings.

Isabel realized she could tell her family how much she missed them and ask them to take turns helping her or simply sitting in the kitchen and talking with her. The family could also schedule at least one unhurried meal together each week. Imagining this additional connection, Isabel thought she could limit herself to one glass of wine a few nights each week.

Finally, we can also use the habit diagnostic to identify obstacles and resistance with implementing a new habit. Here, the question to ask is, "What's getting in my way?" Since the inner critic may be lurking nearby, you might choose to write in your nondominant hand, which helps to dampen it.

For instance, I once trained "Matt," a young Marine who lived in the barracks. Matt wanted to be getting more sleep, but he often found himself playing video games late into the night. The next day, he'd have to pound at least three Monster drinks to make it to his unit's formation in time. Although he had a prescription for sleeping pills, they weren't helping—in large part because Matt was extremely anxious about his inability to fall asleep, which led him to avoid getting in bed. One day, while doing the Contact Points Exercise, Matt realized that he'd never really considered his barracks room to be "home," even though he'd been living there for years. His insight was that if he could personalize his room—and especially make his bed more welcoming—he'd be less likely to procrastinate about climbing in. That weekend, Matt bought a new pillow, a comforter, and fine-thread-count sheets. That night, he slept like a baby. By the time he shared this story with me, he'd been sleeping at least two hours longer each night.

As Matt's example demonstrates, only *his* inner wisdom could have pointed him in that direction—it would have never occurred to me to suggest that Matt buy new bedding for his insomnia!

There is no one-size-fits-all approach to habit change. As you read my suggestions for five window-widening habits, listen to your intuition. You have the insight and guidance you need within you. Change is always an opportunity to discover and become more fully our authentic selves.

AWARENESS AND REFLECTION PRACTICES

The first window-widening habit is developing daily awareness and reflection practices. Set the conscious intention to take periodic breaks from the fast pace of your life and return your attention to the stillness that's always there in the background.

By slowing down, we can see our lives with more clarity. We can also recommit to our intentions and structure our daily lives to pare back distractive energies—so that we can focus on the important rather than the immediate.

Through awareness and reflection practices, we're less likely to deny or hide from important truths about ourselves, our relationships, and our work. By creating this space for truth to arise, we can make better choices.

I've already introduced several awareness practices, such as the Contact Points Exercise, G&R, and paying attention to the three components of emotions in your field of awareness. Others include awareness of breathing and various forms of meditation. Awareness can also be cultivated through movement, such as tai chi, walking meditation, yoga, or martial arts. By paying attention to your moment-to-moment experience in an intentional way, you create the space for creativity, joy, compassion, gratitude, equanimity, calm, and insight to arise naturally.

Reflection practices include writing in a journal, reading a scriptural passage or poem and reflecting on its message, making a gratitude list of the many blessings in your life, or revisiting the day's lessons before bed. My favorite reflection practice is writing in a journal, stream-of-consciousness style. I find it helps me more fully understand different facets of current life situations.

Journal writing can help us become aware of and explore the reasons behind recurring patterns and avoidance mechanisms. As Chapter 14 suggested, it's helpful for getting in touch with our emotions—as well as clarifying what we need and want in a particular situation. In turn, this helps us refine our intentions and our responses. Furthermore, you can use your journal to monitor your progress with habit changes. It can be extremely instructive to go back after a month (or a year) and reread your journal. You may be surprised by the insights or patterns that jump forth when you read the entries from some distance. Thus, it's an excellent way to track your progress and growth.

I recommend establishing at least one awareness and one reflection practice you engage in regularly. You might use an awareness practice in the morning and a reflection practice in the evening, or vice versa. Or you can use these practices during many transition points throughout the day—such as when you arrive home after work or right before you go to sleep.

I especially recommend establishing one period of practice, for fifteen to thirty minutes, first thing in the morning. After a night's rest and before the day's busyness begins, the mind tends to be the most receptive to mind fitness exercises, meditation, tai chi, journaling, reflecting on a scriptural passage, or even simply enjoying a quiet cup of tea on the back porch. When we practice in the morning, we often find that it has a positive effect on our entire day—bringing more clarity, intentionality, and focus to our activities and helping us access more peace and calm throughout the day. We're less likely to be thrown off our purpose by external conditions and distractions—and less sidetracked by counterintentional or self-defeating activities.

DIET

Seventy percent of our immune system is located in and around our digestive system, in what has come to be known as the *microbiome*, the trillions of microscopic gut flora in our digestive tract. These bacteria ward off infection and food poisoning, produce vitamins, improve nutrient absorption, and manufacture antibiotics to the harmful bacteria.

Our microbiome can become imbalanced by a lack of helpful bacteria and/or too many harmful bacteria, fungi (like candida yeast), or parasites. Microbiome imbalances have been linked with a wide range of diseases—including arthritis, osteoporosis, irritable bowel syndrome, food sensitivities and allergies, chronic fatigue syndrome, chronic pain, autoimmune diseases, headaches and migraines, Alzheimer's disease, and colon and breast cancer.

Why are so many different diseases linked with the health of our microbiome? The microorganisms in the microbiome have epigenetic powers! They can turn gene expression on or off. For this reason, *the health of your digestive system is a major determinant of whether you'll develop a disease for which you have a genetic proclivity*. To put that differently, one of the best ways to protect yourself from developing diseases that run in your family is choosing to eat a healthy diet.

Our intestinal tract also has 100 million nerves—way more than our spinal cord—leading some researchers to call it the body's second brain. Our digestive systems manufacture just as many neurotransmitters as our brains—*including 95 percent of our serotonin*. Imbalances in serotonin have been shown to play a big role in depression, anxiety, insomnia, irritable bowel syndrome, and migraine headaches.

What all this means is that improving the health of our digestive systems—balancing the microbiome and healing gut permeability ("leaky gut")—is an effective way to resolve many physical, cognitive, and emotional symptoms. Improving our diet helps us lose weight, clear brain fog, boost our energy, and mend our immune system. For this reason, improving our diet and healing our digestive system is the second window-widening habit.

I *know* this works from firsthand experience. Learning about my food sensitivities and giving those foods up, healing my leaky gut, ridding my body of candida, adding probiotics, and overhauling my diet—these were among the *most* important changes I made on my path to physical and psychological recovery. My new diet cleared up *so many* symptoms, allowing me to terminate medications initially prescribed during my time in the Army. With fewer medications, my liver wasn't so overloaded, so my natural detoxification systems could reboot—improving my energy levels and mood tremendously. I also lost weight, simply because my tissues had been swollen from eating foods to which I was allergic.

It's helpful to avoid—or limit consumption of—the things that contribute to microbiome imbalances, including sugar, toxic chemicals, pesticides, processed foods, trans fats, foods with low nutritional value or that cause allergic reactions, oral contraceptives, steroids, pain medications, and antibiotics.

It's also important to repopulate your microbiome with nutritious foods and live probiotics, the helpful bacteria we need. You might take a probiotic supplement and/or consume probiotic foods and beverages—which are raw and fermented ("cultured")—including Korean kimchi, unpasteurized sauerkraut, cultured vegetables, coconut or milk kefir, unsweetened raw yogurt, raw apple cider vinegar, and kombucha. Most doctors now recommend consuming probiotic foods and/or supplements every day. At the very least, consume probiotics for several weeks after completing a course of antibiotics.

If you have symptoms associated with inflammation, you also might try an *anti-inflammatory diet*. A classic way to start is by eliminating for six weeks the foods that people are most frequently allergic or sensitive to—wheat, gluten, soy, nuts, certain grains (millet, corn), dairy, nightshade vegetables (tomatoes, eggplants, peppers, potatoes), and yeasted products (except for raw apple cider vinegar). When you start to introduce foods back in, test your sensitivity by only adding one new item every few days.

Signs of food sensitivities include brain fog, poor concentration, sluggishness, depression, fatigue, gas and bloating, diarrhea, sneezing, itchiness, nasal congestion, headaches and migraines, skin rashes, and snoring. Definitely keep a log of what you eat and any symptoms, which may not show up until the next day.

If you often feel bloating, cramping, gas, indigestion, or heartburn—or have problems with diarrhea or constipation—you can also become more conscious of how you combine your foods. In general, it's easier for our body to digest food when we partner leafy greens and nonstarchy vegetables with *either* (1) animal protein, eggs, dairy, and nuts, *or* (2) grains, pasta, bread, beans, legumes, and starchy vegetables (potatoes, corn, and squashes). Probiotic foods can also assist digestion.

Most American meals don't follow these food combining principles, since they usually mix animal protein with grains, breads, or potatoes. That's partly why many people feel better when they follow a low-carb diet, like Paleo, because these diets tend to follow the food combining principles that support easy digestion.

In addition, the quality of our food really matters. Toxins in our food accumulate in our bodies—including herbicides, pesticides, and growth hormones/antibiotics fed to livestock that become our sources of meat, eggs, and dairy. These toxins contribute to inflammation, obesity, cellular damage, immune dysfunction, and hormonal imbalances. Thus, as much as possible, try to eat organic: It may be more expensive in the short term, but it will definitely pay off over the long term. At least choose to prioritize organic, free-range, and grass-fed meat, poultry, dairy, and pastured eggs. There are some fruits and vegetables with high pesticide residues—the so-called Dirty Dozen—that we should always aim to eat organic, too.

We could all benefit from drinking more water. Often when we think we're hungry, we're actually just dehydrated. If you get tired of plain water, you can jazz it up with a splash of lemon juice or raw apple cider vinegar—both of which curb sugar cravings, help us detoxify, and stimulate digestion. You can also drink caffeine-free herbal teas. Aim for thirty-two to sixty-four ounces of water (without carbonation) each day.

Finally, reducing caffeine consumption can help. Although caffeine has been shown to improve memory and reduce the risk of dementia, diabetes, and stroke, *it also causes the release of stress hormones.* More importantly, caffeine messes with the microbiome and decreases serotonin levels (95 percent is produced in the gut, remember?). That's why caffeine doesn't mix well with antidepressants, ADHD drugs, and antianxiety medications. Combining caffeine with these drugs may undermine their effectiveness and increase insomnia, panic attacks, anxiety, and irritability. You could replace coffee, caffeinated sodas, Red Bull, and Monster drinks with green tea, black tea, or half-caf coffee. As you decrease caffeine consumption, taper down over a few days to decrease withdrawal headaches.

With any diet changes, it's important to talk with your doctor. I recommend you seek out a functional medicine or integrative medicine practitioner. Not only are these doctors usually the most familiar with diagnosing and treating gut permeability and toxin issues, but they can also guide you to manage symptoms with natural remedies instead of prescriptions—thereby lightening the load on your liver and other internal detoxification systems.

There are also several books where you can learn more. I recommend Dr. Elizabeth Lipski's *Digestive Wellness*, Dr. Gary Kaplan's *Total Recovery*, Dr. Aviva Romm's *The Adrenal Thyroid Revolution*, Annemarie Colbin's *Food and Healing*, Donna Gates's *The Body Ecology Diet*, Dr. Ritchie Shoemaker's *Surviving Mold*, Dr. Neil Nathan's *Mold and Mycotoxins*, and Dr. Dale Bredesen's *The End of Alzheimer's*. The tests, diet shifts, and supplements Bredesen recommends help with a *range* of inflammatory, hormonal, or toxic imbalances, not just cognitive decline.

SLEEP

The third window-widening habit is getting enough sleep, aiming for *eight hours of sleep each night.* We can't make up for weekday sleep deprivation just by sleeping in on weekends, as Chapter 9 explained.

A restful night's sleep boosts mental clarity and supports recovery and healing. With enough high-quality sleep, the PSNS can handle all those "long-term projects" like digestion, tissue repair, toxin elimination, healing, and growth. It can concentrate on eliminating accumulated toxins and reducing inflammation, thereby reducing chronic pain and our risk of inflammatory diseases. Some of this repair and recovery also happens in the brain—pruning synapses, consolidating memories, and clearing amyloid, the plaques associated with dementia.

If you can't sleep eight hours regularly with the suggestions I offer in this section, it's important to get evaluated for any underlying medical issues. You'll be amazed by how much better you feel physically, emotionally, and mentally when you wake up truly rested.

Ideally, we can get eight uninterrupted hours a night without relying on sleeping pills. Many sleep medications actually backfire, interrupting our sleep. They can also compromise cognitive functioning and increase the risk of cancer and premature death. Often sleep aids simply mask underlying dysregulation. In the process, we block the important recovery and healing that needs to happen during sleep.

Therefore, it can be extremely beneficial to wean yourself off these medications, with your doctor's help. Your doctor can help you decide whether to supplement melatonin, which regulates the circadian rhythm and plays a role in our detoxification systems. Many people find they sleep better and awaken more replenished with supplemental melatonin; although it's not addictive, skipping some nights can help the body continue to make its own.

Alternatively, especially if you wake up with your mind racing or ruminating during the night, your doctor may recommend tryptophan or 5-hydroxytryptophan (5-HTP)—but you should definitely avoid these if you're taking an antidepressant. Supplementing with melatonin, tryptophan, or 5-HTP is generally safer than using sleep medications and/or benzodiazepines; however, check with your doctor to select the right dosage and avoid any harmful drug interactions.

About 5 percent of the population has sleep apnea, but many people don't know it. Smoking and being overweight are the biggest risk factors, and it's somewhat more common among men than women. If you snore and never feel rested, this could be your problem. With sleep apnea, people stop breathing periodically during the night, so their body becomes oxygen-deprived.

Understandably, given how important oxygen is for survival, the survival brain feels threatened and turns on stress activation—all night long! Thus, it's not surprising that sleep apnea is linked with PTSD, anxiety disorders, depression, hypertension, chronic pain, type 2 diabetes, and other inflammatory diseases.

Try the following suggestions for improving your sleep hygiene. First, try to go to bed and get up at the same time every day. Ideally, you'll get to bed before eleven p.m., and wake up no later than seven a.m. Try to spend time outside during daylight each day, and avoid napping as much as possible. If you have insomnia, it can also help to reset your morning cortisol response by getting access to bright light first thing in the morning, such as with a light box.

If you have a job that involves shift work or much jet lag, you'll naturally have more challenges with your circadian rhythm. Although you can bank some sleep on your days off, that may not be enough. Be sure to follow the tips in this section, time-shifted to match your schedule. It's also tempting overnight to snack on sweets and junk food. These foods late at night can really whack your circadian rhythm and cortisol cycle. Instead, choose nutrient-dense foods with lots of healthy fats, protein, and resistant starch, which also satisfy the *feeling* of snacking. You might try almonds or walnuts, organic beef or turkey jerky, pumpernickel toast with smoked salmon and/ or slices of avocado, or carrot sticks dipped in hummus.

Second, try to get regular exercise, which has been linked empirically with better sleep. It's important to finish working out at least three hours before bedtime, however, because exercise causes a surge in stress hormones that interferes with falling asleep. The exception to this rule, of course, is quiet stretching, tai chi, or some soothing styles of yoga.

Third, you can shift your diet to support your body winding down for sleep. Avoid caffeine after two p.m.—or after noon, if you have insomnia or you're sensitive. This includes chocolate, sweetened tea, coffee, green and black tea, Monster drinks, Red Bulls, and most soft drinks. Ideally, you'll eat your heavy meal earlier in the day, but regardless, avoid eating and consuming alcohol within three hours of bedtime to minimize a late-evening insulin spike. Although many people drink alcohol to fall asleep, it can lead to fragmented sleep and poor sleep quality. You might also limit fluids two hours before bedtime to decrease overnight bathroom trips.

Fourth, aim to finish working, consuming the news, and engaging in stimulating conversations or arguments a few hours before bedtime. To transition toward rest and recovery, you can use G&R, gentle yoga, or stretching. Some people find that writing in their journal at bedtime can help them let go of the day. Write down anything that might keep you awake, fuel rumination and anxiety, or weigh on you.

Fifth, turn off *all* electronics—including your phone, computer, television, and other devices—at least one hour before bedtime. Melatonin gets suppressed by light, especially the blue light emitted by most electronics. That's why, if you're feeling wired and then choose to unwind by playing video games, watching television, surfing the web, checking your phone, or reading on a tablet—you're actually making it harder to get to sleep. The blue light will make you more wired. Instead, you might read an actual book, stretch, do some restorative yoga, meditate, listen to a guided relaxation exercise or soothing music, or soak in the tub. One of my favorite bedtime activities, especially in the winter, is a hot bath with two cups of Epsom salts and twelve drops of lavender essential oil.

If you find yourself having difficulty falling asleep, don't lie in bed tossing and turning. It will only make things worse! Instead, try getting up and doing something relaxing until you feel sleepy again.

Finally, if you wake up in the middle of the night with nightmares, rumination, anxiety, or distressing thoughts, it can help to leave the bed, move to a chair, and engage in G&R. Especially after nightmares, your mind-body system is activated—so use this opportunity to discharge that stress activation and support recovery. After one cycle of G&R, you could try stretching, yoga, the Contact Points Exercise, deep breathing, meditation, or journaling.

However, if you experience middle-of-the-night awakening often, you should also check with your doctor. This has many potential causes, including hormonal imbalances, menopause, depression, anxiety disorders, and gastroesophageal reflux disease (GERD). It's important to address any underlying causes directly.

EXERCISE

The fourth window-widening habit is exercise. Our bodies were made to move! Since the benefits of exercise are amply covered in other places, I only want to mention several specific ways exercise helps us widen our windows.

First, it allows us to discharge stress activation—especially when we finish a workout with G&R—over time, helping us reduce our allostatic load. Second, it improves our sleep quality. Third, exercise regulates our metabolism, decreasing our risk of insulin resistance, metabolic syndrome, and type 2 diabetes. Fourth, it strengthens the immune system, which could help us fight off infections. Finally, exercise downregulates the microglia to reduce inflammation, decreasing our risk of inflammatory diseases, such as chronic pain, depression, irritable bowel syndrome, autoimmune diseases, arthritis, and Alzheimer's.

The optimal exercise program for physical and cognitive health combines three things.

First, we need *aerobic exercise*, to work our cardiovascular system and increase our stamina. You might try jogging, brisk walking, hiking, spinning, dancing, rowing, swimming, jumping rope, walking flights of stairs, or playing sports. Aim to elevate your heart rate with one of these aerobic activities at least thirty minutes each time, at least three times a week.

Alternatively, you could try *high-intensity interval training (HIIT)*, which lasts from ten to fifteen minutes each session. Not only is HIIT extremely efficient at elevating your heart rate, but you can no longer rely on the old excuse that you don't have enough time to exercise.

Second, we need *weight training*, to boost our metabolism, increase our strength, and strengthen our bone density, thereby reducing our risk of osteoporosis and fractures. You might lift weights in the gym, hire a personal trainer, or attend an exercise class or boot camp that uses equipment to create a weight-bearing load for your muscles.

If you're apprehensive about lifting in the gym, you could create your own weight training setup at home pretty inexpensively. All you need is a large exercise ball, a range of free weights, and a book like Steven Stiefel's *Weights on the Ball Workbook*.

Aim to target each muscle group with weight training at least twice a week, although we can work our core muscles every day. Remember, too, that muscle weighs more than fat—so you may gain weight, but you'll definitely lose inches.

Third, we need *stretching*, *yoga*, or *tai chi* to increase our flexibility and dissipate tension in the body. If you're uneasy about attending yoga classes, try an app or website that lets you stream them. By gently stretching the places where we habitually carry tension—especially the jaw, neck, shoulders, back, and pelvic region—we can keep our bodies in better biomechanical alignment and reduce chronic pain. Stretching is also an important way to warm up and cool down from aerobic exercise and weight training.

Across these three categories, aim for forty-five to sixty minutes of activity at least four times a week. Schedule the time into your weekly plan, and find activities you naturally enjoy. Use the habit diagnostic if you're noticing barriers to exercise.

As with diet, it's important to check with your doctor before starting a new exercise regimen. If you take beta-blockers, for example, it'll be hard to raise your heart rate, so your doctor can help you choose the best exercises.

SOCIAL CONNECTIONS

Americans are spending less and less time connecting with other people— with profound implications for our health and well-being. Research shows that a lack of social connections is roughly twice as dangerous to our health as obesity. It carries a risk of premature death on par with smoking. Having few or low-quality social ties has also been linked with cardiovascular disease, high blood pressure, repeated heart attacks, autoimmune disorders, cancer, and slowed wound healing.

Conversely, the most consistent predictor of happiness in many empirical studies is whether someone has strong personal relationships and social support—a finding that cuts across age, gender, ethnicity, and socioeconomic status. For instance, having a friend whom we see on most days has been shown to have the same impact on our well-being as making an extra $100,000 per year.

Over the last fifty years, there's been a significant decline in American social connectivity. Today, we're less likely to be married, we participate in fewer social groups, we volunteer less, and we entertain others in our homes less often. Television, the Internet, and social media have displaced activities that nourish our relationships. Television is America's number one leisure pastime, yet research shows that the more we watch, the less likely we are to volunteer or spend time with people in our social networks. Likewise, heavy Internet consumers report decreased communication with their families, decreased interaction with their offline social networks, and increases in depression and loneliness.

These trends especially affect young people. Many Americans, especially millennials and young people, are replacing real-world interactions with their smartphones. The more time teenagers spend looking at screens, the more likely they are to report symptoms of depression. At the same time, the number of teens who regularly hang out with their friends dropped by more than 40 percent from 2000 to 2015.

Social isolation—having few friends and social interactions—has grown dramatically in the United States over the last twenty years. According to a recent large-scale survey by the healthcare provider Cigna, nearly half of Americans say they sometimes or always feel alone and "left out." Roughly one in seven Americans report that *zero* people know them well. Nearly half of all meals eaten in the United States are now eaten alone. The average American has forty minutes of interpersonal interaction (outside work) each day—that's 243 hours each year, compared with 1,095 hours each year watching television.

In contrast to social isolation, *loneliness* occurs when we subjectively feel isolated—such as when there's a discrepancy between the number and quality of social interactions we *want*, versus the interactions we're *actually having*.

Social isolation and loneliness don't necessarily correlate. Some people truly enjoy being alone and prefer a secluded lifestyle. Thus, they might be socially isolated but not lonely. Conversely, we can feel lonely when we're surrounded by many people—especially if the interactions we're having feel disconnected, superficial, inauthentic, or not emotionally rewarding. In fact, one recent study showed that most lonely people were married, living with others, and not clinically depressed. Loneliness also seems common among teenagers and young adults. In the Cigna survey, for instance, loneliness among Americans was worse in each successive generation—with Gen Z reporting the most loneliness.

Research has shown that loneliness increases stress hormones, decreases cognitive functioning, and increases inflammation, raising our risk of inflammatory diseases, such as chronic pain, depression, heart disease, arthritis, type 2 diabetes, and dementia.

With this context in mind, spend some time reflecting on your social network. Aim for *at least a handful* of people whom you feel you could confide in and/or ask for help. If you don't have a few people, it's time to grow your social support network.

First, identify your social and relational needs—what feels meaningful, supportive, and nourishing for you. It's important to recognize that different relationships can help you fill different needs. In fact, expecting only one person, such as your spouse, to meet all of them puts too much burden on that relationship.

It can be difficult to ask for help and lean on others for support, especially if we see ourselves as self-sufficient. However, when we keep our vulnerabilities and pain to ourselves, the people around us may assume we don't need help. Then, when we really do, we might feel unable to reach out and ask for it.

If you find yourself resonating with this dynamic, you might start by cultivating relationships where you *offer* support, rather than receive it. You could help an elderly neighbor by mowing their lawn or driving them to an appointment. Do volunteer work to support a cause that feels meaningful. Start a book group. Helping other people can boost resilience just as much as receiving support. It also gives meaning and purpose to our lives.

If you discover that your social network needs bolstering, set some *concrete* intentions to round out this part of your life—and then build these intentions into your resilient planning. Here are a few ideas to get you started:

- Aim to have *at least* one authentic, connected interaction—in person, by phone, or by Skype—each day. *Texts, emails, and social media don't count.*
- Evaluate your existing circle of acquaintances, and pick someone with whom you'd like to deepen your connection. Over the next month, schedule at least two social events with this person.

- If you've recently moved to a new location, or have few acquaintances, you could take a class, volunteer, participate in a meet-up group related to your favorite hobby, or join a church or spiritual group.

- At least once a week, meet a work colleague for lunch, coffee, or a power walk. Bonus points for meeting a colleague you don't normally spend time with. When you're with these colleagues, share something authentic about yourself—and show them a side of you that's unrelated to your work identity. It doesn't have to be earth-shatteringly deep and vulnerable.

- Schedule at least one deep, heart-to-heart phone call or meeting with a confidant each week, such as a close friend, family member, or significant other. If you have children, you might schedule these events after they're in bed or hire a babysitter.

- Schedule one event each week where your whole family can spend quality time together, such as a leisurely meal, a hike in nature, or another fun activity.

- Finally, friends can also be furry, winged, or scaly! If you have a pet, spend some time each day interacting and playing with them.

By setting concrete intentions to build our network of social support, we weave the web of resources that we can call on when we're stressed or need help. Attending to our relationships every day helps us naturally feel more connected with and nourished by the world around us.

RESILIENT LIVING

The lifestyle choices in these five window-widening habits—reflection and awareness practices, a healthy diet, restful sleep, adequate exercise, and supportive relationships—help us reduce our allostatic load. More importantly, *these five window-widening habits have epigenetic powers, protecting us against genetic vulnerabilities to disease.* As Chapter 3 explained, our DNA is changed by whatever we experience frequently—and we certainly experience habits frequently. Epigenetic changes can even be passed on to our children, possibly over many generations.

Window-widening habits that proactively build adaptive capacity—such as stamina, strength, mental flexibility, health, and supportive relationships—provide us greater internal and external resources for coping flexibly during stress and recovering afterward.

How we choose to structure and live each day *is* our life. Regardless of what happens, by tending to what is "up to us" with window-widening habits, we establish the building blocks for resilience, health, creativity, joy, and well-being. We build agency into the very structure of our lives.

CHAPTER EIGHTEEN

Widening the Collective Window

In 2017, after significant procrastination, I had gum graft surgery. I've had a lot of dental trauma in my life, including with a previous gum graft. During that first surgery, the periodontist had peeled back the skin on the roof of my mouth to harvest graft tissue; before he could sew the site closed, however, I had an adverse drug reaction and vomited. As a result, that site became badly infected for several weeks.

Understandably, I wasn't in a hurry to try this again.

This time, I went to a different periodontist. The roof of my mouth and I were delighted to discover that I was a candidate for donor tissue grafting.

On the day of my appointment, I arrived in a rested and well-regulated place. During the procedure, I kept my attention on my contact points to help my survival brain feel stable and safe, while letting other sensations wash through awareness. I stayed in alignment with reality, without any struggle at all, until the *thought* arose that he had cut loose and was now peeling back more than half of my left gum and cheek, to stretch it up and over the donor tissue like a blanket. This mental image of what he was actually doing freaked me out.

With that thought, my survival brain neurocepted danger and turned on stress arousal. Immediately, I observed a wave of nausea and a spike in heart rate, rapidly followed by the closing-in sensations that usually precede blacking out.

In that exact moment, before I could say anything, the periodontist put down his tools, shoved some gauze into my cheek, and said, "I think it's time we take a little break and regroup." With care, he squeezed my shoulder and pushed back his stool.

I focused on my contact points and did G&R. After discharging my stress activation, as I started to feel more stable, I asked him, "How did you know that I needed a break just then?"

"Well, the work I'm doing is extremely detailed. Every millimeter counts, so it's important that I stay focused and take occasional breaks," he said. "But in that moment, I started to feel a little nauseated and dizzy, so I sensed you needed a break."

How did he *do* that?

By paying attention during the procedure, he picked up on our *resonance*, which is a preverbal, automatic, experientially-based connection that occurs between all mammals. Resonance is what allows for *empathy*, the effortless ability to mirror inside ourselves what someone else is feeling, without actually thinking about it. When we're aware of our resonance with others, it's like being inside their skin, while staying inside our own.

Our mind-body systems are constantly resonating with and being influenced by environmental stimuli. That's why it can feel grounding to spend time in nature, while it's activating to watch a horror movie.

Because we're social animals, the most influential environmental stimuli we encounter are other people. Earlier chapters explained how our mind-body systems and our neurobiological wiring, even the width of our window, are indelibly influenced by the people around us—how loneliness, harassment, and tension in interpersonal relationships can increase our stress, while feeling connected with and supported by others can decrease it.

Before concluding, I want to view this interdependence from the opposite direction: how the width of our individual window can influence the world around us. As Chapter 2 pointed out, American culture tends to emphasize the individual as autonomous and to portray success or failure as entirely the result of individual effort. But we can affect each other's resilience and level of regulation through how we interact with each other.

WIRED TO CONNECT

We convey our stress activation and dysregulation—or conversely, our regulation—to other people, and they to us, through several facets of our neurobiology. The first is *stress contagion*, which Chapters 6 and 9 explored. Remember that our survival brain notices the physical sensations of stress in our body, which leads it to neurocept danger and increase stress arousal. Likewise, our survival brain picks up on the physical sensations of stress activation in *others*—especially in people with whom we have attachment bonds or relationships involving power differences, such as boss-subordinate or teacher-student. Stress contagion works through resonance between different people's nervous systems and stress hormone levels.

A similar dynamic exists with *emotion contagion*, as anyone who's ever teared up during a sad movie knows. Especially when we're not paying attention to this consciously, our own mood can be powerfully shaped by *others'* emotions. Researchers who study emotion contagion have found that people tend to synchronize their facial expressions, voices, postures, movements, and emotional behaviors with those of the people around them. This automatic, moment-to-moment mimicry not only affects our emotional state but also plays a major role in social interactions. As with stress contagion, we're most likely to "catch" emotions from others with whom we have relationships involving attachment bonds or power.

In the last fifteen years, neuroscientists have discovered several neurobiological structures that help explain how stress and emotion contagion work. This research shows how much we're biologically wired to be deeply interconnected with each other. A few highlights suggest just how much of our brain is devoted to helping us be more socially connected.

First, when we experience distress from social pain, from, say, the result of social exclusion or separation from a loved one, we activate *the same pain distress network* in the brain as we do when we experience physical pain. This network includes the dorsal ACC and anterior insula, two brain regions you may remember from Chapter 12 that play a role in regulating stress arousal and emotions—and that MMFT has been shown to modulate.

Why does our brain treat physical and social/emotional pain the same way? Just as physical pain helps us take actions to keep our bodies safe, we have inborn neurobiological mechanisms that motivate us to stay connected with others. By making threats to our social connections truly painful, these neurobiological structures prompt us to seek out and offer social support. This makes sense evolutionarily, because human infants and young children would die without continuous support from their caregivers.

We're also motivated to stay connected to others through endorphins and oxytocin, two of the "feel-good" neurotransmitters. As Chapter 7 explained, we release endorphins when we're *being cared for by others*, including when we "make up" after a fight. In contrast, oxytocin increases our willingness *to care for others*. As Chapter 4 explained, oxytocin keeps the nervous system in well-being mode and out of defensive mode, which supports our social engagement and attachment systems. Oxytocin is released in huge quantities during childbirth and breastfeeding, to help mothers bond with and care for their offspring. It also dampens the empathetic distress we feel when we encounter someone in pain. This dampening effect helps us approach and support them; in fact, the more we do, the more oxytocin we release.

Second, our brains have a system of *mirror neurons*, which help us understand the intentions, behaviors, and emotions of others. This is not a thinking brain activity. It's an automatic, prereflective form of imitation. For instance, picking up a cup ourselves and seeing another person picking up a cup have the same effect on the brain's mirror system. The brain fires the same way in both situations, just more intensely when we act ourselves. Through this inner simulation, we experience something known as *motor resonance*, which helps give us an embodied sense of someone else's feelings and intentions.

Mirror neurons stimulate us to imitate and thus learn from others, even playing a role in language acquisition. The mirror system plays an important role in social engagement, helping us unconsciously coordinate gestures, facial expressions, and eye contact. It helps us create shared meaning during a conversation, even during a string of incomplete sentence fragments. Not surprisingly, it also plays an important role in emotion contagion and empathy.

The mirror system may also be implicated in imitative violence. For instance, exposing children to a violent television or film clip increases their odds of aggression soon after. One longitudinal study with a thousand children in New York State showed how heavy exposure to media violence in early childhood was linked with aggressive and antisocial behavior ten years later, after high school graduation. Another cross-national study examined imitative violence induced by watching media violence. Although there were some cultural differences across nations, the basic finding held across all countries: Watching media violence was significantly linked with imitative violence later. Together, these studies suggest a level of neurobiological automaticity resulting from social influences that undermines our conventional wisdom about individual autonomy.

Stress and emotion contagion, the brain's pain distress network, the social bonding neurotransmitters, and mirror neurons can be found in other mammals, too. However, we humans have evolved some additional structures, unique to us, that further ensure our ability to live harmoniously in social groups.

For example, our thinking brain has a *mentalizing system* that helps us perceive and understand others' actions, including their higher-level intentions and mental states, and that complements the more motor-oriented mirror system. Likewise, there's a part of the thinking brain that contributes to self-control—including top-down regulation of stress, emotions, impulses, and cravings—called the *ventrolateral prefrontal cortex*. This brain region also helps shape our behavior so that we'll comply with our social group's norms and values.

A similar dynamic exists with the part of the thinking brain that controls our conceptual sense of self, called the *medial prefrontal cortex (MPFC)*. This is the brain region that fires when we think about "who we are." Interestingly, however, it's also the brain region that fires when we hear persuasive messages from others. For instance, in one experiment that tried to get students to use sunscreen, the more a student's MPFC fired in response to the pro-sunscreen persuasive messages, the more they increased their actual sunscreen usage later on—regardless of what they told the researchers they consciously planned to do.

With all of this in mind, we can appreciate how much our culture's myths of individualism, explored in Chapter 2, are actually at odds with the truth of our neurobiology. Together, these neurobiological structures show how much we're wired to be interconnected with, and influenced by, the social environment around us. This includes friends, family, teachers, coaches, bosses, leaders, celebrities, the media, advertising, social media, and even movies and video games.

For this reason, as we widen our own window and make shifts inside ourselves, we also help shift the social environment for others around us. Thus, one of the biggest gifts we bring to the world can be our own presence and self-regulation.

MIND-BODY SYSTEMS RESONATING TOGETHER

This science of our interconnectedness explains why being in contact with activated, anxious, or irritated people will increase our own stress arousal—if we're not mindful of staying grounded and regulated. Conversely, we can appreciate how being around well-regulated people helps us ground and downregulate *ourselves*. We can also recognize why being in certain social environments—such as traffic jams, cocktail parties where insecure colleagues are clamoring to impress everyone, or lunch dates with a dysregulated friend or family member—take such a toll on us. We can see how even reading or watching the news, especially about a member of our identity group being threatened, bullied, mistreated, or abused, can spur a stressed or traumatized response in us, too.

As Chapter 6 explained, even securely attached individuals are likely to exhibit insecure defensive and relational strategies occasionally—especially when they're stressed, dysregulated, outside their window, or getting triggered by someone else. If we can bring awareness to these indicators, it presents us with a choice point to stop the conversation or argument, take a break, and facilitate some recovery to bring the ventral PSNS fully back online.

In an activated state, our ability to detect positive social cues from others around us is compromised. We're more likely to misinterpret neutral stimuli as threatening and neurocept visceral signals that we're not safe—even if we actually are—amplifying our stress arousal further. The result is that we will likely fall back from the constructive and cooperative strategies associated with social engagement, to fight, flight, and freeze.

Of course, social communication involves at least two people. When we and the person we are interacting with both have our social engagement systems fully online, then connected, constructive, and cooperative social communication is possible.

If the person we're speaking with is dysregulated, however, then our efforts to engage with them are likely to be met by insecure relational and defensive strategies—such as criticism, sarcasm, blame, indifference, withdrawal, or aggression. *Responses like these demonstrate that the other person is activated, possibly even outside their window.*

In this place, their thinking brain functions will be operating in a degraded manner, so "trying to reason with them" is unlikely to achieve the outcome we're seeking. Since their ability to detect our positive social cues is compromised, they're more likely to misinterpret what we're saying and get defensive. Unless they can bring their ventral PSNS back online, it may be difficult to have a constructive conversation. The most important thing is to remember that *their response isn't personal*—it's just an indication of what's happening inside their mind-body system right now.

Because of stress and emotion contagion, without some conscious intention and effort to stay regulated ourselves, their insecure response could easily trigger *our own* survival brain to turn on stress arousal. One way to stay regulated when we're on the receiving end of an insecure defensive and relational strategy is to direct our attention to our contact points. This helps cue our survival brain that we're still grounded and safe, decreasing the likelihood that it will succumb to contagion from the other person's stress activation or emotional intensity.

If we can stay regulated in the face of their dysregulation, our own regulation will resonate back to them—thereby helping their survival brain and nervous system downregulate and access their social engagement system. Conversely, if we *can't* stay regulated in the face of their dysregulation, that's an indication that it's time to take a break, before we say or do something that we might later regret.

When interacting with someone in a highly activated state, we may also help them downregulate simply by meeting their stress arousal and emotional intensity with awareness and nonjudgmental curiosity. We might ask them what they're feeling, and why. *This strategy will work only if we're inside our window, with both our thinking brain and survival brain genuinely interested in listening to and acknowledging the other person's answer.* If not, their response is more likely to trigger in us either thinking brain override or survival brain hijacking—which will only make their survival brain feel more insecure and amp up their arousal even further.

However, if we're inside our window and *can* bring awareness and non-judgmental curiosity to the interaction, we can help set conditions for things to shift. Remember that the survival brain always amps up stress arousal and emotional intensity whenever it perceives that its message is not getting through. By inquiring about the other person's feelings *without judgment, shame, or blame*, we can help *both* thinking brains in the conversation understand what's going on. Allowing both parties' wholeness into the conversation can often be the first step toward creating a connected, constructive interaction.

Since we're all only human, we'll all experience times when we've gone past our stress capacity threshold and then act out in ways that may be frightening or disrespectful to others. Even in these situations, however, we can choose to employ *interactive repair*. It takes wisdom and courage to take responsibility when we've hurt loved ones or colleagues, apologize for the harm we may have caused, and then actively seek to repair the relationship. Even if it's not easy, this simple skill can be strengthened with practice. With each cycle of interactive repair, we strengthen our ventral PSNS circuit—which improves our capacity not only for social engagement but also for recovery. Interactive repair helps us widen our window.

Stress and emotion contagion also affect our decision making. Chapter 14 discussed *internal urgency*, when we feel pressured to speak or act *right now*. If you find yourself feeling pressured to make an immediate decision, notice first if the internal urgency results from others' activation. Just like that old quip, "Your lack of prior planning does not make this my emergency," we can choose to stay grounded and hold a boundary against absorbing their activation and getting spun up ourselves. Rather, if we can stay present and regulated, we can help others downregulate.

Even if the situation is truly urgent—meaning the urgency is dictated by the situation at hand—it's helpful to take a few moments to reach for the most regulated, present, and stable state we can before making any decisions. Even thirty seconds of the Contact Points Exercise can make a world of difference. We may still not be completely regulated, but we'll certainly make better choices than if we hadn't taken that moment. Unless we're literally in an immediate life-or-death situation, we can always take a few moments to down-regulate to some degree. Bathroom breaks are great for this!

Since humans don't finish wiring many thinking brain functions and the capacity for social engagement and self-regulation until their early thirties, youth are especially susceptible to being activated by other people's dysregulation. As Chapters 6 and 7 explored, parents create the social environment that shapes the neurobiological development of their children—profoundly affecting the initial width of their children's windows and setting their mind-body systems on life-long trajectories. Accordingly, parents have a special responsibility to fully recover from the chronic stress and trauma they've experienced, so that they can bring the widest possible window to nurturing and interacting with their children and teenagers.

By extension, people working in certain professional roles have an outsized influence in conveying their own levels of regulation or dysregulation to others. Teachers, therapists, doctors, ministers, coaches, and mentors may all serve as unconscious attachment figures for the survival brains of their students, clients, patients, and congregation members. In these relationships, the width of a professional's window will likely have tremendous impact on the people they serve and lead. Many people come to these service professionals in particular when they're suffering or dysregulated, so it's imperative that these people bring the widest possible window to assisting others. They have an extra responsibility for using the tools in this book to keep themselves as grounded, present, and regulated as possible.

LEADING DURING STRESS, TRAUMA, UNCERTAINTY, AND CHANGE

Leaders send especially strong ripples into the social environment. As the dominant player in the social tribe, the leader sets the social and emotional tone for the entire group, often called the "command climate." Thus, they can have an extraordinarily powerful effect on others' levels of regulation or dysregulation—arguably holding the greatest influence on the width of collective windows.

I have personally experienced the benefits of working in inclusive, collaborative, and diverse environments led by leaders with wide windows—as well as the costs of working in toxic climates where discrimination, harassment, and microaggressions were allowed to fester. Indeed, my passion for training people to better regulate their stress stems in part from having personally faced violence and trauma within toxic, exclusionary organizations. It's not an accident that conflict avoidance, low morale, impulsive and reactive decision making, and unethical or transgressive behavior abound in organizations where the leaders themselves are dysregulated or stressed out.

Leaders strongly influence how their subordinates will respond during stress, uncertainty, and change, in at least two ways. From a thinking brain perspective, leaders can affect how their subordinates will interpret and make sense of stressful and traumatic experiences.

From a survival brain perspective, however, leaders are unconscious attachment figures. Even a tough leader can be loved, when their subordinates neurocept someone who is protective and fair. When leaders are perceived as competent, honest, trustworthy, and attuned to the physical, intellectual, spiritual, social, and emotional needs of their subordinates, they earn their subordinates' trust and, by extension, boost their resilience.

Such attunement allows followers to feel that their leader is providing the "safe home base" typical of secure attachment. In turn, followers feel comfortable exploring, learning, innovating, making mistakes, and growing. Having a secure base helps followers know that they can take risks, speak their minds, participate fully in group decisions, and confront difficulties. Just like our early attachment figures, leaders with wide windows can help us cultivate the traits that widen our individual windows, as well as the group's collective window. A leader who is regulated, rested, and inside their window can convey a calming and creative influence that helps the entire group access curiosity, situational awareness, creative problem solving, improvisation, and connection with others.

In contrast, if the leader is dysregulated, their stress arousal will spread through the group. The leader may convey hyperaroused states, such as anger and fear, or hypoaroused states, such as apathy, despair, victimhood, and powerlessness. Such contagion increases the likelihood that *all* group members will resort to insecure defensive and relational strategies, including violence, conflict avoidance, gossip, defensiveness, disrespect, bigotry, lying, apathy, withdrawal, and indecision. They are also more likely to engage in unethical and transgressive behavior. The leader's dysregulated state can erode the entire group's ability to cooperate, adapt, and learn.

Especially during periods of uncertainty and change, people want their leaders to make things better. They look to their leaders to orient them, help them face their confusion and pain, and find a path forward. They want reassurance that despite the turbulence and difficulty, they're still moving in the right direction.

Yet certain characteristics of stressful and uncertain environments work against leaders being effective "saviors." Stress tends to increase everyone's negativity bias and narrow everyone's perceptions. Stress also tends to narrow the leaders' decision making, so that they rely on fewer inputs and include fewer perspectives. They're more susceptible to biases, especially favoring information that confirms their viewpoint. Dysregulated leaders are also more likely to withdraw, limit information flow, and involve fewer people in their decisions. To manage their own anxiety, they're more likely to engage in micromanagement and other rigid control structures. These behaviors usually exacerbate feelings of apathy, powerlessness, and resentment in their followers.

Many senior leaders whom I've taught complain that time pressures make it impossible for them to create the necessary space to notice cues from their survival brains, listen to their inner wisdom, and keep their mind-body systems inside their windows. Hard-charging high-achievers, many of these leaders get results by pushing themselves and their people hard, and they've been rewarded for it. They admit to overriding their own limits routinely—by not getting adequate sleep and artificially mobilizing energy through stress reaction cycle habits—in order to provide an example for their followers. These choices only narrow their windows further.

As a result, their decision making suffers. They focus disproportionately on putting out short-term fires while neglecting longer-term initiatives. This attentional focus cascades down throughout their organization, as everyone follows their lead to prioritize the immediate over the important.

Alternatively, they say that they're exhausted and burned out from meeting the needs of their organization. They admit to ignoring the need for "white space" in their schedules and deprioritizing their own self-care.

However, once they understand windows—especially the critical role of the leader's window in providing the secure base for the rest of their organization—they come to understand that leader self-care is a *non-negotiable* ingredient for helping their organizations thrive.

Simply put, *leaders who prioritize self-care to stay regulated are better leaders.* By getting adequate sleep, engaging in consistent physical and mind fitness regimens, investing in close relationships, and attending to their own intellectual, emotional, physical, and spiritual needs, these leaders have wider windows for skillfully navigating the challenges they and their organizations face.

They're more effective and deliberate about setting intentions and manifesting their personal and professional visions. They're better skilled at helping their groups see the big picture and create a shared purpose. They're more likely to empower their groups to experiment with new ways of doing business, nonjudgmentally assess the results, and learn and adapt together. They're more likely to allow events to unfold and then respond effectively in the moment, rather than try to shoehorn inappropriate contingency plans or scripts.

These leaders are more likely to see and acknowledge the whole picture clearly—even unpleasant aspects—and tell the truth about it. They're more likely to do the right thing, even when it's difficult or unpopular. They're more likely to see their own and their subordinates' strengths, weaknesses, and limits clearly—which helps them build teams with complementary skills. They're less defensive and more open to feedback. They're more likely to nourish a thriving, inclusive, diverse workplace filled with respect, a sense of humor, connection, and creativity. They're more likely to create the space for the group to process both thinking brain and survival brain responses to challenges and change effectively.

Perhaps most importantly, leaders who prioritize self-care to stay regulated themselves signal to their followers that self-regulation is a critical, non-negotiable aspect of the mission. In fact, *if only one person in the entire organization could engage in consistent window-widening habits, ideally that one person would be the leader.* When they actually walk their talk about "resilience," these leaders help break down societal mixed messaging and its toxic consequences. They help enact new social values. As a result, the entire group prioritizes such behavior and widens the collective window.

WIDENING THE COLLECTIVE WINDOW

There's a maxim in the military: When the enemy is unknown, you need a bigger reserve. Having a large reserve allows you to marshal necessary forces to counterattack or to exploit unexpected opportunities.

This maxim applies more generally to us today. Especially during times of uncertainty, chaos, and change, we need a bigger reserve—in the form of individual and collective adaptive capacity. Since we can never anticipate and prevent all unwanted events from occurring, our best response lies in developing our capacity to meet whatever arises with resilience and resourcefulness.

As I hope this book has made clear, the wider our individual window, the easier it is to flow flexibly with adversity and unexpected events. The easier it is to keep our social engagement system online, so that we can connect with and support each other. The easier it is to keep thinking brain functions online, so that we can assess the situation with clarity, solve problems creatively, and make ethical and effective decisions. The easier it is to access agency, even during extremely stressful situations. And, of course, the more likely it is during challenging events that we'll experience stress instead of trauma.

Just as we can widen our individual window for more resilient responses, regardless of what occurs, we can also build adaptive capacity to widen our collective window.

Our species' dominance may be attributed to our ability to think and act socially—to imagine creative solutions to seemingly insurmountable challenges and then work together cooperatively to manifest these ideas. Thus, *one critical aspect of collective adaptive capacity is the strength of our relationships—* in our families, workplaces, schools, and communities, as well as between our nation and other nations. By nurturing our relationships during times of relative stability, we can rely on them during times of challenge and crisis.

Most features of adaptive capacity—such as creativity, improvisation, adaptability, and connection with others—can be deliberately cultivated through consistent, disciplined practice. Highlighting adaptive capacity in music, for instance, jazz musicians develop a common knowledge base about chord progressions and basic melodies, and through practice they hone their ability to read each other. Then, during performances, they can build on these foundations and their connections to improvise.

Innovative organizations that design new technologies or products often have highly disciplined practices that allow for better codification, replication, and generation of new knowledge. Likewise, high-reliability organizations consistently practice certain scenarios—ideally, not just to get better at those drills but to develop the group's ability to think quickly, adapt, and cooperate during stress.

Indeed, none of the concepts and tools I've shared in this book will be available to you (and your organization) when you're highly activated if you haven't been practicing them consistently when you're not. There are no shortcuts to a wider window. We can't take a pill or cram at a weekend retreat.

That's why the warrior traditions exist in the first place—to give us practices that we can train consistently in order to develop adaptive capacities that'll be available to us during crisis. Rather than binge or bust, it's much better to practice every day—even ten minutes. With consistent practice, we can train ourselves to show up to life with a wide window, trusting that we'll be able to greet whatever we encounter with wisdom and courage.

Practice has value independent of outcome. It's a way of life, not some task with a clear payoff. What a person or group practices reveals what they believe is important—and worth spending time improving.

We know what happens when enough people take up a cause or a habit as a practice: Cultural norms change. In the twentieth century, we saw this with the institutionalization of physical fitness. The sharp decline in smoking. The civil rights movement. Through intentionality and consistent practice, social groups *can* act their way into becoming something better—more inclusive, regulated, courageous, and wise.

With this in mind, using collective intentionality and consistent practice, we can heal our current divisiveness and build collective adaptive capacity to widen our collective window.

We live in an era of profound social, cultural, political, economic, and even planetary instability and change. Economic dislocations from growing domestic financial inequality and the globalized market. Profound environmental shifts from climate change, which have intensified our nation's droughts and increased the frequency of extreme weather events. Exclusionary immigration policies designed to help some Americans feel more secure while other Americans feel more threatened. Increasing cultural and social polarization, as social movements like Black Lives Matter and #MeToo raise public awareness about long-standing patterns of racial inequality, gender discrimination, and sexual harassment—and as counter-movements mobilize in backlash.

Of course our collective survival brains are likely to feel threatened. Survival brains *always* find change threatening, even changes we might collectively evaluate as "positive." Furthermore, *the narrower our individual and collective windows, the more intolerant of this turbulent moment's uncertainty we will be.* The more likely that we'll collectively dig in our heels to change—trying to perpetuate and rely on previous solutions, frameworks, power structures, and ways of understanding the world that reality has outgrown. Perhaps most importantly, *the narrower our windows, the more physiological and emotional arousal we'll experience at having these previous solutions, structures, and plans interrupted when reality inconsiderately refuses to align with them.*

Chapter 15 explored situations that need to shift and sometimes even fall apart. These situations get perpetuated by the opposing energies of overriding some limits while simultaneously falling short of others through limiting beliefs. We see a variant of these dynamics playing out in the United States today, as proponents of an increasingly unstable status quo double down on existing structures, while proponents of change speak out collectively in new ways.

Both groups' survival brains *feel* threatened, although the actual threat they face may not be equal.

As Chapter 2 explored, we've collectively conditioned powerful cultural patterns of disowning pain and trauma. One of the most common of these cultural patterns is for men to *externalize* and inflict their disowned pain onto others, disproportionately engaging in adrenaline-seeking behavior and perpetrating our society's violence against women, children, marginalized groups, and each other. Conversely, women are culturally conditioned to *internalize* their pain, overriding it for "peace at any price" and manifesting internalizing disorders disproportionately experienced in our society by women, including impostor syndrome, depression, anxiety, eating disorders, and autoimmune diseases.

Today, these long-standing cultural patterns for how we collectively relate to our stress and trauma—as well as its toxic effects—are being brought into collective awareness for reconsideration, reconditioning, and change. In the past, these cultural patterns have helped perpetuate sexism, heterosexism, racism, bigotry, and deeply entrenched economic inequalities in our nation. Now, we face a choice, as individuals and as a society: We could use this opening to acknowledge and heal the stress, trauma, and pain that we've previously disowned in our society. Or we could continue to perpetuate conditions for more individuals to be (re)traumatized and to strengthen cultural pressures for Americans to continue disowning the problems we have created for ourselves and others.

There is no doubt that our nation's collective window is narrow(ed) today. We see it in the incivility, deepening distrust, and polarization of our civic discourse—as leaders and followers both "fall back" from cooperative and constructive social engagement to insecure defensive and relational strategies (fight, flight, and freeze). Many groups feel under assault, with 64 percent of Americans in a recent Pew poll saying they believe that their group has been losing most of the time. When a group feels threatened, it's easier for them to dehumanize people in other groups, setting the psychological stage for violence. Leaders can stoke this siege mentality by deliberately choosing to whip their supporters into a dysregulated frenzy, using partisan dog whistles and "alternative facts." For instance, during Justice Brett Kavanaugh's Supreme Court nomination hearings, President Donald Trump galvanized his base by ridiculing and making false personal attacks on Christine Blasey Ford—a dysregulated fight response—which was spread via emotion contagion throughout his cheering crowd.

As this book has explained, discrimination, prejudice, and harassment don't have to be personally experienced to create toxic effects in our mind-body systems. Unless we're mindful, our survival brains will experience a surge of stress arousal while reading or watching the news about events where other members of our identity group are being marginalized. Our survival brains can also neurocept danger and trigger stress arousal while remembering or anticipating events when we are marginalized ourselves. Thus, in our nation's current culture of divisiveness, most survival brains are easily triggered from everyday relational trauma—turning our systems on without ever turning them off and thereby narrowing our collective window.

We see the narrow(ed) collective window in how many Americans today seek out information and experiences that confirm what they already believe, and avoid information and experiences that don't. As Chapter 17 explored, by the time we're adults, we tend to ignore, forget, reinterpret, or try to discredit information that isn't consistent with our internal structures; that's why we seek out like-minded people and information sources. One unfortunate downside of the digital age, however, is that we've created a world in which it is so much easier for us to live in separate siloes. This gets exacerbated by our culture of immediacy, where the unending fire hose of data impedes most Americans' ability to digest information thoughtfully and feeds disinformation and conspiracy theories. Moreover, after decades of political gerrymandering, we've also created a siloed political geography. In turn, we've degraded our collective resilience for listening nonjudgmentally and communicating respectfully across differences. Not surprisingly, therefore, a Pew poll in August 2018 found that 78 percent of Americans say that Democrats and Republicans disagree not only on "plans and policies" but on "basic facts."

We see the narrow(ed) collective window in our degraded thinking brain functions, such as in our distorted perceptions of others' behavior and our inability to downregulate our own stress and emotions. When we're stressed, our attention gets biased by negativity, and we're more likely to misinterpret neutral cues as threatening and/or negative. In turn, we're more likely to get defensive and react with fight, flight, or freeze responses. For instance, we've seen this dynamic in responses to Colin Kaepernick, the American football star who started the "take a knee" national anthem protest against police brutality. In fact, taking a knee is actually a neutral, nonviolent act to draw attention to aspects of disowned trauma—but in our collectively activated state, many survival brains have interpreted it as disrespectful or threatening.

We see the narrow(ed) collective window in our collective denial and disregard for limits and boundaries—not just for our own, but for those of other nations and even the planet. We continue to promote myths of American exceptionalism and implement America First policies, even as the biggest potential threats to our nation would be most effectively addressed by transnational cooperation. We consume most of the planet's resources, a stance at odds with the welfare of others and the planet itself. A 2018 Gallup poll shows that most Americans believe climate change won't affect them personally; only 45 percent think that global warming will pose a serious threat in their lifetime—even as a recent United Nations report projects that the worst effects will start occurring by 2040. Avoiding climate change's expected damage will require cooperating with other nations to transform the global economy at a historically unprecedented scale and speed. Instead, the United States today promotes coal, which releases more greenhouse gases than any other energy source; produces the most oil it has since the 1970s; and still buys relatively fuel-inefficient trucks and SUVs at exceptional rates. In 2017, 65 percent of new vehicle sales in the United States were trucks and SUVs. In our collectively stressed state, we're focusing on the immediate while denying and ignoring what's most important—having a viable planet where we, and the generations that follow, can live safely.

Similarly, to perpetuate our nation's longest period at war, we've mostly relied on our all-volunteer force, which disproportionately draws from evangelical, rural, minority, and lower socioeconomic groups. Americans serving in the AVF are also significantly more likely to have experienced ACEs than their civilian counterparts, as Chapter 7 explored. With 80 percent of U.S. military service-members related to someone else who's served, in effect we've created a separate warrior caste that unduly bears the costs of our nation's overextended foreign policy—and sets the stage for intergenerational trauma in military-connected families.

Highlighting not just our nation's civil-military gap but also our collective disconnection from *actual* limits, about two thirds of American civilians in 2011 said the disproportionate burden shouldered by U.S. troops since 9/11 is "just part of being in the military." Conversely, 84 percent of post-9/11 veterans feel that civilians don't understand the problems they and their families face. Iraq and Afghanistan veterans are nearly twice as likely as veterans of other wars to say that readjusting to civilian life was difficult. As a society, we've squared the circle of the Forever War by overriding the limits of our military service-members, with the longer-term costs on their physical and mental health just beginning to come due. At the same time, these conflicts have cost more than $2 trillion, financed on the nation's credit card. With neither a draft nor war taxation, we've eliminated some critical accountability links between the American public and our nation's foreign policy. We've also perpetuated a democratic society where the rights and responsibilities of citizenship are increasingly disconnected, which undermines our collective window.

We see the narrow(ed) collective window in the large allostatic loads that many Americans outside the military carry, as well. As of 2010, 70 percent of Americans were overweight or obese. Forty million Americans have a chronic sleep disorder. Collectively, we experience many conditions associated with chronic inflammation—as well as dysregulated endocrine and immune systems and nonfunctioning vagal brakes—including high blood pressure (103 million), chronic pain (100 million), cardiovascular disease (60 million), allergies (50 million), rheumatoid arthritis (50 million), autoimmune diseases (24 million), depression (21 million), and diabetes (14 million). As Chapter 1 explored, we've also seen an increase in mental health problems, especially anxiety (40 million), depression (21 million), substance abuse, and a skyrocketing suicide rate. In 2017, U.S. deaths from alcohol, drugs, and suicide reached their highest levels since the federal government started collecting this type of mortality data in 1999. We now have the world's second-highest per capita mass shooting rate—behind only Yemen, a borderline failed state at civil war—and the world's highest drug-death rate, with the opioid epidemic causing two thirds of such deaths.

Finally, we see the narrow(ed) collective window in the stress reaction cycle habits that many Americans rely on. In addition to externalized violence and addictions, many of us mismanage our stress by not getting adequate sleep and exercise, eating poorly, relying on a range of substances to up- and downregulate our stress, and compulsively engaging in self-harming or adrenaline-seeking behaviors. We increasingly cope by distracting and numbing ourselves with television, movies, video games, the Internet, and social media—with detrimental effects on our social interactions, as Chapter 17 explained.

Of these coping mechanisms, our electronics addiction stands out as perhaps the most worrying, because it works against our capacity to deepen the relationships that we need to widen our collective window. Indeed, in one recent study, video gamers playing a *nonviolent* immersive game showed significant reductions in pain sensitivity, both in themselves and in resonance with others. Compared to a control group completing a non-electronic puzzle, immersive video gamers could retrieve significantly more paper clips from icy water. Viewing pictures of other people experiencing pain or pleasure, the video gamers also rated other people's pain more indifferently (i.e., lower). This study suggests how much our social wiring may be blunted by our technology.

The big picture is overwhelming, and it's easy to feel helpless. The problems seem so intractable, while we each seem so insignificant and powerless.

So what can we do to help our nation build adaptive capacity and widen our collective window?

First, we can't appreciate the whole situation when we collectively continue to deny, disown, suppress, compartmentalize, and write off the parts of reality that are inconvenient, painful, or traumatic. The more we can *collectively* tolerate the whole picture—even its most unpleasant parts—the more clarity we can achieve about the actual situation, and the more we can cooperatively create a viable pathway forward.

As painful as letting in the truth can be, in the long term it's much more painful to live in denial—projecting a make-believe world that fits our comfort level or crying "fake news" when the facts don't align with our preferences.

As Chapter 15 explored, the greater our collective denial—and the longer we artificially prop up an increasingly shaky status quo that actually needs to be transformed—the greater the tension that builds from the ever-widening gap between reality and our collective thinking brain agenda. Even more importantly, *the greater the course correction that's eventually needed to release that tension to bring the situation back into alignment with reality.* We saw this dynamic with the implosion of the Warsaw Pact at the end of the Cold War, after decades of denial.

Second, we can remember that military maxim—and focus on building a bigger reserve, our individual *and* collective adaptive capacity. We can never anticipate, predict, and prevent all unwanted events from occurring. The best we can do is build the nation's ability to respond resiliently, regardless of what happens in the future.

To build this collective adaptive capacity, what could each of us as citizens help create—and as voters, advocate for as we elect and hold our leaders accountable?

We could nurture the relationships we'd need during crisis and recommit to building honesty, trust, and respectful communication in our communities. We could invest in affordable healthcare coverage for *all* Americans, to help everyone lower their allostatic load and build a resilient mind-body system. We could address income inequality, so more Americans could afford to get sick and cope with other unexpected emergencies, and first responders and public servants could afford to live in the communities they serve. We could teach pain management skills and fund treatment through drug addiction and recovery programs, to help end the devastation of the opioid epidemic on our communities and businesses.

We could invest in the safety and resilience of our nation's infrastructure, to dampen any effects of future catastrophic weather or potential cyberattacks. We could lower our staggering level of national debt, so that we'll have the financial resources we need to respond to global economic instability in the future. The national debt now tops $21 trillion for the first time, while U.S. household indebtedness, now at $13.5 trillion, has been growing steadily since 2012.

We could tend respectfully to our alliances, our trade relationships, our international partnerships, and the global institutions that we'll need for addressing transnational challenges effectively. We could address our addiction to fossil fuels and proactively protect the environment, to minimize the effects of climate change and provide responsible stewardship of our planet's constrained resources.

Perhaps most importantly, we could educate and train our children in ways that could help them develop wide windows for the many challenges ahead.

WE ALWAYS HAVE CHOICE

We create ourselves individually and collectively through what we choose to notice about ourselves and our surroundings. To break out of our conditioned filters, we must first be willing to let go of our thinking brain's false sense of certainty. We can't learn and change unless we're aware.

Awareness and nonjudgmental curiosity extend beyond our conditioning, which is why they can serve as alchemical agents of healing and transformation. They're always available to us. Since they belong to neither the thinking brain nor the survival brain, they can help us create an allied relationship between them. Even when our body is experiencing intense pain, emotions, or stress, if we can stay inside our bodies with these qualities online, we *can* make profound shifts to guide our mind-body system toward recovery, healing, and change.

Even more importantly, *anything conditioned can be unlearned, reconditioned, and extinguished.* Although our conditioned structures and programming have powerful inertia, we always have choice. To exercise that choice, we must first be willing to turn toward our habits, emotions, pain, addictions, and vulnerabilities without denying them.

All we can ever control is our choice in this moment: Whether we direct our attention consciously—or whether we let it get hijacked unconsciously by habit, impulse, emotion, or stress. Whether a thinking brain and survival brain alliance helps us choose the most appropriate course of action— or whether we let thinking brain override or survival brain hijacking drive our behavior. Whether we nurture our relationships through social engagement, attunement, and interactive repair—or whether we default to insecure relational strategies. Whether our choice embodies our highest potential and serves the greatest good—or whether we choose to let short-term gratification or narrow self-interest win out.

Over time, as enough moments pass in which we choose to pay attention, access presence and self-regulation, and make skillful choices, we *can* recondition our neurobiological structures. We can transform the entire landscape of our lives, our relationships, and our communities. Although this is not a linear progression, the shifts do become apparent over time.

As recent neuroscience research shows, we each contribute to the collective window. Are we helping to widen or narrow it? Are we adding regulation, presence, creativity, wisdom, courage, and connection with others? Or are we adding dysregulation, fear, anger, confusion, violence, denial, and discord?

We can't go back and magically undo the many assaults that our mind-body systems have experienced until now. All we can do is start where we are right now.

We can choose to break the cycle of intergenerational trauma, by healing our addictions and internal divisions, rewiring our insecure defensive and relational strategies, and cultivating a wider window. We can turn off detrimental epigenetic changes through new window-widening habits, ensuring that they don't get passed to the next generation. We can help our children develop the widest possible windows, to set them on a lifelong trajectory toward resilience.

We can choose to decrease our allostatic load by getting enough sleep and exercise and balancing our microbiome with a healthy diet, thereby reducing our susceptibility to many diseases. We can prioritize what really matters with resilient planning. We can proactively cultivate stillness each day, creating the space for insight, inspiration, creativity, love, compassion, and joy to shine through.

We can choose to investigate our unskillful habits, self-defeating behaviors, externalized violence, and addictions with nonjudgmental curiosity, understand what's driving them, and then choose to replace them with more skillful choices.

We can choose to nurture our relationships—in our families, workplaces, and communities—so that we strengthen the relational web that we'll need to rely on during chaos and crisis. We can practice interactive repair. We can seek out people with whom we don't agree and make a sincere effort to listen nonjudgmentally and understand their perspective. We can show up to each encounter with another person with presence, self-regulation, respect, and kindness.

We can choose to embody the best aspects of being human, inspiring the people we come into contact with that not only is meaningful change possible but that wholeness is our human birthright.

These are the things that are up to us.

APPENDIX

The recommendations made in this book provide an introduction to the concepts and skills taught in MMFT; however, they are not the full MMFT course. Furthermore, they are not meant to replace formal medical or psychiatric treatment. Individuals with medical problems and/or with intense symptoms of dysregulation should consult their doctors about the appropriateness of following the recommendations in this book. They should also discuss appropriate modifications relevant to their unique circumstances.

If you're confronting profound dysregulation, I believe that working with a therapist trained in body-based techniques while you begin your mind fitness practice is essential. They can help you pace your survival brain's bottom-up processing, so that it happens gradually and safely. I *strongly* recommend that you seek out a trained professional to help you navigate your way through the process, so that you don't inadvertently flood your system, retraumatize your survival brain, and exacerbate your dysregulation. You can find a certified practitioner for Somatic Experiencing at https://sepractitioner.membergrove.com, and a certified practitioner for sensorimotor psychotherapy at https://www.sensorimotorpsychotherapy.org/referral.html.

I wrote this book with the intention of sharing some basic practices and techniques for working skillfully with *your own* stress activation and symptoms of dysregulation. However, this book is *not* designed to provide you with the necessary training or any competency to teach MMFT skills to others. Because of the deep physiological and psychological processes that can be affected by these practices and techniques, it's important that you seek additional training and certification before you attempt to teach MMFT or any other mindfulness-based training program to others. So as not to inadvertently cause harm, this warning is especially important if you intend to teach or work with individuals in high-stress environments and/or with histories of chronic stress and/or trauma.

In fact, to teach or work with others most effectively—especially with people who are coping with profound dysregulation—requires that we have already deeply engaged that process for ourselves, with intensive practice dealing with the range of experiences that can come up in the mind-body system. There is no substitute for deepening one's own mind fitness practice and widening one's own window.

Basic Guidelines for Mind Fitness Practice

If you can, find a quiet place where you won't be interrupted or distracted by lots of activity or noise. Especially when you're experiencing a lot of stress activation, it can also be especially helpful to sit in a stable chair, with your back facing a solid wall rather than a door, window, or open space. This can help your survival brain feel more stable and secure.

Practicing in the same place each day can help you build the habit of practicing. As you begin to associate a particular space with practice, you'll find that returning to that space for your practice period will help you call to mind the qualities of mind fitness, especially awareness and nonjudgmental curiosity.

Practicing at the same time each day can also help build the habit. Many people find the morning to be a particularly effective time for practice, since the mind tends to be more receptive to practice after the night's rest and before the day's busyness has begun. Others find it effective to practice right after physical exercise. Research suggests that exercise encourages neuroplastic changes in the brain, so practicing mind fitness exercises right after cardiovascular exercise may be especially effective. Finally, many people find it helpful to practice during transition periods throughout the day, to provide some reset and recovery before engaging in the next activity, such as upon awakening, after lunch, after exercise, right after work, and at bedtime.

It's best to use a timer when you practice, to help you disengage from any thoughts about the passing time and to stand your ground against any temptations to check your watch and/or end your practice session early. It helps your thinking brain and survival brain relax into practice. Even better, the timer helps you commit fully to your intended period of practice and supports you in navigating skillfully through any resistance you may experience.

The greatest support for developing the qualities of a fit mind is regular practice. This momentum can help you generate the motivation to practice, even on days when you don't feel like it. It's much better to practice even five minutes of the Contact Points Exercise once each day than to binge and bust with practice.

Even if you don't think that a particular exercise is working for you, you will benefit from sticking with it. Sticking with it, even when you find it unpleasant, helps break down our habitual pattern of pushing away experiences we don't like. However, if you find even the Contact Points Exercise significantly distressing, it's best to back off and seek some help from a therapist, preferably someone trained in body-based trauma techniques (such as sensorimotor psychotherapy or Somatic Experiencing).

Using Mind Fitness Practice to Cultivate Wisdom and Courage

As Chapter 11 explained, wisdom is the ability to see the whole picture, nothing left out, just as it is—and then choose the most appropriate response in light of this information. To rouse wisdom during a period of mind fitness practice, we only need ask: *What's happening right now?*

Courage is the ability to tolerate our experience without needing it to be different. If we can first allow reality to be exactly as it is, then we're more likely to act skillfully to change those aspects of reality that we can. To rouse courage during a period of mind fitness practice, we only need ask: *Can I be with this experience, just as it is, without needing it to be different?*

For example, on days when you're feeling resistance to practicing, you can let the resistance become part of your practice session—and cultivate wisdom and courage along the way.

To do this, you might bring awareness and nonjudgmental curiosity to the resistance you are experiencing. Investigate: How does the resistance manifest as stress activation and/or physical sensations in your body? What is your current energy level? Are there any places of tension or tightness in the body? How does the resistance manifest in your thinking brain as thoughts, storylines, or "excuses" for why you can't practice right now? Do you notice any emotions or impulses co-arising with the resistance? For instance, you may notice anxiety, time pressure, and planning about other things that you need to get done. Or you may notice boredom and the desire to watch television. See if you can spend a few minutes investigating, with nonjudgmental curiosity, how the resistance is manifesting right now. As you do, you'll be rousing wisdom.

Then, as an advanced move, see if you can be with the resistance just as it is, without needing it to be different. You might allow the thoughts and sensations of the resistance to simply move through your mind-body system, while also noticing the felt sense of the support and stability provided by the chair and floor beneath you. If the resistance has a lot of stress activation associated with it, you can transition into one cycle of G&R. If not, you can transition into a few more minutes of the Contact Points Exercise. As you follow these advanced moves, you'll be rousing courage.

You also will have completely circumvented your resistance and completed your daily mind fitness practice!

Suggested Sequence for Mind Fitness Exercises

<u>Week 1:</u> Contact Points Exercise for five minutes each session, preferably two to three times each day.

<u>Week 2:</u> Contact Points Exercise for eight to ten minutes each session, preferably twice a day.

<u>Week 3:</u> Contact Points Exercise for ten to fifteen minutes each session, preferably twice a day.

This week, if you already have an established mindfulness practice that predates reading this book, you can also start using the G&R exercise after mild or moderate stress activation, such as after cardiovascular exercise, a wave of anxious planning, a nightmare, or an argument with a loved one.

<u>Week 4:</u> Contact Points Exercise for fifteen minutes each session, preferably twice a day.

This week, everyone can start using the G&R exercise after mild or moderate stress activation, such as after cardiovascular exercise, a wave of anxious planning, a nightmare, or an argument with a loved one.

<u>After a Month of Consistent Practice:</u> You can increase your practice sessions to twenty to thirty minutes. You can also start incorporating the other techniques from this book that build on G&R, such as with chronic pain and intense emotions (summarized shortly).

You may also want to incorporate other forms of mindfulness practice, such as awareness of breathing or mindful walking. During practice, if you notice yourself experiencing a wave of stress activation, such as panic or anger, you can always transition into one cycle of G&R to discharge that activation. Then you can easily return to the other form of mindfulness practice.

Remember that daily practice has value independent of any beneficial outcomes that you may observe. This is the reason why the warrior traditions evolved in the first place—to give us practices that we can train consistently in order to develop adaptive capacities that'll be available to us during crisis and challenge. None of the concepts and tools I've shared in this book will be available to you when you're highly activated if you haven't been practicing them consistently when you're not. There are no shortcuts to a wider window.

Summary of Contact Points Exercise (for complete instructions, see Chapter 12)

Find a comfortable place to sit, preferably in a chair with your back toward a solid wall, rather than toward a door, window, or open space. Sit with your feet shoulder-width apart and flat on the ground. If it feels comfortable to you, close your eyes; if not, direct your gaze at the ground in front of you. Sit so that your spine is both upright yet relaxed.

Allow yourself to notice the feeling of being supported by the chair and ground. You're aiming for the *felt sense* of this support, in your body, rather than *thinking about* or analyzing this support. Notice the contact between the backs of your legs and butt and the chair, and between the soles of your feet and the floor. If you have trouble feeling your feet, you can gently wiggle your toes or press your feet into the ground.

As you notice this support from your surroundings, briefly scan your body for any places holding tension or tightness. In particular, check your brow, jaw, neck, and shoulders. Without trying to make anything particular happen, see if by bringing attention to those places, the tension shifts. It may or may not, but either outcome is perfectly fine.

Now, bring your attention back to the *physical sensations of contact* between your body and your surroundings. You may notice things like pressure, hardness, softness, heat, coolness, tingling, numbness, sweatiness, or dampness. Pay attention to these sensations of contact at three different places: (1) between your legs, butt, and lower back and the chair; (2) between your feet and the ground; and (3) where your hands are touching your legs or touching each other.

After you investigate the sensations of contact at each point, select the place where you most strongly notice the sensations of contact. *This one contact point will now be your target object of attention.* If you find that you aren't noticing any sensations at all, try taking off your shoes and socks, sitting on a hard surface, or running your hands slowly along your thighs.

Once you've selected your contact point, direct and sustain your attention there. Notice the sensations at this contact point in great detail. Without thinking about it, simply notice and investigate the physical sensations with nonjudgmental curiosity.

If you notice your attention wandering off, simply recognize that it's wandered and gently and nonjudgmentally redirect it back to the sensations of contact. Each time you redirect your attention back to the sensations of contact, you're breaking up your old mental canyons and building attentional control. Think of it as a neuroplastic rep.

To conclude the exercise, widen your attention to take in your whole body seated in the chair. Notice if anything has changed in your mind-body system from having done this exercise. With nonjudgmental curiosity, investigate: Is the body more relaxed or more agitated? Is there more or less muscle tension? Is your energy level higher or lower? Are you sleepier or more alert? Is the mind more focused or more distracted? Is it calmer or more anxious and irritated? You may notice shifts in your mind-body system after practice, or you may not. Either way is perfectly fine. The goal is simply noticing the state of your mind-body system right now.

Summary of the Ground and Release Exercise
(for complete instructions, see Chapter 13)

To use G&R, say that you've just had an argument with a loved one, or you've just awoken from a nightmare, and now you're experiencing stress activation.

Common Symptoms of Stress Activation

• Faster breathing	• Reduced visual field/tunnel vision
• Difficulty breathing	• Hair standing on end
• Tightness in chest or belly	• Losing bladder/bowel control
• Faster heart rate/pounding heart	• Racing thoughts
• Nausea	• Anxious thoughts
• Butterflies in stomach	• Rumination/looping thoughts
• Dry mouth	• Anxiety or panic
• Clenched jaw	• Impatience, irritation, or rage
• Pale and cold skin	• Sadness
• Sweaty palms	• Shame
• Sweating	• Overwhelm
• Hunched or collapsed body posture	• Restlessness/fidgeting
• Dizziness	

Move yourself to a quiet place where you can be alone. Think about structuring the external environment to facilitate your survival brain neurocepting safety. For example, you should find a comfortable and safe place to sit, preferably with your back toward a solid wall, rather than toward a door, window, or open space.

With interoceptive awareness and nonjudgmental curiosity online, bring your attention to your symptoms of stress activation. Once you've noticed the physiological component of your stress activation, consciously recognize that you're activated. It can be helpful to note this to yourself, as in "Oh, I'm activated right now." However, you *don't* need to make an exhaustive list of all of your symptoms of activation. You also *don't* need to focus on the activation in great detail, which would only amplify your stress activation. Instead, simply notice from a more global perspective what's happening in the body.

If you notice restlessness, such as when your legs are bouncing or you feel like you're ready to jump out of your skin, see if you can sit perfectly still. When you do, the stress activation that had been coursing through you as fidgeting energy will almost certainly shift to become autonomic arousal, the form the activation needs to take in order to be discharged.

After consciously recognizing that you're activated, redirect your attention to a place in the body where you feel the most solid, stable, grounded, and strong. This is usually at a contact point—with your butt, your lower back, or the backs of your legs touching the chair, with your feet touching the floor, or perhaps with your hands touching each other or your legs. If you cannot feel contact with the chair or floor, consciously push your butt into the chair and your feet into the floor until you feel that support.

Notice the sense of being grounded and supported by the chair and floor. Just as in the Contact Points Exercise, you're aiming for the *felt sense* of this support in your body, rather than your thinking brain trying to *think about* or *analyze* this support. For instance, you may notice pressure, hardness, softness, dampness, heat, tingling, or coolness at these places of contact. Keep your awareness on your contact point or another place in the body that feels most stable, comfortable, safe, and grounded.

Of course, your attention may get pulled back to the physical sensations of stress activation, or back to the storyline, image, flashback, emotion, or thoughts that are fueling the activation. When this happens, keep redirecting your attention back to the contact point or other solid place in the body. Keep redirecting your attention in this way until you feel more relaxed or more settled, or until you notice one of the signs of release.

Signs of Nervous System Discharge and Recovery

• Shaking/trembling	• Crying
• Twitching	• Laughing/giggling
• Slower, deeper breathing	• Yawning
• Slower heart rate	• Sighing
• Relaxation in the chest or belly	• Stomach gurgling
• Tingling/buzzing	• Burping
• Waves of warmth/heat	• Farting
• Chills	• Coughing (phlegm)
• Flushed skin/sweating	• Itching

As you notice symptoms of release, there is no need to try to control or stop them. As you pay attention to the signs of release, you may notice that they amplify. If you can tolerate it, this is perfectly fine. For instance, if you bring your attention to yawning, you're likely to experience several more yawns in quick succession. Or if you focus your attention on trembling in your hands, the trembling will likely intensify.

Once you notice some sensations of release, you can shuttle your attention between these sensations and your chosen contact point. Shuttling in this way may help you feel more stable while the release is going on. If the release is quite intense, or if you find shuttling your attention to be uncomfortable, then simply keep your attention on your contact points. It is possible for release to feel strange, unfamiliar, or even frightening. Remind yourself that discharging stress activation is a natural part of the recovery process. If you find that you're getting caught up in the narrative, images, or thoughts that triggered the activation, redirect your attention to the sensations at your contact points.

You can stay with the release process as long as you are noticing signs of release. Once the signs of release stop, you have completed one cycle of recovery. *When we use G&R, it's important to complete only one cycle of recovery and then stop the exercise.*

If you find your mind-body system cycling back into activation again, or if you never experience any signs of release, redirect your attention away from the activation. Instead, keep bringing your attention back to sensations at your contact points. Sustain your attention there. You can also open your eyes; move your head and neck; notice sights, sounds, and smells around you; and name to yourself what you're noticing. As you orient to your surroundings through one of these senses, also notice the felt sense of being grounded and supported by the chair and floor. Keep directing your attention in these ways that support the survival brain from cycling back into more stress activation until you feel more relaxed.

If you've tried all of these steps and still not experienced any signs of release, it's likely that you're experiencing stress arousal outside your window. In this situation, the best way to expend some of the excess stress activation is to engage in cardiovascular exercise for at least fifteen to twenty minutes during which you are slightly out of breath. We each differ in terms of how much and what kind of exercise we need to attain this state. Then, during the cooldown after your workout, you can try G&R again.

Working Skillfully with Thoughts

When we first begin mind fitness practice, we quickly observe that we get lost in thought frequently. This is completely normal; it's just what thinking brains do. The goal of mind fitness practice is not to try to *stop* the thinking. The goal is to *notice* the thinking.

Try not to make your thoughts the enemy. If you try to suppress your thoughts, you will quickly become exhausted and discouraged. In fact, you may notice that trying to suppress thoughts will actually cause thoughts to proliferate!

Instead, when you notice that you've been lost in thought, simply acknowledge nonjudgmentally that your mind had wandered. Notice that although you were lost in thought just moments ago, you are paying attention again now. Simply notice the thoughts as observable events moving through your field of awareness.

It can be helpful to give thoughts a label to describe the kind of thinking that you notice, such as planning, worrying, fantasizing, remembering, comparing, narrating, or anticipating. With such labels, you can begin to observe some of your thinking brain's default habits. Then see if you can disengage your energy and attention from the thoughts. Each time you notice that your mind has wandered and you choose to redirect your attention back to the target object of the exercise, you are strengthening attentional control. Mind fitness practice is all about beginning again, noticing that the mind has wandered and gently guiding it back, over and over.

Working Skillfully with Emotions

When an intense emotion arises, see if you can acknowledge its presence with a simple label. Emotions provide a great opportunity to cultivate wisdom and courage; you can begin by asking those two questions: *What's happening right now? Can I be with this experience just as it is, without needing it to be different?*

Each emotion has its own "tells," which we can learn to identify. With awareness and nonjudgmental curiosity, you can investigate the three components of each emotion: (1) particular physical sensations, body postures, and arousal patterns; (2) a particular flavor in the mind; and (3) a particular "voice" and belief structure, which usually manifests as storylines, narratives, and thought patterns. When we disaggregate our emotions into these components, we obtain three different doorways into recognizing an emotion's presence.

The sooner we notice an emotion's presence at the mild end of the spectrum of intensity and support its release, the more likely that we'll safeguard our ability to work with it skillfully and choose the most effective response. Whenever we experience emotional intensity outside our window—manifesting as either thinking brain override or survival brain hijacking—we can always find agency by disentangling the emotion into its three components.

As much as possible, disengage from the content of any thoughts or storylines associated with the emotion and simply notice the thoughts as observable events. Especially when we're stressed, tired, or outside our window, thinking brain–dominant techniques, such as cognitive reappraisal, positive thinking, or accessing gratitude, tend to fall short. That's because our executive functioning capacity may be too depleted or degraded to support such cognitive effort.

Therefore, especially when you're outside your window, try to disengage from the content of your thoughts and postpone any decision making, problem solving, and planning until you're back inside your window.

Focus instead on physical sensations, grounded in the present moment. Allow the emotional wave to wash through your awareness while tracking physical sensations. You might use G&R to discharge and release the emotional arousal. Alternatively, you could shuttle your attention between sensations at your contact points and the sensations associated with the emotion moving through your body. As much as possible, give the emotion permission to fully blossom in your field of awareness and then wash through. With practice, we can learn to ride a wave of emotion out—allowing it to pass through our mind-body system while neither suppressing nor indulging in it. In the process, we gain access to the information the survival brain is trying to convey.

Sometimes an emotional wave *is* too intense to bear—in which case, skillful avoidance may be the right choice for a while. That's when it could be helpful to call a friend for some support, walk in nature, do some gentle stretching, or soak in the tub. This is *not* the time to vent or journal about the emotion—such activities will only feed its intensity. When we encounter emotional intensity that seems disproportionate to the current situation, it's often linked to emotional patterns from earlier in our life. If you find yourself experiencing intolerable emotional intensity frequently, it can also help to work with a therapist.

Especially if you have difficulty identifying emotions, tend to get caught in the content of distressing thoughts, rely on emotion suppression, or have beliefs that deny certain emotional experiences, it can be helpful to journal. Ideally, you might journal once you're back inside your window, after the emotional intensity has subsided. For instance, when you notice that you are unsettled or upset, you could write about an unresolved incident. As you notice the thoughts, beliefs, and physical sensations associated with emotions, you might put down your pen and seamlessly transition into a cycle of G&R, to support the movement of the emotional wave through your body. Continue to alternate between journaling and G&R, until you feel like the wave has completely passed.

Working Skillfully with Intense Physical Sensations and Chronic Pain

During mind fitness practice, it is helpful to find a posture that balances being alert with being at ease. In addition, keeping the body still and resisting the urge to move can help calm the mind.

When you sit without moving, however, you will almost certainly experience uncomfortable physical sensations occasionally. When they become strong enough, it can become difficult to pay attention to the target object of the exercise.

For the first few weeks of practice, I recommend that when an uncomfortable sensation arises, you deliberately disengage your attention from it and redirect your attention to sensations at your contact points.

After you've developed a baseline of attentional control, however, physical discomfort can become another opportunity for cultivating wisdom and courage. First, acknowledge that the uncomfortable sensation is keeping you from staying focused on your target object of attention. Then allow your attention to go to the sensation. Ask: *What's happening right now?* As you bring awareness and nonjudgmental curiosity to the sensation, see if you can distinguish between your thoughts and feelings about the sensation and the bare experience of the sensation itself.

For instance, you might have the thought, *This itch is driving me nuts; I need to scratch it*. Recognize this as a thought. Then, with nonjudgmental curiosity, explore the itching sensation directly. Is it constant or changing? Is it one solid sensation or a combination of a variety of sensations? See if you can track the sensation over time to observe how it changes and eventually disappears. When the sensation ends or becomes less intense, you can return your attention to the original target object of the exercise.

While you observe and track uncomfortable sensations in the body, you can also ask, *Can I be with this experience, just as it is, without needing it to be different?* Many times, the answer to this question is yes and we have the opportunity to cultivate courage. Other times, however, the answer to this question becomes no—the sensation becomes too intense to simply observe it. When this occurs, you might acknowledge how unpleasant the sensation is, as well as your desire to move. Then, very slowly and mindfully, move to relieve the discomfort. Notice the sense of relief you experience after you move. Then you can return your attention to the target object of the exercise. As you continue practicing, you can also notice how long the relief from moving actually lasts.

Beyond the intense sensations that can develop when we sit still without moving, many people experience chronic pain. As Chapter 14 explored, there are additional tools that you can use when you are experiencing chronic pain and muscle tension.

First, if you are experiencing chronic pain, to work with pain mindfully, you first need to develop attentional control, such as by using the Contact Points Exercise. While initially cultivating attentional control, it's important to practice in a posture that is relatively pain-free. For instance, many people with back pain find that practicing the Contact Points Exercise while lying down, with the knees bent and together, minimizes pain during practice. Returning your attention again and again to sensations at the contact points helps cue the survival brain that you're safe, stable, and settled. As the survival brain neurocepts safety, it can help the pain subside.

Recognize that pain will be worse when you're outside your window, such as when you're sleep-deprived, overwhelmed, or experiencing intense emotions. *Chronic pain is a symptom of dysregulation.* Thus, by using tools for discharging stress activation and recovering back inside your window—such as getting more sleep and exercise, eating well, and using G&R—you'll also expedite pain reduction.

Second, you can modify G&R for working with places of tension and pain. Say, for instance, you're experiencing neck pain. You could practice rolling your neck *very slowly*, coming right up to the place where you feel it "catch." It's important to stretch slowly enough that you don't inadvertently override this place. As you stretch right up to the edge of the painful place, hold still and focus your attention there. As you hold, the pain may intensify momentarily, but then it will release. You can also use these same principles while pressing a tennis or golf ball against a particularly sore or tender spot. If you can tolerate the discomfort for about ninety seconds, it usually releases.

You can also use this G&R stretching variant to supplement body-based trauma therapies and/or bodywork.

Third, once you've developed attentional control, you can begin to work directly with the pain, building your capacity to tolerate painful sensations. To be clear, *this exercise is best done when you're feeling relatively rested, regulated, and balanced,* such as first thing in the morning.

Begin by paying attention to a neutral target object, such as sounds or sensations at your contact points. Then, you can direct your attention to the *edge* of the place where you feel pain. It's not helpful to take your attention right smack into the middle of the most painful place—this will only trigger more stress arousal and pain.

As you pay attention to sensations at the *edge* of the pain, it's important to distinguish between the *actual physical sensations*—such as throbbing, burning, stabbing, or heat—and *your thinking brain's narrative* about the pain. As much as possible, disengage your attention from any thoughts about the pain; keep your attention focused on the sensations.

When paying attention to the edge of the pain becomes too much, simply redirect your attention back to the neutral target object. While you practice, you can shuttle your attention back and forth between the neutral target object and the edge of the pain many times.

As you pay attention at the edges of the pain, you may notice signs of stress activation, such as increased heart rate, shallow breathing, nausea, dry mouth, sweaty palms, or chest constriction. Indeed, chronic pain often relates to unresolved memory capsules and incomplete defensive strategies that you've not yet discharged. Sometimes when we pay attention at the edge of the pain, we tap into one of these. No problem. If this happens, *redirect your attention away from the pain and back to the sensations at your contact points—followed by one cycle of G&R.*

If this happens frequently, however, I *strongly* recommend that you find a therapist trained in body-based techniques, such as sensorimotor psychotherapy or Somatic Experiencing. They can help you move through and discharge these memory capsules safely—and more efficiently—than you could on your own.

Since emotional and physical pain use the same neural circuits, you could also work with the underlying emotions directly, using the tools in the previous section for emotions. You could also journal about your pain, asking the pain to share its message: Write down a question, and then let "the pain" write the answer using your nondominant hand, which tends to sideline your inner critic. As you do this, it may bring up emotions and then you can seamlessly transition into one cycle of G&R.

Finally, you might journal about any limiting beliefs about the pain—such as assumptions about what you can and can't do physically—and then test these beliefs with small experiments. Are these beliefs really true? In the process, you may move things from the "can't" to the "can" category, minimizing the pain's negative repercussions in your daily life.

Summary of the Window-Widening Habits

The Principles of Resilient Planning—Planning 2.0 (see Chapter 16)

1. Set aside time when you're in a regulated, rested, and unhurried place to plan for the upcoming week, before the week has started.

2. Get clear about your intentions in the physical, emotional, spiritual, and intellectual domains.

3. Ensure that your weekly plan matches your intentions and goals for the week to the space-time continuum.

4. Schedule first the tasks for meeting long-term professional and personal goals, including time for window-widening habits and time off for recovery. Aim for at least one day each week without any work, errands, or household tasks.

5. Ideally, schedule some time for window-widening activities and long-term goals each day.

6. Build plenty of white space into the schedule for unexpected challenges and opportunities.

7. Build a few hours into each week's plan for attending to "squeaky wheels" in your environment.

8. Group tasks of a similar energy level together, and as much as possible, match tasks of a similar energy level to the energy that you typically have at different times throughout the day.

Awareness and Reflection Practices (see Chapter 17)

1. Aim for fifteen to thirty minutes of practice first thing in the morning, as much as possible.

2. Aim for at least one awareness and one reflective practice you engage in regularly, such as one in the morning and one in the evening.

Healthy Diet (see Chapter 17)

1. Avoid—or limit consumption of—foods and substances that lead to microbiome imbalances.

2. Repopulate your microbiome with nutritious foods and live probiotics; you might take a probiotic supplement and/or consume probiotic foods and beverages, ideally every day.

3. If you have symptoms associated with inflammation, such as chronic pain, you also might try an anti-inflammatory diet, eliminating for six weeks the foods that people are typically sensitive or allergic to. Be sure to keep a symptom diary as you start to add foods back in.

4. If you often feel bloating, cramping, gas, indigestion, or heartburn—or have problems with diarrhea or constipation—pay attention to food-combining principles. It's easier for our body to digest food when we partner leafy greens and nonstarchy vegetables with *either* (1) animal protein, eggs, dairy, and nuts, *or* (2) grains, pasta, bread, beans, legumes, and starchy vegetables (potatoes, corn, squashes).

5. As much as possible, try to eat organic—especially for meat, poultry, dairy, pastured eggs, and the fruits and vegetables on the Dirty Dozen list.

6. Aim for thirty-two to sixty-four ounces of caffeine-free herbal tea and/or water (without carbonation) each day. You can jazz it up with a splash of lemon juice or raw apple cider vinegar.

7. Examine your caffeine intake. Especially if you have difficulties with sleeping or if you currently take antidepressants, ADHD drugs, or anti-anxiety medications, consider decreasing (or eliminating) caffeine consumption.

Sleep (see Chapter 17)

1. Aim to go to bed and get up at the same time every day, preferably before eleven p.m. and waking up before seven a.m.

2. Limit late-night snacking. If you work overnight shifts, try to avoid junk food. Instead, choose nutrient-dense foods with lots of healthy fats, protein, and resistant starch.

3. Get regular exercise but try to finish cardiovascular workouts at least three hours before bedtime—unless you are already experiencing hyperarousal, in which case twenty minutes of cardio followed by G&R can help you prepare for bed.

4. Avoid caffeine after two p.m., and avoid eating and drinking alcohol for at least three hours before you go to bed.

5. Aim to finish working, consuming the news, and engaging in stimulating conversations or arguments a few hours before bedtime. Aim to turn off all electronics within one hour before bedtime.

6. If you wake up in the night with nightmares or anxious rumination, get out of bed and do one cycle of G&R.

Exercise (see Chapter 17)

1. Aim to elevate your heart rate with aerobic activities at least thirty minutes each time, at least three times a week.

2. Aim to target each muscle group with weight training at least twice a week, although you can work core muscles every day.

3. Add in regular stretching, yoga, or tai chi to increase flexibility and dissipate tension in the body.

4. Across these three categories, aim for forty-five to sixty minutes of activity at least four times a week.

Social Connections (see Chapter 17)

1. Aim for at least a handful of people that you feel you could confide in and/or ask for help.

2. If you don't have a few people, it's time to grow your social support network, using the concrete tips in Chapter 17 that you can build into your weekly plan.

3. If you see yourself as self-sufficient, or you find it difficult to ask for help and lean on others for support, you might start by cultivating relationships where you offer support, rather than receive it. Helping other people can boost resilience just as much as receiving support.

ACKNOWLEDGMENTS

My heart is filled with gratitude as I reflect on the interconnected web that helped me birth this book. I've chosen not to include ranks and titles, because I am thanking people.

In part because this project grew from my own journey to wholeness, there are literally thousands of people who contributed to its manifestation, both directly and indirectly—such as through the roles they played in my life experiences with stress and trauma or, alternatively, with healing, growth, and connection. Space will not permit me to thank all of these people by name. Each one was a teacher to me, and I was indelibly touched by our contact. Without all of those experiences, this particular book would not have been possible. I appreciate the part they played in helping this book eventually come to be.

This book has been strongly shaped by my students, both in my Georgetown decision-making courses and at other MMFT teaching venues. For me, teaching is truly the most joyful form of learning. While for privacy's sake I cannot name them here, I have been deeply honored by the opportunity to witness—and help to facilitate—their window-widening journeys. I am grateful for their courage, curiosity, honesty, and commitment to growth, which inspires me to distill and share what's been so freely given to me. I thank them for their countless questions, which often prompted new ways to explain particular concepts or skills; and for the generous sharing of their stories, many of which help illuminate material in this book. I also thank the many participants at the various seminars, workshops, conferences, and media interviews where I have presented portions of this work over the years. The book benefited greatly from the comments and questions that these presentations elicited.

I'm deeply grateful to John Schaldach for his essential contributions to MMFT. His commitment to practice, savvy with all things technical, and willingness to help me digest the insights that emerged from the MMFT pilot study—and then generate with me the revised, manualized curriculum that resulted—was critical to MMFT's creation.

I thank my MMFT teaching colleagues who each dedicated significant time and resources to becoming MMFT Trainers. Their commitment to widening their own windows, while learning how to and then helping others do the same, is inspiring. I thank Luann Barndt, Meg Campbell-Dowling, Jeanne Cummings, Kathleen Cutshall, Mark Davies, Janet Durfee, Christine Frazita, Jane Grafton, Vajra Grinelli, Michael Hayduk, Ninette Hupp, Solwazi Johnson, Wynne Kinder, Sam Levy, Raz Mason, Colleen Mizuki, Elizabeth Mumford, Pat Roach, Tuere Sala, Jim Saveland, John Schaldach, Jared Smyser, Erin Treat, and Judith Vanderryn. I am especially indebted to Colleen and Erin for their efforts with training other trainers; and to John, Colleen, and Solwazi for serving as trainers in the MMFT research studies.

I feel deep gratitude for the many generous friends and colleagues who played a part in the nonprofit Mind Fitness Training Institute. In addition to the trainers already named, I thank Joe Burton, Mirabai Bush, Randi Cohen, Milica Cosic, Carrie

Getsinger, Dan Edelman, Bob Gallucci, Esra Hudson, Marcia Johnston, Brian Kelly, Shannon King, Fred Krawchuk, Trevor Messersmith, Whitney Poulin, Andy Powell, Tim Rosenberg, Holly Roth, Kristin Siebenacher, Jennifer Sims, Robert Skidmore, Mark Williams, and Cristin Zeiser. I am particularly indebted to our longest-serving volunteer executive board, including James Gimian, Robert Moser, Tammy Schultz, and Alan Schwartz, and our pro bono counsel Rob Begland. There really aren't words to express their exceptional contributions.

I also feel deep gratitude for the generous individuals who shared their wisdom, connections, reputations, and support in so many ways while serving on MFTI's Advisory Board, including Richard Davidson, Chuck Hagel, Richard Hearney, Judith Richards Hope, Jack Kornfield, Peter Levine, Richard Strozzi-Heckler, Tim Ryan, Loree Sutton, and Bessel van der Kolk. I'm fortunate to have such a wise and courageous group of friends and mentors. MFTI was also blessed by the generosity of the 1440 Foundation and countless donors, who helped support our work over the years.

Several individuals and organizations were critical in facilitating research about MMFT's efficacy. None of this could have happened without the 2008 pilot study, promoted by three things: my research collaboration with Amishi Jha; the intrepid participation of U.S. Marines from the 4th Air Naval Gunfire Liaison Company (AN-GLICO); and John and Tussi Kluge, who not only helped fund the pilot study when our initial funding fell through but also graciously provided a place for me to live during part of the study—and lead the Marines in a day of silent practice—at their Palm Beach home.

Since then, countless other people helped support MMFT research in various ways. Special thanks to participants in the research studies: the 4th ANGLICO Marines in the pilot study; soldiers in the U.S. Army's 25th Infantry Division in the 2010 study; Marines in the 1st Marine Expeditionary Force in the 2011 study; and Marines attending a course at the U.S. Marine Corps' School of Infantry–West in the 2013–2014 study. I also thank Amy Adler, Mark Bates, Kelly Bickel, Jeff Bearor, Mike Brumage, Marion Cain, Carl Castro, Kaye Coker, Jeff Davis, Chris Demuro, Frank DiGiovanni, Janet Hawkins, Mylene Huyhn, Tom Jones, Ryan Keating, Doug King, Kenn Knarr, Paul Lester, Clarke Lethin, Pat Martin, Bill McNulty, Nisha Money, Scott Naumann, Sam Newland, Eric Schoomaker, Jason Spitaletta, Peter Squire, Doug Todd, Jim Toth, Al Vigilante, and Stephen Xenakis. I am especially grateful for the pivotal support and guidance from Joseph Dunford, Rich Hearney, Dave Hodne, Walt Piatt, Tim Ryan, Mel Spiese, and Loree Sutton.

I have learned so much from my research collaboration with neuroscientists, clinicians, and stress researchers on these studies. I thank Sara Algoe, Toby Elliman, Lori Haase, Amishi Jha, Chris Johnson, Anastasia Kiyonaga, Brian Lakey, Tom Minor, Alexandra Morrison, Suzanne Parker, Martin Paulus, Traci Plumb, Nina Rostrup, Sarah Stearlace, Nate Thom, and Tony Zanesco.

Financial support for MMFT-related research came from the U.S. Department of Defense Centers of Excellence for Psychological Health and Traumatic Brain Injury;

the U.S. Army Medical Research and Materiel Command; the U.S. Office of Naval Research; the U.S. Naval Health Research Center; the U.S. Department of the Navy Bureau of Medicine and Surgery; and the John Kluge Foundation. I was also blessed with financial support for time away from teaching to conduct MMFT research and/or work on this book from Georgetown University; the Smith Richardson Foundation; and the Woodrow Wilson International Center for Scholars, where I spent a deeply enriching fellowship year among friends. Georgetown's Security Studies Program also provided financial support for research assistance, from Carrie Getsinger, Haotian Qi, Mart Stewart-Smith, and Lindsay Windsor.

There aren't really words to express my gratitude for the many amazing teachers, healers, clinicians, and fellow travelers on the path who either helped me widen my own window or supported my training, in various ways, in the two lineages from which MMFT springs. I feel truly blessed by many friendships among this group. In addition to people I've already named who belong here—you know who you are—I also thank Adyashanti, Ariyañani, Guy Armstrong, Pascal Auclair, Lois Bass, Lynn Bourbeau, Sarah Bowen, Barry Boyce, Tara Brach, Lynda Bradley, Jessica Briscoe-Coleman, Willoughby Britton, Jud Brewer, Barb Cargill, Berns Galloway, Joseph Goldstein, Bree Greenberg-Benjamin, Alicen Halquist, Dan Harris, Steve Hoskinson, Jeremy Hunter, Phyllis Jacobson-Kram, Gary Kaplan, Sara Lazar, Erika LeBaron, Laurie Leitch, Susan Lemak, Brian LeSage, Narayan Liebenson, Elaine Miller-Karas, Nancy Napier, Mariellen O'Hara, Gerry Piaget, Kristin Quigley, Evan Rabinowitz, Reginald Ray, Sharon Salzberg, Saki Santorelli, Cliff Saron, Naomi Schwiesow, Martin Skopp, Rodney Smith, Sayadaw U Indaka, Virañani, Joe Weston, and Carol Wilson.

I am especially grateful to my friends and mentors Jon Kabat-Zinn and Bessel van der Kolk. Jon pushed me early on to articulate the ethical framework upon which MMFT is built, gave me a chance to learn from his teaching genius when we taught together, and served as wise counsel during several growthful periods along the way. In addition to serving on MFTI's Advisory Board, Bessel generously shared clinical insights that helped me hone my understanding of the role of mindfulness in trauma extinction and offered to write the book's foreword.

I am grateful for my connection with the National Center for Faculty Development and Diversity, which helped me overcome my binge-or-bust writing habit (and leave keyboard vomiting behind forever!). I thank the whole NCFDD family, especially my fellow coaches and participants in my small groups, who inspire me to maintain my daily writing. I'm especially thankful for my current book-writing coach group— Tamara Beauboeuf, Marta Robertson, and Ilona Yim—who held my hand and offered excellent advice through this book's ups and downs.

I've also been blessed by generous friends and colleagues who offered detailed feedback on earlier draft chapters or related manuscripts. I thank Ty Flinton, Jim Gimian, Bree Greenberg-Benjamin, LeNaya Hazel, Bruce Hoffman, Charles King, Laurie Leitch, Rose McDermott, Colleen Mizuki, Elizabeth Mumford, Mary Stewart-Smith, Loree Sutton, Ariane Tabatabai, Kenton Thibaut, Kate Hendricks Thomas, Claire

Wings, Jennifer Woolard, and Ilona Yim. I'm deeply thankful for Stephanie Tade's kindness and the special role she played early in this book's conceptualization. I've also honed this book's argument greatly through my enjoyable research and writing collaboration with Kelsey Larsen on several related projects.

Special thanks to Caroline Sutton, my editor at Avery, and the whole Avery team for their enthusiastic support for this project. Caroline's insightful guidance, candor, and patience during our journey together were pivotal to this book becoming what it has. In addition, the Cosmos surely gifted me with a marvelous agent, and an even better friend, in Lauren Sharp. I'm deeply grateful to Lauren and the whole Aevitas team for believing in this book, helping me find its voice and supporting me through its many twists and turns. Finally, words cannot express the appreciation I feel for the editing brilliance and sense of humor of my dear friend Beth Blaufuss, who deftly helped me love this book into being and arrive somewhere near word count.

I'm blessed with a true wealth of friends, neighbors, special beings, and special places that have had my (and Chloe's) back and supported my heart as this book was being born, especially Alan, Ari, Audrey, Beth, Betsi, Bob, Bree, Bruce, Claire, Dennis, Ellis, Jennifer, Jessica, Joe, Judy, Kelsey, Kenton, Kristin, Liz, Loree, Marty, Mayra, Meredith, Milica, Naomi, Phyllis, Rhonda, Rob, Rose, and Tammy. I'm deeply grateful for my connection to Blue Ridge Farm—where some of this book's most joyful writing happened—and for the deep support from the plants, animals, and people who live there, particularly Leslie and Michael. Special thanks to Chloe, my loyal friend, for generating one of my favorite stories in this book and teaching me so much about trauma extinction in animals—and by extension, in humans.

Finally, I want to express my love to my family, to whom this book is dedicated—my parents, Cissie and Deane, and my sisters, Alison and Karalyn. Although I'm sad that Alzheimer's prevents Cissie from understanding that this book is finally finished, I'm truly grateful for the journey that we've all shared together as a family. With deep respect and gratitude, I bow to both of my family lineages—the warriors and the intergenerational trauma—and release them to the wind.

NOTES

CHAPTER 1

12 **Our brain was designed:** Pat Ogden, Kekuni Minton, and Clare Pain, *Trauma and the Body: A Sensorimotor Approach to Psychotherapy* (New York: Norton, 2006), chap. 1.

13 **One of the survival brain's most important functions:** Stephen W. Porges, *The Polyvagal Theory: Neurophysiological Foundations of Emotions, Attachment, Communication, and Self-Regulation* (New York: Norton, 2011), chap. 1.

14 **Trauma is especially likely to result:** Robert C. Scaer, *The Trauma Spectrum: Hidden Wounds and Human Resiliency* (New York: Norton, 2005), 205; Bruce S. McEwen and Elizabeth Norton Lasley, *The End of Stress as We Know It* (Washington, D.C.: Joseph Henry, 2002), chap. 1.

16 **59 percent of American adults:** American Psychological Association, "Stress in America: State of Our Nation" (2017), 1.

16 **Beyond these human conflicts:** G. Ceballos, P. R. Ehrlich, and R. Dirzo, "Biological Annihilation via the Ongoing Sixth Mass Extinction Signaled by Vertebrate Population Losses and Declines," *Proceedings of the National Academy of Sciences* 114, no. 30 (2017): E6089–E6096.

17 **The United States experienced more than 1,500 mass shootings:** German Lopez, "America's Unique Gun Violence Problem, Explained in 17 Maps and Charts," *Vox* (November 5, 2017), www.vox.com/policy-and-politics/2017/10/2/16399418/us-gun -violence-statistics-maps-charts; Nurith Aizenman, "Gun Violence: Comparing the U.S. to Other Countries," NPR, *Morning Edition* (November 6, 2017), www.npr.org /sections/goatsandsoda/2017/11/06/562323131/gun-violence-in-US.

17 **Only Yemen:** Max Fisher and Josh Keller, "Only One Thing Explains Mass Shootings in the United States," *New York Times*, November 8, 2017, A15.

17 **U.S. residents experienced 5.7 million violent victimizations:** Rachel E. Morgan and Grace Kena, "Criminal Victimization, 2016," Bureau of Justice Statistics (Department of Justice, 2017).

17 **the world's highest incarceration rate:** Peter Wagner and Bernadette Rabuy, "Mass Incarceration: The Whole Pie 2017" (Prison Policy Initiative, 2017), www.prisonpolicy .org/reports/pie2017.html; "World Prison Populations," news.bbc.co.uk/2/shared/spl/hi /uk/06/prisons/html/nn2page1.stm.

17 **Between 4 to 6 percent of men:** D. G. Kilpatrick et al., "National Estimates of Exposure to Traumatic Events and PTSD Prevalence Using DSM-IV and DSM-5 Criteria," *Journal of Traumatic Stress* 26, no. 5 (2013): 537–547; J. J. Fulton et al., "The Prevalence of Posttraumatic Stress Disorder in Operation Enduring Freedom/Operation Iraqi Freedom (OEF/OIF) Veterans: A Meta-Analysis," *Journal of Anxiety Disorders* 31 (2015): 98–107; B. P. Dohrenwend et al., "The Psychological Risks of Vietnam for U.S. Veterans: A Revisit with New Data and Methods," *Science* 313, no. 5789 (2006): 979–982; H. S. Resnick et al., "Prevalence of Civilian Trauma and Posttraumatic Stress Disorder in a Representative National Sample of Women," *Journal of Consulting and Clinical Psychology* 61, no. 6 (1993): 984–991; Sandra L. Bloom and Michael Reichert, *Bearing Witness: Violence and Collective Responsibility* (Binghamton, NY: Haworth, 1998), chap. 1.

17 **When someone has PTSD and one of these other conditions:** J. R. Cougle, H. Resnick, and D. G. Kilpatrick, "PTSD, Depression, and Their Comorbidity in Relation to Suicidality: Cross-Sectional and Prospective Analyses of a National Probability Sample of Women," *Depression and Anxiety* 26, no. 12 (2009): 1151–1157; I. R. Galatzer-Levy et al., "Patterns of Lifetime PTSD Comorbidity: A Latent Class Analysis," *Depression and Anxiety* 30, no. 5 (2013): 489–496.

18 **About one quarter of American adults:** R. C. Kessler et al., "Lifetime Prevalence and Age-of-Onset Distributions of Mental Disorders in the World Health Organization's

World Mental Health Survey Initiative," *World Psychiatry* 6, no. 3 (2007): 168–176; W. C. Reeves et al., "Mental Illness Surveillance among Adults in the United States," *Morbidity and Mortality Weekly Report* 60, no. 3 (2001): 1–32; M. K. Nock et al., "Cross-National Analysis of the Associations among Mental Disorders and Suicidal Behavior: Findings from the WHO World Mental Health Surveys," *PLOS Medicine* 6, no. 8 (2009), e1000123.

18 **Today, the lifetime rate of major depression:** Kessler et al., "Lifetime Prevalence and Age-of-Onset Distributions"; R. C. Kessler et al., "Anxious and Non-Anxious Major Depressive Disorder in the World Health Organization World Mental Health Surveys," *Epidemiology and Psychiatric Sciences* 24, no. 3 (2015): 210–226; L. J. Andrade et al., "The Epidemiology of Major Depressive Episodes: Results from the International Consortium of Psychiatric Epidemiology (ICPE) Surveys," *International Journal of Methods in Psychiatric Research* 12 (2003): 3–21; J. M. Twenge et al., "Birth Cohort Increases in Psychopathology among Young Americans, 1938–2007: A Cross-Temporal Meta-Analysis of the MMPI," *Clinical Psychology Review* 30 (2010): 145–154; Nock et al., "Cross-National Analysis"; Scott Stossel, *My Age of Anxiety: Fear, Hope, Dread, and the Search for Peace of Mind* (New York: Vintage, 2014), 213; S. Lee et al., "Lifetime Prevalence and Inter-Cohort Variation in DSM-IV Disorders in Metropolitan China," *Psychological Medicine* 37 (2007): 61–71; B. H. Hidaka, "Depression as a Disease of Modernity: Explanations for Increasing Prevalence," *Journal of Affective Disorders* 140, no. 3 (2012): 205–214.

18 **in large part because they seek relief for symptoms:** P. S. Wang et al., "Twelve Month Use of Mental Health Services in the United States," *Archives of General Psychiatry* 62, no. 6 (2005): 629–640; Jean M. Twenge, *Generation Me: Why Today's Young Americans Are More Confident, Assertive, Entitled—and More Miserable Than Ever Before, Revised and Updated* (New York: Atria, 2014), 140–143; Stossel, *My Age of Anxiety*, 300–301; R. C. Kessler et al., "Prevalence, Severity, and Comorbidity of Twelve-Month DSM-IV Disorders in the National Comorbidity Survey Replication (NCS-R)," *Archives of General Psychiatry* 62, no. 6 (2005): 617–627; Kessler et al., "Anxious and Non-Anxious Major Depressive Disorder"; R. C. Kessler et al., "Lifetime Prevalence and Age-of-Onset Distributions of DSM-IV Disorders in the National Comorbidity Survey Replication (NCS-R)," *Archives of General Psychiatry* 62, no. 6 (2005): 593–602; Benoit Denizet-Lewis, "The Kids Who Can't," *New York Times Magazine*, October 15, 2017.

18 **One third of Americans have abused or been dependent on alcohol:** Steven M. Southwick and Dennis S. Charney, *Resilience: The Science of Mastering Life's Greatest Challenges* (Cambridge, UK: Cambridge University Press, 2012), 14; L. Saad, "Few Americans Meet Exercise Targets: Self-Reported Rates of Physical Exercise Show Little Change since 2001," January 1, 2008, www.gallup.com/poll/103492/few-americans-meet-exercise-targets.aspx; D. S. Hasin, F. S. Stinson, E. Ogburn, and B. F. Grant, "Prevalence, Correlates, Disability, and Comorbidity of DSM-IV Alcohol Abuse and Dependence in the United States: Results from the National Epidemiologic Survey on Alcohol and Related Conditions," *Archives of General Psychiatry* 64, no. 7 (2007): 830–842.

18 **since 2000, emergency room visits:** Gabrielle Glaser, "America, It's Time to Talk about Your Drinking," *New York Times*, December 31, 2017.

18 **we consume 75 percent of the world's prescriptions:** Stossel, *My Age of Anxiety*, 176, 97; Benedict Carey and Robert Gebeloff, "The Murky Perils of Quitting Antidepressants after Years of Use," *New York Times*, April 8, 2018; National Institute on Drug Abuse, "Popping Pills: Prescription Drug Abuse in America," www.drugabuse.gov/related-topics /trends-statistics/infographics/popping-pills-prescription-drug-abuse-in-america; U.S. Substance Abuse and Mental Health Services Administration, "Results from the 2011 National Survey on Drug Use and Health: Summary of National Findings," NSDUH Series H-44 (Rockville, MD: Substance Abuse and Mental Health Services Administration, 2012); National Center for Health Statistics, "Health, United States, 2013: With Special Feature on Prescription Drugs" (Hyattsville, MD: U.S. Government

Printing Office, 2014); United Nations Office on Drugs and Crime, "World Drug Report" (New York: United Nations, 2011).

19 **the United States also has the world's highest drug-death rate:** Amanda Erickson, "Opioid Abuse in the U.S. Is So Bad It's Lowering Life Expectancy. Why Hasn't the Epidemic Hit Other Countries?," *Washington Post*, December 28, 2017; Lenny Bernstein, "U.S. Life Expectancy Declines Again, a Dismal Trend Not Seen since World War I," *Washington Post*, November 29, 2018; Josh Katz, "Just How Bad Is the Drug Overdose Epidemic?" *New York Times*, October 26, 2017, www.nytimes.com/interactive/2017/04/14 /upshot/drug-overdose-epidemic-you-draw-it.html.

19 **suicide has attracted attention:** Dan Keating and Lenny Bernstein, "U.S. Suicide Rate Has Risen Sharply in the 21st Century," *Washington Post*, April 22, 2016; Adeel Hassan, "Deaths from Drugs and Suicide Reach a Record in U.S.," *New York Times*, March 7, 2019; Bernstein, "U.S. Life Expectancy Declines Again"; Anne Case and Angus Deaton, "Mortality and Morbidity in the 21st Century," *Brookings Papers on Economic Activity* (Spring 2017): 397–476.

19 **Over the last decade, hospital admissions for suicidal teenagers:** Denizet-Lewis, "The Kids Who Can't"; Jamie Ducharme, "More Than 90% of Generation Z Is Stressed Out. And Gun Violence Is Partly to Blame," *Time*, October 30, 2018.

19 **American adults also report subjectively feeling more stressed and more anxious:** American Psychological Association, "Stress in America" (2010), 5–12; Stossel, *My Age of Anxiety*, 300; Seth Stephens-Davidowitz, "Fifty States of Anxiety," *New York Times*, August 7, 2016.

20 **Consider these U.S. lifestyle indicators:** American Psychological Association, "Stress in America;" American Psychological Association, "Stress in America: State of Our Nation"; Southwick and Charney, *Resilience*, 14; National Sleep Foundation, "Sleep in America Poll: Summary of Findings" (2009); Hidaka, "Depression as a Disease of Modernity."

20 **To explain these rising anxiety and depression rates:** Hidaka, "Depression as a Disease of Modernity"; Sebastian Junger, *Tribe: On Homecoming and Belonging* (New York: Twelve Books, 2016), 18–23; Lee et al., "Lifetime Prevalence and Inter-Cohort Variation in DSM-IV Disorders in Metropolitan China"; W. A. Vega et al., "12-Month Prevalence of DSM-III-R Psychiatric Disorders among Mexican Americans: Nativity, Social Assimilation, and Age Determinants," *Journal of Nervous and Mental Disease* 192 (2004): 532–541; J. Colla et al., "Depression and Modernization," *Social Psychiatry and Psychiatric Epidemiology* 41, no. 4 (2006): 271–279; J. Peen et al., "The Current Status of Urban-Rural Differences in Psychiatric Disorders," *Acta Psychiatrica Scandinavica* 121 (2010): 84–93.

23 **we can regulate our stress levels upward or downward:** Ogden et al., *Trauma and the Body*, chap. 2.

CHAPTER 2

30 **being "gritty" . . . is a good predictor of eventual *external* success:** A. Duckworth et al., "Grit: The Perseverance and Passion for Long-Term Goals," *Journal of Personality and Social Psychology* 92, no. 6 (2007): 1087–1101; L. Eskreis-Winkler et al., "The Grit Effect: Predicting Retention in the Military, the Workplace, School and Marriage," *Frontiers in Personality Science and Individual Differences* 5, no. 36 (2014): 1–12; Angela L. Duckworth, *Grit: The Power of Passion and Perseverance* (New York: Scribner, 2016).

34 **Child-centered, labor-intensive, and financially expensive parenting:** Claire Cain Miller, "Stress, Exhaustion and Guilt: Modern Parenting," *New York Times*, December 25, 2018.

35 **we're collectively reluctant to accept poverty, abuse, discrimination:** Terrence Real, *I Don't Want to Talk about It: Overcoming the Secret Legacy of Male Depression* (New York: Simon and Schuster, 1998), 104–107.

36 **Poor Americans . . . are roughly five times as likely to report "fair" or "poor" health:** Steven H. Woolf et al., *How Are Income and Wealth Linked to Health and Longevity?* (Urban Institute, 2015), www.urban.org/sites/default/files/publication/49116/2000178 -How-are-Income-and-Wealth-Linked-to-Health-and-Longevity.pdf; J. S. Schiller, J. W. Lucas, and J. A. Peregoy, "Summary Health Statistics for U.S. Adults: National Health Interview Survey, 2011," *Vital Health and Statistics* 10 (2012): 256; A. Case and A. Deaton, "Mortality and Morbidity in the 21st Century," *Brookings Papers on Economic Activity* (Spring 2017): 397–476.

36 **perceived sexism has been linked empirically with depression:** R. J. Hurst and D. Beesley, "Perceived Sexism, Self-Silencing, and Psychological Distress in College Women," *Sex Roles* 68, no. 5 (2013): 311–320; A. N. Zucker and L. J. Landry, "Embodied Discrimination: The Relation of Sexism and Distress to Women's Drinking and Smoking Behaviors," *Sex Roles* 56, no. 3–4 (2007): 193–203; J. K. Swim et al., "Everyday Sexism: Evidence for Its Incidence, Nature, and Psychological Impact from Three Daily Diary Studies," *Journal of Social Issues* 57 (2001): 31–53; E. A. Klonoff, H. Landrine, and R. Campbell, "Sexist Discrimination May Account for Well-Known Gender Differences in Psychiatric Symptoms," *Psychology of Women Quarterly* 24 (2000): 93–99.

36 **In experimental research . . . researchers examined women's stress hormones:** S. S. M. Townsend et al., "From 'In the Air' to 'Under the Skin': Cortisol Responses to Social Identity Threat," *Personality and Social Psychology Bulletin* 37, no. 2 (2011): 151–164.

36 **Heterosexism . . . has been shown to contribute to psychological distress:** S. E. Bonds, "Shame Due to Heterosexism, Self-Esteem and Perceived Stress: Correlates of Psychological Quality of Life in a Lesbian, Gay and Bisexual Sample," master's thesis (University of North Texas, 2015); A. L. Roberts et al., "Pervasive Trauma Exposure among U.S. Sexual Orientation Minority Adults and Risk of Posttraumatic Stress Disorder," *American Journal of Public Health* 100 (2010): 2433–2441; M. R. Woodford et al., "Contemporary Heterosexism on Campus and Psychological Distress among LGBQ Students: The Mediating Role of Self-Acceptance," *American Journal of Orthopsychiatry* 84, no. 5 (2014): 519–529; K. T. Straub, A. A. McConnell, and T. L. Messman-Moore, "Internalized Heterosexism and Posttraumatic Stress Disorder Symptoms: The Mediating Role of Shame Proneness among Trauma-Exposed Sexual Minority Women," *Psychology of Sexual Orientation and Gender Diversity* 5, no. 1 (2018): 99–108.

37 **sexual minorities who experience workplace heterosexism:** C. R. Waldo, "Working in a Majority Context: A Structural Model of Heterosexism as Minority Stress in the Workplace," *Journal of Counseling Psychology* 46, no. 2 (1999): 218–232.

37 **Racism is significantly associated with poorer health . . . [and] economic injustice:** Y. Paradies et al., "Racism as a Determinant of Health: A Systematic Review and Meta-Analysis," *PLOS One* 10, no. 9 (2015): e0138511; Paul F. Campos, "White Economic Privilege Is Alive and Well," *New York Times,* July 30, 2017.

37 **Researchers examined stress arousal among Latinas:** P. J. Sawyer et al., "Discrimination and the Stress Response: Psychological and Physiological Consequences of Anticipating Prejudice in Interethnic Interactions," *American Journal of Public Health* 102, no. 5 (2012): 1020–1026.

38 **write down five adjectives that describe how you see yourself:** Rodney Smith, *Awakening: A Paradigm Shift of the Heart* (Boston: Shambala, 2014), 66–67.

40 **girls and women are taught . . . that anger is not a suitable emotion:** Leslie Jamison, "I Used to Insist I Didn't Get Angry. Not Anymore," *New York Times Magazine,* January 17, 2018.

40 **boys and men are taught to be competitive and aggressive:** Real, *I Don't Want to Talk about It*; Jackson Katz, *Macho Paradox: Why Some Men Hurt Women and How All Men Can Help* (Naperville, IL: Sourcebooks, 2006); Sandra L. Bloom and Michael Reichert, *Bearing Witness: Violence and Collective Responsibility* (Binghamton, NY: Haworth, 1998), 34–37; H. Braswell and H. I. Kushner, "Suicide, Social Integration, and Masculinity in

the U.S. Military," *Social Science and Medicine* 74, no. 4 (2012): 530–536; G. Green et al., "Exploring the Ambiguities of Masculinity in Accounts of Emotional Distress in the Military among Young Ex-Servicemen," *Social Science and Medicine* 71, no. 8 (2010): 1480–1488; R. P. Auerbach, J. R. Z. Abela, and M. R. Ho, "Responding to Symptoms of Depression and Anxiety: Emotion Regulation, Neuroticism, and Engagement in Risky Behaviors," *Behaviour Research and Therapy* 45, no. 9 (2007): 2182–2191; A. L. Teten et al., "Intimate Partner Aggression Perpetrated and Sustained by Male Afghanistan, Iraq, and Vietnam Veterans with and without Posttraumatic Stress Disorder," *Journal of Interpersonal Violence* 25, no. 9 (2010): 1612–1630; E. B. Elbogen et al., "Violent Behavior and Post-Traumatic Stress Disorder in U.S. Iraq and Afghanistan Veterans," *British Journal of Psychiatry* 204, no. 5 (2014): 368–375; American Psychological Association, "APA Guidelines for Psychological Practice with Boys and Men" (August 2018), www .apa.org/about/policy/psychological-practice-boys-men-guidelines.pdf.

41 **Terry Real calls this "the great divide":** Terrence Real, "The Long Shadow of Patriarchy: Couples Therapy in the Age of Trump," *Psychotherapy Networker* (September /October 2017), www.psychotherapynetworker.org/magazine/article/1112/the-long -shadow-of-patriarchy.

41 **A large body of empirical research . . . links somatization:** B. A. van der Kolk et al., "Dissociation, Somatization, and Affect Dysregulation: The Complexity of Adaptation to Trauma," *American Journal of Psychiatry* 153, no. 7 (1996): 83–93; C. W. Hoge et al., "Association of Posttraumatic Stress Disorder with Somatic Symptoms, Health Care Visits, and Absenteeism among Iraq War Veterans," *American Journal of Psychiatry* 164, no. 1 (2007): 150–153; M. M. Lilly et al., "Gender and PTSD: What Can We Learn from Female Police Officers?," *Journal of Anxiety Disorders* 23, no. 6 (2009): 767–774; M. A. Hom et al., "The Association between Sleep Disturbances and Depression among Firefighters: Emotion Dysregulation as an Explanatory Factor," *Journal of Clinical Sleep Medicine* 12, no. 2 (2016): 235–245; W. D. S. Killgore et al., "The Effects of Prior Combat Experience on the Expression of Somatic and Affective Symptoms in Deploying Soldiers," *Journal of Psychosomatic Research* 60, no. 4 (2006): 379–385; K. B. Koh et al., "The Relation between Anger Expression, Depression, and Somatic Symptoms in Depressive Disorders and Somatoform Disorders," *Journal of Clinical Psychiatry* 66, no. 4 (2005): 485–491; Y. I. Nillni et al., "Deployment Stressors and Physical Health among OEF/OIF Veterans: The Role of PTSD," *Health Psychology* 33, no. 11 (2014): 1281–1287; J. C. Shipherd et al., "Sexual Harassment in the Marines, Posttraumatic Stress Symptoms, and Perceived Health: Evidence for Sex Differences," *Journal of Traumatic Stress* 22, no. 1 (2009): 3–10; G. J. G. Asmundson, K. D. Wright, and M. B. Stein, "Pain and PTSD Symptoms in Female Veterans," *European Journal of Pain* 8, no. 4 (2004): 345–350.

42 **Somatization has also been linked with suicidal behavior:** R. A. Bernert et al., "Sleep Disturbances as an Evidence-Based Suicide Risk Factor," *Current Psychiatry Reports* 17, no. 3 (2015): 1–9; E. B. Elbogen et al., "Risk Factors for Concurrent Suicidal Ideation and Violent Impulses in Military Veterans," *Psychological Assessment* 30, no. 4 (2017): 425–435; V. Vargas de Barros et al., "Mental Health Conditions, Individual and Job Characteristics and Sleep Disturbances among Firefighters," *Journal of Health Psychology* 18, no. 3 (2012): 350–358; J. D. Ribeiro et al., "Sleep Problems Outperform Depression and Hopelessness as Cross-Sectional and Longitudinal Predictors of Suicidal Ideation and Behavior in Young Adults in the Military," *Journal of Affective Disorders* 136, no. 3 (2012): 743–750; J. D. Ribeiro et al., "An Investigation of the Interactive Effects of the Capability for Suicide and Acute Agitation on Suicidality in a Military Sample," *Depression and Anxiety* 32, no. 1 (2015): 25–31; D. D. Luxton et al., "Prevalence and Impact of Short Sleep Duration in Redeployed OIF Soldiers," *Sleep* 34, no. 9 (2011): 1189–1195.

42 **they may perceive less stigma for seeking help for physical . . . concerns:** Killgore et al., "The Effects of Prior Combat Experience"; Hoge et al., "Association of

Posttraumatic Stress Disorder"; T. M. Greene-Shortridge, T. W. Britt, and C. A. Castro, "The Stigma of Mental Health Problems in the Military," *Military Medicine* 172, no. 2 (2007): 157–161; A. M. Berg et al., "An Exploration of Job Stress and Health in the Norwegian Police Service: A Cross Sectional Study," *Journal of Occupational Medicine and Toxicology* 1, no. 26 (2006): 115–160; I. H. Stanley, M. A. Hom, and T. E. Joiner, "A Systematic Review of Suicidal Thoughts and Behaviors among Police Officers, Firefighters, EMTs, and Paramedics," *Clinical Psychology Review* 44 (2016): 25–44; C. J. Bryan et al., "Understanding and Preventing Military Suicide," *Archives of Suicide Research* 16, no. 2 (2012): 95–110.

42 **a 2003 study with 82nd Airborne Division soldiers:** Killgore et al., "The Effects of Prior Combat Experience."

43 **recent research finds that toxic, unethical workers enjoy longer tenures:** Michael Housman and Dylan Minor, "Toxic Workers: Working Paper 16-057," *Harvard Business School Working Papers* (Cambridge, 2015); L. Pierce and J. A. Snyder, "Unethical Demand and Employee Turnover," *Journal of Business Ethics* 131, no. 4 (2015): 853–869.

43 **David Petraeus . . . once told reporters he "rarely feels stress at all":** Scott Stossel, *My Age of Anxiety: Fear, Hope, Dread, and the Search for Peace of Mind* (New York: Vintage, 2014), 28.

44 **Western civilization is replete with stories of the heroic individual:** Ian Watt, *Myths of Modern Individualism: Faust, Don Quixote, Don Juan, Robinson Crusoe* (Cambridge, UK: Cambridge University Press, 1996); Peter L. Callero, *The Myth of Individualism: How Social Forces Shape Our Lives, Third Edition* (Lanham, MD: Rowman and Littlefield, 2018), 22–23.

45 **This preference for quantified information:** Edward Stewart and Milton Bennett, *American Cultural Patterns: A Cross-Cultural Perspective* (Yarmouth, ME: Intercultural, 1991), 41–42; Thomas S. Kuhn, *The Structure of Scientific Revolutions*, 3rd ed. (Chicago: University of Chicago Press, 1996); Quentin J. Schultze, *Habits of the High-Tech Heart: Living Virtuously in the Information Age* (Grand Rapids, MI: Baker Academic, 2002), 31–37; Antonio Damasio, *Descartes' Error: Emotion, Reason, and the Human Brain* (New York: Penguin, 1994); Morris Berman, *The Reenchantment of the World* (Ithaca, NY: Cornell University Press, 1981), 28.

45 **our cultural preference for knowledge derived from "rational" thought:** Stewart and Bennett, *American Cultural Patterns*, 32–37; Theodore M. Porter, *Trust in Numbers: The Pursuit of Objectivity in Science and Public Life* (Princeton, NJ: Princeton University Press, 1995).

46 **none to date has shown empirical efficacy:** Laura Aiuppa Denning, Marc Meisnere, and Kenneth E. Warner, *Preventing Psychological Disorders in Service Members and Their Families: An Assessment of Programs* (Washington, D.C.: National Academies Press, 2014); S. L. Smith, "Could Comprehensive Soldier Fitness Have Iatrogenic Consequences? A Commentary," *Journal of Behavioral Health Services and Research* 40, no. 2 (2013): 242–246; M. M. Steenkamp, W. P. Nash, and B. T. Litz, "Post-Traumatic Stress Disorder: Review of the Comprehensive Soldier Fitness Program," *American Journal of Preventive Medicine* 44, no. 5 (2013): 507–512; Institute of Medicine, *Preventing Psychological Disorders in Service Members and Their Families: An Assessment of Programs* (Washington, D.C.: National Academies Press, 2014); Roy Eidelson and Stephen Soldz, "Does Comprehensive Soldier Fitness Work: CSF Research Fails the Test," working paper no. 1 (Bala Cynwyd, PA: Coalition for an Ethical Psychology, 2012); C. A. Vaughan et al., *Evaluation of the Operational Stress Control and Readiness (OSCAR) Program* (Santa Monica, CA: RAND, 2015).

47 **they may internalize that ineffectiveness as a "problem" with themselves:** Braswell and Kushner, "Suicide, Social Integration, and Masculinity in the US Military"; K. L. Larsen and E. A. Stanley, "Conclusion: The Way Forward," in *Bulletproofing the Psyche: Preventing Mental Health Problems in Our Military and Veterans*, edited by Kate Hendricks

Thomas and David Albright (Santa Barbara, CA: Praeger, 2018), 233–253; Smith, "Could Comprehensive Soldier Fitness Have Iatrogenic Consequences?"

CHAPTER 3

53 **The first institutionalized PT programs:** Andrew J. Thompson, "Physical Fitness in the United States Marine Corps: History, Current Practices, and Implications for Mission Accomplishment and Human Performance," master's thesis (Naval Postgraduate School, 2005), 8–9; Michael D. Krause, "History of U.S. Army Soldier Physical Fitness" (paper presented at the National Conference on Military Physical Fitness—Proceedings Report, Washington, D.C., 1990); Whitfield B. East, *A Historical Review and Analysis of Army Physical Readiness Training and Assessment* (Fort Leavenworth, KS: Combat Studies Institute Press, 2013), 25–39.

53 **in 1906, the Army mandated servicewide requirements:** Krause, "History of U.S. Army Soldier Physical Fitness"; East, *A Historical Review*, 38–44.

53 **Roosevelt issued an Executive Order:** Krause, "History of U.S. Army Soldier Physical Fitness"; Thompson, "Physical Fitness in the United States Marine Corps," 9–10; East, *A Historical Review*, 42–44.

53 **military training during both wars:** East, *A Historical Review*, 49–72, 79–93; Krause, "History of U.S. Army Soldier Physical Fitness"; Thompson, "Physical Fitness in the United States Marine Corps," 13–17; L. C. Dalleck and L. Kravitz, "The History of Fitness," *Idea, Health, and Fitness Source* (January 2002): 26–30.

54 **This society-wide campaign . . . culminated during the Kennedy administration:** Dalleck and Kravitz, "The History of Fitness"; H. M. Barrow, and J. P. Brown, *Man and Movement: Principles of Physical Education*, 4th ed. (Philadelphia: Lea and Febiger, 1988).

55 **the well-documented theories of neuroplasticity:** Jeffrey M. Schwartz and Sharon Begley, *The Mind and the Brain: Neuroplasticity and the Power of Mental Force* (New York: HarperPerennial, 2003).

55 **Neuroscientists now know that the brain changes throughout our lifetime:** Norman Doidge, *The Brain That Changes Itself* (New York: Penguin, 2007).

56 **the memory and brain structures of London taxi drivers:** E. A. Maguire et al., "Navigation-Related Structural Change in the Hippocampi of Taxi Drivers," *Proceedings of the National Academy of Sciences* 97 (2000): 4398–4403; E. A. Maguire et al., "Navigation Expertise and the Human Hippocampus: A Structural Brain Imaging Analysis," *Hippocampus* 13, no. 2 (2003): 250–259.

56 **areas of the brain may shrink or expand:** Schwartz and Begley, *The Mind and the Brain*.

56 **As science writer Sharon Begley beautifully expresses it:** Sharon Begley, *Train Your Mind, Change Your Brain: How a New Science Reveals Our Extraordinary Potential to Transform Ourselves* (New York: Ballantine, 2007), 8–9.

56 **as worrying becomes a habit:** B. K. Hölzel et al., "Stress Reduction Correlates with Structural Changes in the Amygdala," *Social Cognitive and Affective Neuroscience* 5, no. 1 (2010): 11–17.

57 **prolonged stress, trauma, depression, anxiety, and PTSD are all associated with declines in our cognitive performance:** B. P. Marx, S. Doron-Lamarca, S. P. Proctor, and J. J. Vasterling, "The Influence of Pre-Deployment Neurocognitive Functioning on Post-Deployment PTSD Symptom Outcomes among Iraq-Deployed Army Soldiers," *Journal of the International Neuropsychological Society* 15, no. 6 (2009): 840–852; S. Kuhlman, M. Piel, and O. T. Wolf, "Impaired Memory Retrieval after Psychological Stress in Healthy Young Men," *Journal of Neuroscience* 25, no. 11 (2005): 2977–2982; J. Douglas Bremner, *Does Stress Damage the Brain? Understanding Trauma-Related Disorders from a Mind-Body Perspective* (New York: Norton, 2005).

58 **brain imaging studies show that people diagnosed with PTSD:** J. D. Bremner et al., "MRI-Based Measurement of Hippocampal Volume in Patients with Combat-Related

Posttraumatic Stress Disorder," *American Journal of Psychiatry* 152 (1995): 973–981; J. D. Bremner et al., "Magnetic Resonance Imaging–Based Measurement of Hippocampal Volume in Posttraumatic Stress Disorder Related to Childhood Physical and Sexual Abuse—A Preliminary Report," *Biological Psychiatry* 41 (1997): 23–32; R. K. Pitman, "Hippocampal Diminution in PTSD: More (or Less?) Than Meets the Eye," *Hippocampus* 11, no. 73–74 (2001): 82–174; R. J. Lindauer et al., "Smaller Hippocampal Volume in Dutch Police Officers with Posttraumatic Stress Disorder," *Biological Psychiatry* 56 (2004): 356–363; M. Vythilingam et al., "Smaller Head of the Hippocampus in Gulf War–Related Posttraumatic Stress Disorder," *Psychiatry Research: Neuroimaging* 139 (2005): 89–99.

58 **one large study of U.S. Army soldiers:** J. J. Vasterling et al., "Neuropsychological Outcomes of Army Personnel Following Deployment to the Iraq War," *Journal of the American Medical Association* 296, no. 5 (2006): 519–529.

58 **troops who participated in MMFT before their combat deployments:** A. P. Jha et al., "Examining the Protective Effects of Mindfulness Training on Working Memory Capacity and Affective Experience," *Emotion* 10, no. 1 (2010): 54–64; A. P. Jha et al., "Minds 'at Attention': Mindfulness Training Curbs Attentional Lapses in Military Cohorts," *PLOS One* 10, no. 2 (2015): e0116889; A. P. Jha, A. B. Morrison, S. C. Parker, and E. A. Stanley, "Practice Is Protective: Mindfulness Training Promotes Cognitive Resilience in High-Stress Cohorts," *Mindfulness* 8, no. 1 (2017): 46–58; A. P. Jha et al., "Short-Form Mindfulness Training Protects against Working-Memory Degradation over High-Demand Intervals," *Journal of Cognitive Enhancement* 1, no. 2 (2017): 154–171; L. Haase et al., "Mindfulness-Based Training Attenuates Insula Response to an Aversive Interoceptive Challenge," *Social Cognitive and Affective Neuroscience* 11, no. 1 (2016): 182–190; D. C. Johnson et al., "Modifying Resilience Mechanisms in at-Risk Individuals: A Controlled Study of Mindfulness Training in Marines Preparing for Deployment," *American Journal of Psychiatry* 171, no. 8 (2014): 844–853.

58 **forty years of empirical scientific research have documented a range of neuroplastic benefits:** Jon Kabat-Zinn, *Full Catastrophe Living (Revised Edition): Using the Wisdom of Your Body and Mind to Face Stress, Pain, and Illness* (New York: Bantam, 2013); Begley, *Train Your Mind, Change Your Brain.*

59 **Research with rodents shows that voluntary exercise:** Begley, *Train Your Mind, Change Your Brain,* 67–68; Gretchen Reynolds, "Jogging Your Brain," *New York Times Magazine,* April 22, 2012; M. S. Nokia et al., "Physical Exercise Increases Adult Hippocampal Neurogenesis in Male Rats Provided It Is Aerobic and Sustained," *Journal of Physiology* 594, no. 7 (2016): 1855–1873.

59 **Humans experience these neuroplastic benefits from voluntary physical exercise:** C. H. Hillman, K. I. Erickson, and A. F. Kramer, "Be Smart, Exercise Your Heart: Exercise Effects on Brain and Cognition," *National Review of Neuroscience* 9 (2008): 58–65; H. A. Slagter, R. J. Davidson, and A. Lutz, "Mental Training as a Tool in the Neuroscientific Study of Brain and Cognitive Plasticity," *Frontiers in Human Neuroscience* 5, no. 1 (2011): 1–12; A. F. Kramer and K. I. Erickson, "Capitalizing on Cortical Plasticity: Influence of Physical Activity on Cognition and Brain Function," *Trends in Cognitive Sciences* 11, no. 8 (2007): 342–348; Reynolds, "Jogging Your Brain"; Gretchen Reynolds, "A for Effort," *New York Times Magazine,* June 18, 2017; A. Z. Burzynska et al., "Physical Activity Is Linked to Greater Moment-to-Moment Variability in Spontaneous Brain Activity in Older Adults," *PLOS One* 10, no. 8 (2015): e0134819; K. I. Erickson, R. L. Leckie, and A. M. Weinstein, "Physical Activity, Fitness, and Gray Matter Volume," *Neurobiology of Aging* 35, Suppl. 2 (2014): S20–S28.

60 **mind-wandering is linked with attentional lapses and declines in performance:** D. Stawarczyk et al., "Mind-Wandering: Phenomenology and Function as Assessed with a Novel Experience Sampling Method," *Acta Psychologica* 136, no. 3 (2011): 370–381; J. Smallwood et al., "Subjective Experience and the Attentional Lapse: Task Engagement

and Disengagement during Sustained Attention," *Consciousness and Cognition* 13, no. 4 (2004): 657–690; J. Smallwood et al., "Going AWOL in the Brain: Mind Wandering Reduces Cortical Analysis of External Events," *Journal of Cognitive Neuroscience* 20, no. 3 (2008): 458–469; J. C. McVay and M. J. Kane, "Conducting the Train of Thought: Working Memory Capacity, Goal Neglect, and Mind Wandering in an Executive-Control Task," *Journal of Experimental Psychology: Learning, Memory, and Cognition* 35, no. 1 (2009): 196–204; J. W. Kam et al., "Mind Wandering and the Adaptive Control of Attentional Resources," *Journal of Cognitive Neuroscience* 25, no. 6 (2013): 952–960; M. J. Kane et al., "For Whom the Mind Wanders, and When: An Experience-Sampling Study of Working Memory and Executive Control in Daily Life," *Psychological Science* 18, no. 7 (2007): 614–621; B. Baird et al., "Unnoticed Intrusions: Dissociations of Meta-Consciousness in Thought Suppression," *Consciousness and Cognition* 22, no. 3 (2013): 1003–1012; M. K. T. Takarangi, D. Strange, and D. S. Lindsay, "Self-Report May Underestimate Trauma Intrusions," *Consciousness and Cognition* 27 (2014): 297–305; M. Bastian and J. Sackur, "Mind Wandering at the Fingertips: Automatic Parsing of Subjective States Based on Response Time Variability," *Frontiers in Psychology* 4 (2013): 10.3389; J. Smallwood and J. W. Schooler, "The Restless Mind," *Psychological Bulletin* 132, no. 6 (2006): 946–958; J. W. Kam and T. C. Handy, "The Neurocognitive Consequences of the Wandering Mind: A Mechanistic Account of Sensory-Motor Decoupling," *Frontiers in Psychology* 4 (2013): 10.3389.

60 **Their study of 2,250 American adults:** M. A. Killingsworth and D. T. Gilbert, "A Wandering Mind Is an Unhappy Mind," *Science* 330, no. 6006 (2010): 932.

61 **negative moods are known to cause mind-wandering:** J. Smallwood et al., "Shifting Moods, Wandering Minds: Negative Moods Lead the Mind to Wander," *Emotion* 9, no. 2 (2009): 271–276.

61 **college students using Instant Messenger:** L. L. Bowman et al., "Can Students Really Multitask? An Experimental Study of Instant Messaging While Reading," *Computers and Education* 54, no. 4 (2010): 927–931.

61 **an experiment tracked the work patterns of twenty-seven Microsoft employees:** S. T. Iqbal and E. Horvitz, "Disruption and Recovery of Computing Tasks: Field Study, Analysis, and Directions" (paper presented at the Conference on Human Factors in Computing Systems, 2007).

62 **people who interrupt their work flow to answer email:** Gloria Mark, Daniela Gudith, and Ulrich Klocke, "The Cost of Interrupted Work: More Speed and Stress" (unpublished paper, 2008).

62 **One observational study of fifty-six thousand drivers:** D. L. Strayer and J. M. Watson, "Supertaskers and the Multitasking Brain," *Scientific American Mind* 23, no. 1 (2012): 22–29.

62 **Most people check their smartphones:** Jane E. Brody, "Hooked on Our Smartphones: Curbing Our Digital Dependence," *New York Times*, January 10, 2017.

62 **frequent multitaskers tend to be more impulsive:** Strayer and Watson, "Supertaskers and the Multitasking Brain"; D. M. Sanbonmatsu et al., "Who Multi-Tasks and Why? Multi-Tasking Ability, Perceived Multi-Tasking Ability, Impulsivity, and Sensation Seeking," *PLOS One* 8, no. 1 (2013): e54402; Matt Richtel, "Hooked on Gadgets, and Paying a Mental Price," *New York Times*, June 7, 2010; Dario D. Salvucci and Niels A. Taatgen, *The Multitasking Mind* (Oxford, UK: Oxford University Press, 2011).

62 **One study assessed how much people multitask and then tested their performance:** E. Ophir, C. Nass, and A. D. Wagner, "Cognitive Control in Media Multitaskers," *Proceedings of the National Academy of Sciences* 106, no. 37 (2009): 15583–15587.

63 **A related study asked three hundred participants to rate their multitasking frequency:** Sanbonmatsu et al., "Who Multi-Tasks and Why?"

63 **heavy multitaskers had less dense gray matter in the anterior cingulate cortex:** K. K. Loh and R. Kanai, "Higher Media Multi-Tasking Activity Is Associated with

Smaller Gray-Matter Density in the Anterior Cingulate Cortex," *PLOS One* 9, no. 9 (2014): e106698.

64 **Throughout our life, we may accumulate epigenetic changes in both directions:** The science of gene expression is complicated, but for an introduction to epigenetics, see Robert M. Sapolsky, *Behave: The Biology of Humans at Our Best and Worst* (New York: Penguin, 2017), chaps. 7–8; Bruce E. Wexler, *Brain and Culture: Neurobiology, Ideology, and Social Change* (Cambridge, MA: MIT Press, 2006), chap. 3; Bessel A. van der Kolk, *The Body Keeps the Score: Brain, Mind, and Body in the Healing of Trauma* (New York: Penguin, 2015), chap. 10; Richard J. Davidson and Sharon Begley, *The Emotional Life of Your Brain: How Its Unique Patterns Affect the Way You Think, Feel, and Live—and How You Can Change Them* (New York: Plume, 2012), chap. 5.

64 **chronic sleep deprivation and shift work can adversely affect the genes regulating the circadian rhythm:** S. M. James et al., "Shift Work: Disrupted Circadian Rhythms and Sleep—Implications for Health and Well-Being," *Current Sleep Medicine Reports* 3, no. 2 (2017): 104–112.

65 **people who died by suicide and had been abused during childhood showed distinct epigenetic changes:** P. O. McGowan et al., "Epigenetic Regulation of the Glucocorticoid Receptor in Human Brain Associates with Childhood Abuse," *Nature Neuroscience* 12, no. 3 (2009): 342–348.

65 **Another study investigated people who'd experienced trauma in childhood versus adulthood:** The study included 169 people with a history of least two types of trauma besides child abuse, who were into three groups: 108 who did not currently have PTSD (trauma-exposed controls), 32 with current PTSD and a history of childhood abuse, and 29 with current PTSD but no childhood abuse history. See D. Mehta et al., "Childhood Maltreatment Is Associated with Distinct Genomic and Epigenetic Profiles in Posttraumatic Stress Disorder," *Proceedings of the National Academy of Sciences* 110, no. 20 (2013): 8302–8307.

65 **early life adversity can leave lasting neuroplastic and epigenetic changes on our mind-body systems:** M. Szyf, P. O. McGowan, and M. J. Meaney, "The Social Environment and the Epigenome," *Environmental and Molecular Mutagenesis* 49, no. 1 (2008): 46–60; M. Gunnar and K. Quevedo, "The Neurobiology of Stress and Development," *Annual Review of Psychology* 58 (2007): 145–173; B. Labonté et al., "Genome-Wide Epigenetic Regulation by Early-Life Trauma," *Archives of General Psychiatry* 69, no. 7 (2012): 722–731; S. Galea, M. Uddin, and K. Koenen, "The Urban Environment and Mental Disorders: Epigenetic Links," *Epigenetics* 6, no. 4 (2011): 400–404; M. Uddin et al., "Epigenetic and Immune Function Profiles Associated with Posttraumatic Stress Disorder," *Proceedings of the National Academy of Sciences* 107, no. 20 (2010): 9470–9475; M. Uddin et al., "Epigenetic and Inflammatory Marker Profiles Associated with Depression in a Community-Based Epidemiologic Sample," *Psychological Medicine* 41, no. 5 (2011): 997–1007.

65 **Chronic stress arousal affects the programming of . . . macrophages:** G. E. Miller, E. Chen, and K. J. Parker, "Psychological Stress in Childhood and Susceptibility to the Chronic Diseases of Aging: Moving toward a Model of Behavioral and Biological Mechanisms," *Psychological Bulletin* 137, no. 6 (2011): 959–997; G. E. Miller and E. Chen, "Harsh Family Climate in Early Life Presages the Emergence of a Proinflammatory Phenotype in Adolescence," *Psychological Science* 21, no. 6 (2010): 848–856; N. Slopen et al., "Early Origins of Inflammation: An Examination of Prenatal and Childhood Social Adversity in a Prospective Cohort Study," *Psychoneuroendocrinology* 51 (2015): 403–413; C. P. Fagundes, R. Glaser, and J. K. Kiecolt-Glaser, "Stressful Early Life Experiences and Immune Dysregulation across the Lifespan," *Brain, Behavior, and Immunity* 27, no. 1 (2013): 8–12; R. Nusslock and G. E. Miller, "Early-Life Adversity and Physical and Emotional Health across the Lifespan: A Neuroimmune Network Hypothesis," *Biological*

Psychiatry 80, no. 1 (2016): 23–32; Gary Kaplan and Donna Beech, *Total Recovery: Solving the Mystery of Chronic Pain and Depression* (New York: Rodale, 2014), chaps. 4, 7.

65 Macrophages . . . are responsible for recognizing and destroying the "bad guys": G. Arango Duque and A. Descoteaux, "Macrophage Cytokines: Involvement in Immunity and Infectious Diseases," *Frontiers in Immunology* 5, no. 491 (2014): 1–12.

66 Gary Kaplan, a physician who clinically treats and writes about these processes: Kaplan and Beech, *Total Recovery*, 170.

66 chronic inflammation may manifest in many different ways: V. H. Perry and C. Holmes, "Microglial Priming in Neurodegenerative Disease," *Nature Reviews Neurology* 10, no. 4 (2014): 217–224; V. H. Perry and J. Teeling, "Microglia and Macrophages of the Central Nervous System: The Contribution of Microglia Priming and Systemic Inflammation to Chronic Neurodegeneration," *Seminars in Immunopathology* 35, no. 5 (2013): 601–612; Kaplan and Beech, *Total Recovery*, chap. 4; C. M. Eklund, "Proinflammatory Cytokines in CRP Baseline Regulation," *Advances in Clinical Chemistry* 48 (2009): 111–136; A. A. Appleton et al., "Divergent Associations of Adaptive and Maladaptive Emotion Regulation Strategies with Inflammation," *Health Psychology* 32, no. 7 (2013): 748–756; Bruce S. McEwen and Elizabeth Norton Lasley, *The End of Stress as We Know It* (Washington, D.C.: Joseph Henry, 2002), chap. 6.

66 Several experiments show that rat pups with attentive mothers: A. Gonzalez et al., "Intergenerational Effects of Complete Maternal Deprivation and Replacement Stimulation on Maternal Behavior and Emotionality in Female Rats," *Developmental Psychobiology* 38, no. 1 (2001): 11–32; R. M. Sapolsky, "Mothering Style and Methylation," *Nature Neuroscience* 7, no. 8 (2004): 791–792; I. C. G. Weaver et al., "Epigenetic Programming by Maternal Behavior," *Nature Neuroscience* 7, no. 8 (2004): 847–854; D. Francis et al., "Nongenomic Transmission across Generations of Maternal Behavior and Stress Responses in the Rat," *Science* 286, no. 5442 (1999): 1155–1158; A. S. Fleming et al., "Mothering Begets Mothering," *Pharmacology Biochemistry and Behavior* 73, no. 1 (2002): 61–75. For reviews of related research, see Wexler, *Brain and Culture*, chap. 3; Sapolsky, *Behave*, 219–221; G. N. Neigh, C. F. Gillespie, and C. B. Nemeroff, "The Neurobiological Toll of Child Abuse and Neglect," *Trauma, Violence and Abuse* 10, no. 4 (2009): 389–410; Gunnar and Quevedo, "The Neurobiology of Stress and Development."

67 one study exposed male mice to traumatic conditions: K. Gapp et al., "Implication of Sperm RNAs in Transgenerational Inheritance of the Effects of Early Trauma in Mice," *Nature Neuroscience* 17, no. 5 (2014): 667–669.

67 experiments with mice help us understand why regular exercise reduces anxiety: T. J. Schoenfeld et al., "Physical Exercise Prevents Stress-Induced Activation of Granule Neurons and Enhances Local Inhibitory Mechanisms in the Dentate Gyrus," *Journal of Neuroscience* 33, no. 18 (2013): 7770–7777.

68 people who reported low mind-wandering . . . had longer telomeres: E. S. Epel et al., "Wandering Minds and Aging Cells," *Clinical Psychological Science* 1, no. 1 (2013): 75–83.

68 studies have shown how mindfulness meditation has a significant buffering effect against inflammation: M. A. Rosenkranz et al., "A Comparison of Mindfulness-Based Stress Reduction and an Active Control in Modulation of Neurogenic Inflammation," *Brain, Behavior, and Immunity* 27 (2013): 174–184; J. D. Creswell et al., "Mindfulness-Based Stress Reduction Training Reduces Loneliness and Pro-Inflammatory Gene Expression in Older Adults: A Small Randomized Controlled Trial," *Brain, Behavior, and Immunity* 26 (2012): 1095–1101. See also J. Kabat-Zinn et al., "Influence of a Mindfulness-Based Stress Reduction Intervention on Rates of Skin Clearing in Patients with Moderate to Severe Psoriasis Undergoing Phototherapy (UVB) and Photochemotherapy (PUVA)," *Psychosomatic Medicine* 60 (1998): 625–632; R. J. Davidson et al., "Alterations in Brain and Immune Function Produced by Mindfulness Meditation," *Psychosomatic Medicine* 65, no. 4 (2003): 564–570.

CHAPTER 4

74 **stress means something harmful—to be avoided, reduced, or managed:** Kelly McGonigal, *The Upside of Stress: Why Stress Is Good for You, and How to Get Good at It* (New York: Avery, 2015), xi.

75 **Stressors can also be classified as acute or chronic, and physical or psychological:** Robert M. Sapolsky, *Why Zebras Don't Get Ulcers*, 3rd ed. (New York: Holt, 2004), chaps. 1, 13; Jon Kabat-Zinn, *Full Catastrophe Living (Revised Edition): Using the Wisdom of Your Body and Mind to Face Stress, Pain, and Illness* (New York: Bantam, 2013), chap. 19.

75 **being able to imagine and prepare for future contingencies is one of humanity's unique talents:** Sapolsky, *Why Zebras Don't Get Ulcers*, 10; Jennifer Kavanagh, *Stress and Performance: A Review of the Literature and Its Applicability to the Military* (Arlington, VA: RAND, 2005), 31.

76 **individuals with Type A personalities:** Kavanagh, *Stress and Performance*, 30.

76 **Our survival brain will perceive greater threat:** S. J. Lupien, "Brains under Stress," *Canadian Journal of Psychiatry* 54, no. 1 (2009): 4–5; Kavanagh, *Stress and Performance*, 32–34; Bruce S. McEwen and Elizabeth Norton Lasley, *The End of Stress as We Know It* (Washington, D.C.: Joseph Henry, 2002), 32; Sapolsky, *Why Zebras Don't Get Ulcers*, chap. 13.

76 **if aspects of the current stressor contain cues or triggers related to past traumatic events:** Robert C. Scaer, *The Trauma Spectrum: Hidden Wounds and Human Resiliency* (New York: Norton, 2005), 62–64; Pat Ogden and Janina Fisher, *Sensorimotor Psychotherapy: Interventions for Trauma and Attachment* (New York: Norton, 2015), 181.

76 **During the Nazi bombings of Britain:** Sapolsky, *Why Zebras Don't Get Ulcers*, 260; Kavanagh, *Stress and Performance*, 32–37.

77 **individuals with a strong internal locus of control experience more stress:** J. Haidt and J. Rodin, "Control and Efficacy as Interdisciplinary Bridges," *Review of General Psychology* 3 (2000): 317–337; Kavanagh, *Stress and Performance*, 30–33; Sapolsky, *Why Zebras Don't Get Ulcers*, chap. 13; McEwen and Lasley, *The End of Stress as We Know It*, 149–152.

78 **Allostasis allows us to vary internal conditions:** McEwen and Lasley, *The End of Stress as We Know It*, 6.

79 **the survival brain . . . directs the endocrine system to release adrenaline:** For an overview of the neurobiology of stress arousal and its various hormonal shifts, see McEwen and Lasley, *The End of Stress as We Know It*, chaps. 2, 5, 6; Sapolsky, *Why Zebras Don't Get Ulcers*, chaps. 1, 2.

79 **stress activation is initially aimed at transporting oxygen and glucose:** McEwen and Lasley, *The End of Stress as We Know It*, chap. 2.

80 **As part of the second wave, the HPA axis activates hormones to mobilize energy:** The hypothalamus releases corticotropin-releasing hormone to stimulate the pituitary gland to produce corticotropin (also known as adreno-corticotropic hormone, or ACTH). ACTH, in turn, stimulates the adrenals to produce cortisol and other glucocorticoids.

81 **sustained stress arousal suppresses immunity:** McEwen and Lasley, *The End of Stress as We Know It*, chap. 6; Sapolsky, *Why Zebras Don't Get Ulcers*, chap. 8, esp. 154–155.

81 **the HPA axis inhibits hormones related to long-term needs:** Stephen W. Porges, *The Polyvagal Theory: Neurophysiological Foundations of Emotions, Attachment, Communication, and Self-Regulation* (New York: Norton, 2011), 174–179, 290–293; Sapolsky, *Why Zebras Don't Get Ulcers*, 30–32.

82 **The ANS has wide-ranging authority:** Sapolsky, *Why Zebras Don't Get Ulcers*, chap. 2; McEwen and Lasley, *The End of Stress as We Know It*, chap. 5; Bessel A. van der Kolk, *The Body Keeps the Score: Brain, Mind, and Body in the Healing of Trauma* (New York: Penguin, 2015), chap. 5; Porges, *The Polyvagal Theory*, chap. 3.

82 **The ANS has two branches:** McEwen and Lasley, *The End of Stress as We Know It*, chap. 5; van der Kolk, *The Body Keeps the Score*, chap. 5.

83 **the ANS also has a feedback loop from the organs back to the survival brain:** McEwen and Lasley, *The End of Stress as We Know It*, 72; Scaer, *The Trauma Spectrum*, 22–25.

83 **Porges . . . developed the polyvagal theory:** Porges, *The Polyvagal Theory*.

83 **if the survival brain perceives that fight-or-flight won't work:** Porges, *The Polyvagal Theory*, chap. 2.

84 **The survival brain's neuroception process automatically determines:** Porges, *The Polyvagal Theory*, chap. 1.

84 **Whether our nervous system is in well-being mode or defensive mode:** Sex is especially complicated—it starts with a delicate balance between the ventral PSNS and SNS for seduction and foreplay, followed by the dorsal PSNS and SNS for erection and visceral arousal, followed by the PSNS shutting off entirely at ejaculation. Because of this delicate balance, premature ejaculation and erectile dysfunction are common male symptoms of nervous system dysregulation and allostatic load. Porges, *The Polyvagal Theory*, 172–179, 81–83, 275–277, 91–93; Sapolsky, *Why Zebras Don't Get Ulcers*, 120–126.

84 **The parasympathetic nervous system (PSNS) is divided into two branches:** The ventral branch of the vagus nerve is myelinated, meaning it has a fatty sheath around the nerve fibers so they can conduct electricity and communicate more quickly with other neurons. All mammals have the myelinated vagus and a ventral PSNS. In contrast, the dorsal branch is unmyelinated. Most vertebrates have an unmyelinated vagus; thus, the dorsal PSNS is shared by mammals, fish, reptiles, and amphibians.

86 **The ventral PSNS controls three functions:** Porges, *The Polyvagal Theory*, 92–93; McEwen and Lasley, *The End of Stress as We Know It*, 78–81.

86 **High HRV . . . means our vagal brake is working efficiently:** McEwen and Lasley, *The End of Stress as We Know It*, chap. 5; Porges, *The Polyvagal Theory*, chap. 4.

86 **the ventral PSNS triggers the release of acetylcholine:** McEwen and Lasley, *The End of Stress as We Know It*, 44–45, 76.

86 **The third function of the ventral PSNS is controlling bodily functions for the social engagement system:** van der Kolk, *The Body Keeps the Score*, 80–82; Porges, *The Polyvagal Theory*, 55, 92–93; Pat Ogden, Kekuni Minton, and Clare Pain, *Trauma and the Body: A Sensorimotor Approach to Psychotherapy* (New York: Norton, 2006), 32.

87 **Because the ventral PSNS is deeply involved with both social engagement and recovery:** McEwen and Lasley, *The End of Stress as We Know It*, chap. 5; van der Kolk, *The Body Keeps the Score*, chap. 5; Porges, *The Polyvagal Theory*, 269–271.

88 **many defensive SNS behaviors will not expend all this mobilized energy:** Scaer, *The Trauma Spectrum*, 43–44; McEwen and Lasley, *The End of Stress as We Know It*, chap. 2.

89 **massive conservation of oxygen and energy:** Porges, *The Polyvagal Theory*, 55, 92–93; McEwen and Lasley, *The End of Stress as We Know It*, 78–79; Ogden et al., *Trauma and the Body*, 94–96; Ogden and Fisher, *Sensorimotor Psychotherapy*, 520–521.

90 **our body is preparing for shutdown:** Ogden and Fisher, *Sensorimotor Psychotherapy*, 520–521; Scaer, *The Trauma Spectrum*, 44–45; Porges, *The Polyvagal Theory*, 177–183; Ogden et al., *Trauma and the Body*, 94–96; van der Kolk, *The Body Keeps the Score*, 82.

90 **Each of these descriptions of freeze:** If you want to see an excellent fictionalized depiction of freeze—one I use when I teach about freeze—I highly recommend the scene from the movie *Saving Private Ryan* when Tom Hanks's character first reaches the Normandy Beach on D-Day. It's a wonderful depiction of his ANS going into and then coming out of freeze.

90 **freeze . . . is a highly activated state:** Scaer, *The Trauma Spectrum*, 45.

91 **the survival brain will once again turn on the ventral PSNS:** Porges, *The Polyvagal Theory*, chap. 4.

91 **When allostasis isn't working properly, however:** Porges, *The Polyvagal Theory*, 58, 69; Ogden and Fisher, *Sensorimotor Psychotherapy*, 519–522.

92 **In these states, they'll not only have difficulty downregulating their stress:** Ogden and Fisher, *Sensorimotor Psychotherapy*, chap. 25; McEwen and Lasley, *The End of Stress as We Know It*, 82–83; van der Kolk, *The Body Keeps the Score*, 79–80.

92 **Some of this default wiring comes from our early social environment:** Ogden and Fisher, *Sensorimotor Psychotherapy*, 519, 21.

CHAPTER 5

97 **Kahneman describes this as System 1 thinking:** For a list of characteristics of System 1 thinking, see Daniel Kahneman, *Thinking, Fast and Slow* (New York: Macmillan, 2011), 105.

97 **Because the survival brain isn't verbal:** Pat Ogden, Kekuni Minton, and Clare Pain, *Trauma and the Body: A Sensorimotor Approach to Psychotherapy* (New York: Norton, 2006), chap. 1; Robert C. Scaer, *The Trauma Spectrum: Hidden Wounds and Human Resiliency* (New York: Norton, 2005), chap. 2; Stephen W. Porges, *The Polyvagal Theory: Neurophysiological Foundations of Emotions, Attachment, Communication, and Self-Regulation* (New York: Norton, 2011), chaps. 1, 3.

97 **Implicit learning is supported by implicit (or nondeclarative) memory:** Scaer, *The Trauma Spectrum*, 40–42.

97 **At moderate stress levels, the amygdala works with the hippocampus:** Scaer, *The Trauma Spectrum*, 40–42; Robert M. Sapolsky, *Why Zebras Don't Get Ulcers*, 3rd ed. (New York: Holt, 2004), chap. 15; J. Douglas Bremner, *Does Stress Damage the Brain? Understanding Trauma-Related Disorders from a Mind-Body Perspective* (New York: Norton, 2005), chap. 4; Bruce S. McEwen and Elizabeth Norton Lasley, *The End of Stress as We Know It* (Washington, D.C.: Joseph Henry, 2002), 108–110.

98 **Implicit memory is not just facts or information:** Sapolsky, *Why Zebras Don't Get Ulcers*, 320–322; Scaer, *The Trauma Spectrum*, 40–42; J. LeDoux, "The Emotional Brain, Fear, and the Amygdala," *Cellular and Molecular Neurobiology* 23, no. 4–5 (2003): 727–738; McEwen and Lasley, *The End of Stress as We Know It*, chap. 7; Bremner, *Does Stress Damage the Brain?*, chap. 4.

99 **The thinking brain . . . engages in "top-down processing":** Paul D. MacLean, *The Triune Brain in Evolution: Role in Paleocerebral Functions* (New York: Plenum, 1990); Ogden et al., *Trauma and the Body*, chap. 1.

99 **System 2 is indeed slow and effortful:** Kahneman, *Thinking, Fast and Slow*, 21; J. Evans, "In Two Minds: Dual-Process Accounts of Reasoning," *Trends in Cognitive Sciences* 7, no. 10 (2003): 454–459.

99 **Executive functioning allows us to focus:** K. N. Ochsner and J. J. Gross, "The Cognitive Control of Emotion," *Trends in Cognitive Sciences* 9, no. 5 (2005): 242–249; J. M. Hinson, T. L. Jameson, and P. Whitney, "Impulsive Decision Making and Working Memory," *Journal of Experimental Psychology: Learning, Memory, and Cognition* 29, no. 2 (2003): 298–306; W. Hofmann, B. J. Schmeichel, and A. D. Baddeley, "Executive Functions and Self-Regulation," *Trends in Cognitive Sciences* 16, no. 3 (2012): 174–180; W. Hofmann et al., "Working Memory Capacity and Self-Regulatory Behavior: Toward an Individual Differences Perspective on Behavior Determination by Automatic versus Controlled Processes," *Journal of Personality and Social Psychology* 95, no. 4 (2008): 962–977; M. L. Pe, F. Raes, and P. Kuppens, "The Cognitive Building Blocks of Emotion Regulation: Ability to Update Working Memory Moderates the Efficacy of Rumination and Reappraisal on Emotion," *PLOS One* 8, no. 7 (2013): e69071; T. F. Heatherton and D. D. Wagner, "Cognitive Neuroscience of Self-Regulation Failure," *Trends in Cognitive Sciences* 15, no. 3 (2011): 132–139.

100 **Executive functioning is like a credit bank:** For a review, see Hofmann et al., "Executive Functions and Self-Regulation." For a review of sleep deprivation's effects on

executive functioning, see L. K. Barger et al., "Neurobehavioral, Health, and Safety Consequences Associated with Shift Work in Safety-Sensitive Professions," *Current Neurology and Neuroscience Reports* 9, no. 2 (2009): 155–164.

100 **Regardless of how executive functioning gets impaired:** Heatherton and Wagner, "Cognitive Neuroscience of Self-Regulation Failure"; Hofmann et al., "Executive Functions and Self-Regulation."

100 **Conscious learning is supported by explicit (or declarative) memory:** Scaer, *The Trauma Spectrum*, 38–39; Bremner, *Does Stress Damage the Brain?*, 45–47.

101 **Explicit memory is influenced by our intelligence:** Scaer, *The Trauma Spectrum*, 38; Pe et al., "The Cognitive Building Blocks of Emotion Regulation"; Hofmann et al., "Working Memory Capacity and Self-Regulatory Behavior."

101 **Mild to moderate stress arousal enhances explicit memory:** Sapolsky, *Why Zebras Don't Get Ulcers*, chap. 10.

101 **the brain is quite glucose-greedy:** McEwen and Lasley, *The End of Stress as We Know It*, 130; E. R. De Kloet et al., "Brain Corticosteroid Receptor Balance in Health and Disease," *Endocrine Review* 19, no. 3 (1998): 269–301.

101 **one study examined Israeli parole board judges:** S. Danziger, J. Levav, and L. Avnaim-Pesso, "Extraneous Factors in Judicial Decisions," *Proceedings of the National Academy of Sciences* 108, no. 17 (2011): 6889–6892. See also M. T. Gailliot and R. F. Baumeister, "The Physiology of Willpower: Linking Blood Glucose to Self-Control," *Personality and Social Psychology Review* 11 (2007): 303–327; J. M. Tyler and K. C. Burns, "After Depletion: The Replenishment of the Self's Regulatory Resources," *Self and Identity* 7 (2008): 305–321.

101 **executive functioning and explicit memory functions may be impaired:** For an excellent review of these dynamics, see McEwen and Lasley, *The End of Stress as We Know It*, chap. 7; Bremner, *Does Stress Damage the Brain?*, chap. 4; Sapolsky, *Why Zebras Don't Get Ulcers*, chap. 10.

102 **the soldiers who deployed to Iraq:** J. J. Vasterling et al., "Neuropsychological Outcomes of Army Personnel Following Deployment to the Iraq War," *Journal of the American Medical Association* 296, no. 5 (2006): 519–529.

102 **Research with high-stress occupations:** B. Vila, G. B. Morrison, and D. J. Kenney, "Improving Shift Schedule and Work-Hour Policies and Practices to Increase Police Officer Performance, Health, and Safety," *Police Quarterly* 5, no. 1 (2002): 4–24; A. Gohar et al., "Working Memory Capacity Is Decreased in Sleep-Deprived Internal Medicine Residents," *Journal of Clinical Sleep Medicine* 5, no. 3 (2009): 191–197; Barger et al., "Neurobehavioral, Health, and Safety Consequences"; M. R. Baumann, C. L. Gohm, and B. L. Bonner, "Phased Training for High-Reliability Occupations: Live-Fire Exercises for Civilian Firefighters," *Human Factors* 53, no. 5 (2011) : 548–557; H. R. Lieberman et al., "Severe Decrements in Cognition Function and Mood Induced by Sleep Loss, Heat, Dehydration, and Undernutrition during Simulated Combat," *Biological Psychiatry* 57, no. 4 (2005): 422–429; H. R. Lieberman et al., "Effects of Caffeine, Sleep Loss, and Stress on Cognitive Performance and Mood during U.S. Navy Seal Training," *Psychopharmacology* 164, no. 3 (2002): 250–261; Jennifer Kavanagh, *Stress and Performance: A Review of the Literature and Its Applicability to the Military* (Arlington, VA: RAND, 2005); C. A. Morgan et al., "Stress-Induced Deficits in Working Memory and Visuo-Constructive Abilities in Special Operations Soldiers," *Biological Psychiatry* 60, no. 7 (2006): 722–729; C. A. Morgan et al., "Accuracy of Eyewitness Memory for Persons Encountered during Exposure to Highly Intense Stress," *International Journal of Law and Psychiatry* 27, no. 3 (2004): 265–279; C. A. Morgan et al., "Symptoms of Dissociation in Humans Experiencing Acute, Uncontrollable Stress: A Prospective Investigation," *American Journal of Psychiatry* 158, no. 8 (2001): 1239–1247; C. A. Morgan et al., "Neuropeptide-Y, Cortisol, and Subjective Distress in Humans Exposed to Acute Stress: Replication and Extension of Previous Report," *Biological Psychiatry* 52, no. 2 (2002): 136–142; C. A. Morgan et al.,

"Relationships among Plasma Dehydroepiandrosterone Sulfate and Cortisol Levels, Symptoms of Dissociation, and Objective Performance in Humans Exposed to Acute Stress," *Archives of General Psychiatry* 61, no. 8 (2004): 819–825; C. A. Morgan et al., "Relationship among Plasma Cortisol, Catecholamines, Neuropeptide Y, and Human Performance during Exposure to Uncontrollable Stress," *Psychosomatic Medicine* 63, no. 3 (2001): 412–422; B. P. Marx, S. Doron-Lamarca, S. P. Proctor, and J. J. Vasterling, "The Influence of Pre-Deployment Neurocognitive Functioning on Post-Deployment PTSD Symptom Outcomes among Iraq-Deployed Army Soldiers," *Journal of the International Neuropsychological Society* 15, no. 6 (2009): 840–852; Vasterling et al., "Neuropsychological Outcomes"; S. Maguen et al., "Description of Risk and Resilience Factors among Military Medical Personnel before Deployment to Iraq," *Military Medicine* 173, no. 1 (2008): 1–9; E. A. Stanley et al., "Mindfulness-Based Mind Fitness Training: A Case Study of a High-Stress Predeployment Military Cohort," *Cognitive and Behavioral Practice* 18, no. 4 (2011): 566–576; A. P. Jha et al., "Examining the Protective Effects of Mindfulness Training on Working Memory Capacity and Affective Experience," *Emotion* 10, no. 1 (2010): 54–64; A. P. Jha et al., "Minds 'at Attention': Mindfulness Training Curbs Attentional Lapses in Military Cohorts," *PLOS One* 10, no. 2 (2015): e0116889; A. P. Jha, A. B. Morrison, S. C. Parker, and E. A. Stanley, "Practice Is Protective: Mindfulness Training Promotes Cognitive Resilience in High-Stress Cohorts," *Mindfulness* 8, no. 1 (2017): 46–58.

102 **Since stress arousal may also occur with symbolic threats:** Scaer, *The Trauma Spectrum*, 62–64; Sapolsky, *Why Zebras Don't Get Ulcers*, chap. 1; Kavanagh, *Stress and Performance*, 31.

102 **Aging may also exacerbate these effects:** S. J. Lupien et al., "Cortisol Levels during Human Aging Predict Hippocampal Atrophy and Memory Deficits," *Nature Neuroscience* 1 (1998): 69–73; Sapolsky, *Why Zebras Don't Get Ulcers*, chap. 10. Fortunately, other research with animals and humans shows that the effects of elevated cortisol likely will not cause permanent damage; when cortisol levels get corrected, the hippocampus begins to resume its normal size. See McEwen and Lasley, *The End of Stress as We Know It*, chap. 7. The relationship between elevated cortisol and PTSD is less conclusive—many people with PTSD have low cortisol levels at times, although their HPA axis remains hypersensitive to being activated—so researchers used to believe hippocampal volume loss from PTSD was irreversible. Yet some research shows that hippocampal volume loss with PTSD may also be reversed. For example, see E. Vermetten et al., "Long-Term Treatment with Paroxetine Increases Verbal Declarative Memory and Hippocampal Volume in Posttraumatic Stress Disorder," *Biological Psychiatry* 54 (2003): 693–702; Bremner, *Does Stress Damage the Brain?*, 60–62, 115–119. Other research posits that the causal relationship goes the other way: that having a small hippocampus and memory problems *before* stressful experiences increases the risk of developing PTSD when exposed to trauma. This could occur after childhood adversity, for example. See Marx et al., "The Influence of Pre-Deployment Neurocognitive Functioning"; R. A. Parslow and A. F. Jorm, "Pretrauma and Posttrauma Neurocognitive Functioning and PTSD Symptoms in a Community Sample of Young Adults," *American Journal of Psychiatry* 164, no. 3 (2007): 509–515; Sapolsky, *Why Zebras Don't Get Ulcers*, chap. 10, esp. 222.

102 **Intercontinental flight attendants with long careers:** K. Cho, "Chronic 'Jet Lag' Produces Temporal Lobe Atrophy and Spatial Cognitive Deficits," *Nature Neuroscience* 4 (2001): 567–568.

102 **People taking prescription steroids:** E. S. Brown and P. A. Chandler, "Mood and Cognitive Changes during Systemic Corticosteroid Therapy," *Primary Care Companion for Journal of Clinical Psychiatry* 3, no. 1 (2001): 17–21; S. J. Lupien and B. S. McEwen, "The Acute Effects of Corticosteroids on Cognition: Integration of Animal and Human Model Studies," *Brain Research Reviews* 24, no. 1 (1997): 1–27.

102 **People with prolonged major depression:** Sapolsky, *Why Zebras Don't Get Ulcers*, 221.

103 People with PTSD after repeated trauma: Bremner, *Does Stress Damage the Brain?*, 60–62, 115–119.

104 optimal performance . . . [is] most likely to occur at moderate stress levels: Bremner, *Does Stress Damage the Brain?*, chap. 4; Kavanagh, *Stress and Performance*, 16–19; McEwen and Lasley, *The End of Stress as We Know It*, chap. 7.

104 our neurobiological window of tolerance: Daniel J. Siegel, *The Developing Mind: How Relationships and the Brain Interact to Shape Who We Are* (New York: Guilford, 1999), 253; Ogden et al., *Trauma and the Body*, chap. 2.

104 Inside our window, we're more likely to engage in accurate neuroception: Porges, *The Polyvagal Theory*, chap. 1; Ogden et al., *Trauma and the Body*, chap. 2; Bremner, *Does Stress Damage the Brain?*, chap. 4; McEwen and Lasley, *The End of Stress as We Know It*, chap. 7; Sapolsky, *Why Zebras Don't Get Ulcers*, chaps. 10, 15.

104 Different tasks need different arousal levels: Kavanagh, *Stress and Performance*, 16–17.

105 Outside our window, the survival brain is . . . driving information search, assessment, and decision making: Porges, *The Polyvagal Theory*, chap. 1; E. A. Stanley, "War Duration and the Micro-Dynamics of Decision-Making under Stress," *Polity* 50, no. 2 (2018): 178–200; J. Renshon, J. J. Lee, and D. Tingley, "Emotions and the Micro-Foundations of Commitment Problems," *International Organization* 71, no. S1 (2017): S189–S218.

105 We focus on information we perceive to be psychologically central: Kavanagh, *Stress and Performance*, 17–18; Karl E. Weick, *Sensemaking in Organizations* (New York: Sage, 1995), 129.

105 we also tend to gather less information: Kahneman, *Thinking, Fast and Slow*; Elizabeth A. Stanley, *Paths to Peace: Domestic Coalition Shifts, War Termination and the Korean War* (Stanford, CA: Stanford University Press, 2009), chap. 2; Kavanagh, *Stress and Performance*, 17–19; Scott Sigmund Gartner, *Strategic Assessment in War* (New Haven, CT: Yale University Press, 1999).

105 we become biased toward the negative: R. F. Baumeister et al., "Bad Is Stronger Than Good," *Review of General Psychology* 5, no. 4 (2001): 323–370; P. Rozin and E. B. Royzman, "Negativity Bias, Negativity Dominance, and Contagion," *Personality and Social Psychology Review* 5, no. 4 (2001): 296–320.

105 A similar narrowing and rigidity occur with decision making: Stanley, "War Duration"; Kavanagh, *Stress and Performance*, 17–20; Kahneman, *Thinking, Fast and Slow*.

107 When the survival brain encodes implicit memories: Scaer, *The Trauma Spectrum*, 58–59, 132–133.

107 until the survival brain, nervous system, and body have an opportunity to finish: Scaer, *The Trauma Spectrum*, chap. 3; Ogdenet al., *Trauma and the Body*, 20–22, 86–87.

107 Without complete recovery, the survival brain remains frozen in time: Scaer, *The Trauma Spectrum*, 42, 95, chap. 3; Ogden et al., *Trauma and the Body*, 34–36, 86–87.

108 The survival brain will continue to rely on such default programming: Scaer, *The Trauma Spectrum*, chap. 3, 42; Ogden et al., *Trauma and the Body*, 18–23, 86–87, 104–105; B. A. van der Kolk, "Clinical Implications of Neuroscience Research in PTSD," *Annals of the New York Academy of Sciences* 1071, no. 1 (2006): 277–293.

108 Memory capsules include sensory input: Scaer, *The Trauma Spectrum*, 59–64; Robert C. Scaer, *The Body Bears the Burden: Trauma, Dissociation, and Disease*, 3rd ed. (New York: Routledge, 2014), 91–95.

109 the thinking brain's explicit memory may have been disrupted or offline: Bessel A. van der Kolk, *The Body Keeps the Score: Brain, Mind, and Body in the Healing of Trauma* (New York: Penguin, 2015), chap. 11; Sapolsky, *Why Zebras Don't Get Ulcers*, 320–323; Scaer, *The Body Bears the Burden*, 91–95.

109 the survival brain and body create symptoms: Scaer, *The Trauma Spectrum*, 62–67.

110 Each time kindling occurs, it "ups the ante": Scaer, *The Trauma Spectrum*, 62–64.

110 unresolved memory capsules unconsciously bias the traumatized person's perceptions: Ogden et al., *Trauma and the Body*, 18–23.

110 symptoms from earlier traumatic events can actually worsen: Scaer, *The Trauma Spectrum*, 62–64; Pat Ogden and Janina Fisher, *Sensorimotor Psychotherapy: Interventions for Trauma and Attachment* (New York: Norton, 2015), 181.

110 integration of information processing between the different parts of the brain: Ogden et al., *Trauma and the Body*, 7.

110 Humans have more and bigger neural circuits: McEwen and Lasley, *The End of Stress as We Know It*, 37–38; Joseph LeDoux, *The Emotional Brain: The Mysterious Underpinnings of Emotional Life* (New York: Touchstone, 1998); M. R. Delgado, A. Olsson, and E. A. Phelps, "Extending Animal Models of Fear Conditioning to Humans," *Biological Psychiatry* 73 (2006): 39–48; A. Feder, E. J. Nestler, and D. S. Charney, "Psychobiology and Molecular Genetics of Resilience," *Nature Reviews Neuroscience* 10 (2009): 446–457; M. R. Delgado et al., "Neural Circuitry Underlying the Regulation of Conditioned Fear and Its Relation to Extinction," *Neuron* 59 (2008): 829–838; M. R. Milad et al., "Thickness of Ventromedial Prefrontal Cortex in Humans Is Correlated with Extinction Memory," *Proceedings of the National Academy of Sciences* 102, no. 30 (2005): 10706–10711; D. Schiller et al., "From Fear to Safety and Back: Reversal of Fear in the Human Brain," *Journal of Neuroscience* 28 (2008): 11517–11525.

111 since the survival brain doesn't support the thinking brain's "reasonable" belief: Ogden et al., *Trauma and the Body*, 10–11.

113 using these techniques alone, it's possible for a traumatized person to rigidly hold: van der Kolk, *The Body Keeps the Score*, 182; van der Kolk, "Clinical Implications of Neuroscience Research in PTSD," 281–282; Ogden et al., *Trauma and the Body*, 23–24, 37.

113 as trauma clinician Pat Ogden and her colleagues explain: Ogden et al., *Trauma and the Body*, 24. Italics in original.

CHAPTER 6

117 The parts of the brain that most distinguish humans: Bruce E. Wexler, *Brain and Culture: Neurobiology, Ideology, and Social Change* (Cambridge, MA: MIT Press, 2006), 36.

117 It begins its development during the last trimester of pregnancy: Stephen W. Porges, *The Polyvagal Theory: Neurophysiological Foundations of Emotions, Attachment, Communication, and Self-Regulation* (New York: Norton, 2011), 122.

118 colic . . . may be a sign that the infant's ventral PSNS circuit is having difficulty learning: Porges, *The Polyvagal Theory*, 77.

118 positive bonding during nursing and early life begins the infant's learning trajectory: Porges, *The Polyvagal Theory*, chap. 8.

118 Premature birth, infant illness, and neglect or abuse can all interrupt: Porges, *The Polyvagal Theory*, 122.

118 In one study, nine-month-old infants: S. W. Porges et al., "Infant Regulation of the Vagal 'Brake' Predicts Child Behavior Problems: A Psychobiological Model of Social Behavior," *Developmental Psychobiology* 29, no. 8 (1996): 697–712; Porges, *The Polyvagal Theory*, chaps. 7, 8.

119 Bowlby, the British psychoanalyst who developed attachment theory: J. Bowlby, "Attachment and Loss: Retrospect and Prospect," *American Journal of Orthopsychiatry* 52, no. 4 (1982): 664–678.

119 As trauma researcher and clinician Bessel van der Kolk puts it: Bessel A. van der Kolk, *The Body Keeps the Score: Brain, Mind, and Body in the Healing of Trauma* (New York: Penguin, 2015), 122.

119 Attachment theory developed in two parallel communities: For more information about differences in how the two communities conceptualize and measure attachment styles, see K. Bartholomew and P. R. Shaver, "Methods of Assessing Adult Attachment:

Do They Converge?," in *Attachment Theory and Close Relationships*, edited by J. A. Simpson and W. S. Rholes (New York: Guilford, 1998), 25–45.

120 **Our primary attachment style usually remains stable throughout our life:** van der Kolk, *The Body Keeps the Score,* chap. 7; Pat Ogden, Kekuni Minton, and Clare Pain, *Trauma and the Body: A Sensorimotor Approach to Psychotherapy* (New York: Norton, 2006), chap. 3; Daniel J. Siegel, *The Developing Mind: How Relationships and the Brain Interact to Shape Who We Are* (New York: Guilford, 1999), chap. 8.

120 **we developed different attachment patterns with each attachment figure:** Ogden et al., *Trauma and the Body*, 46–47.

120 **Empirical studies suggest that about three quarters of adults:** Amir Levine and Rachel S. F. Heller, *Attached: The New Science of Adult Attachment and How It Can Help You Find and Keep Love* (New York: Tarcher/Penguin, 2010), 140. However, the evidence from several scientific studies suggests there's a weak correlation, at best, between attachment style in childhood and adulthood. For a review, see R. C. Fraley, "Attachment Stability from Infancy to Adulthood: Meta-Analysis and Dynamic Modeling of Developmental Mechanisms," *Personality and Social Psychology Review* 6, no. 2 (2002): 123–151; E. Scharfe and K. I. M. Bartholomew, "Reliability and Stability of Adult Attachment Patterns," *Personal Relationships* 1, no. 1 (1994): 23–43.

121 **Attachment styles include emotional communication patterns and relational strategies:** Ogden et al., *Trauma and the Body*, chap. 3; Siegel, *The Developing Mind*, 276–278.

121 **the parents provide a "holding environment" for the baby's needs and growth:** Ogden et al., *Trauma and the Body*, 41–43; Siegel, *The Developing Mind*, 278–283; van der Kolk, *The Body Keeps the Score*, 110–114.

122 **Ideally, the child learns to rely on his parents as a "safe home base":** van der Kolk, *The Body Keeps the Score*, 110–114; Ogden et al., *Trauma and the Body*, 41–48; Siegel, *The Developing Mind*, 278–283.

122 **having what they call "good enough" parents:** E. Z. Tronick, "Emotions and Emotional Communication in Infants," *American Psychologist* 44, no. 2 (1989): 112–119; Ogden et al., *Trauma and the Body*, 46; Siegel, *The Developing Mind*, 291.

123 **With interactive repair, "good enough" parents:** Donald Winnicott argued that most mothers did just fine attuning to their infants; he is responsible for the term "good enough" mother/care-provider. See Donald W. Winnicott, *Primary Maternal Preoccupation* (London: Tavistock, 1956).

123 **by imitating and internalizing his parents' . . . patterns, the child's brain and nervous system get shaped:** Wexler, *Brain and Culture*, 92–121; van der Kolk, *The Body Keeps the Score*, 113–114; Ogden et al., *Trauma and the Body*, 43–48; Siegel, *The Developing Mind*, 282–283.

123 **securely attached adults have learned how to display congruence:** Ogden et al., *Trauma and the Body*, 47–48; Siegel, *The Developing Mind*, 282–283; Levine and Heller, *Attached*, chap. 7.

123 **Between 50 and 63 percent of adults have a secure attachment style:** T. Ein-Dor et al., "The Attachment Paradox: How Can So Many of Us (the Insecure Ones) Have No Adaptive Advantages?," *Perspectives on Psychological Science* 5, no. 2 (2010): 123–141. For instance, one study found 59 percent secure, 25 percent insecure-avoidant, and 11 percent insecure-anxious: K. D. Mickelson, R. C. Kessler, and P. R. Shaver, "Adult Attachment in a Nationally Representative Sample," *Journal of Personality and Social Psychology* 73, no. 5 (1997): 1092–1106. Another study found 63.5 percent secure, 22 percent avoidant, 6 percent anxious, and 9 percent unclassified: X. Meng, C. D'Arcy, and G. C. Adams, "Associations between Adult Attachment Style and Mental Health Care Utilization: Findings from a Large-Scale National Survey," *Psychiatry Research* 229, no. 1 (2015): 454–461. Levine and Heller review several studies to suggest the adult rates as 50 percent secure, 25 percent avoidant, 22 percent anxious, and 3 percent disorganized; see Levine

and Heller, *Attached*, 8. M. Mikulincer and P. R. Shaver, *Attachment in Adulthood: Structure, Dynamics, and Change* (New York: Guilford, 2007), review studies to find that the proportion of secure attachment is lower in more disadvantaged (e.g., poorer, less socially stable) populations.

123 **Older adults, from all socioeconomic backgrounds, may also have lower rates of secure attachment:** For a review of these studies, see C. Magai et al., "Attachment Styles in Older European American and African American Adults," *Journal of Gerontology B: Psychological and Social Sciences* 56, no. 1 (2001): S28–S35.

124 **we are neurobiologically programmed to attach to someone:** van der Kolk, *The Body Keeps the Score*, 115.

124 **Their eventual parenting styles were profoundly influenced by their own initial wiring:** For reviews, see G. N. Neigh, C. F. Gillespie, and C. B. Nemeroff, "The Neurobiological Toll of Child Abuse and Neglect," *Trauma, Violence and Abuse* 10, no. 4 (2009): 389–410; Robert M. Sapolsky, *Behave: The Biology of Humans at Our Best and Worst* (New York: Penguin, 2017), chap. 7.

125 **A child's gender and temperament have little effect on their attachment style:** M. H. van Ijzendoorn, C. Schuengel, and M. J. Bakermans-Kranenburg, "Disorganized Attachment in Early Childhood: Meta-Analysis of Precursors, Concomitants, and Sequelae," *Development and Psychopathology* 11, no. 2 (1999): 225–250.

125 **siblings develop different attachment styles—but birth order does not explain this variation:** M. Kennedy, L. R. Betts, and J. D. M. Underwood, "Moving beyond the Mother-Child Dyad: Exploring the Link between Maternal Sensitivity and Siblings' Attachment Styles," *Journal of Genetic Psychology* 175, no. 3–4 (2014): 287–300; E. M. Leerkes, "Maternal Sensitivity during Distressing Tasks: A Unique Predictor of Attachment Security," *Infant Behavior and Development* 34 (2011): 443–446. Recent research has also found that insecure attachment styles are at least partially associated with variants of particular dopamine and serotonin receptor genes. See R. C. Fraley et al., "Interpersonal and Genetic Origins of Adult Attachment Styles: A Longitudinal Study from Infancy to Early Adulthood," *Journal of Personality and Social Psychology* 104, no. 5 (2013): 817–838; O. Gillath et al., "Genetic Correlates of Adult Attachment Style," *Personality and Social Psychology Bulletin* 34, no. 10 (2008): 1396–1405.

125 **studies found evidence of similar stress hormone levels between a mother and her child:** L. C. Hibel et al., "Intimate Partner Violence Moderates the Association between Mother–Infant Adrenocortical Activity across an Emotional Challenge," *Journal of Family Psychology* 23 (2009): 615–625; L. M. Papp, P. Pendry, and E. K. Adam, "Mother-Adolescent Physiological Synchrony in Naturalistic Settings: Within-Family Cortisol Associations and Moderators," *Journal of Family Psychology* 23 (2009): 882–894; S. R. Williams et al., "Exploring Patterns in Cortisol Synchrony among Anxious and Nonanxious Mother and Child Dyads: A Preliminary Study," *Biological Psychology* 93 (2013): 287–295.

125 **A recent experiment with sixty-nine mothers and their babies:** S. F. Waters, T. V. West, and W. B. Mendes, "Stress Contagion: Physiological Covariation between Mothers and Infants," *Psychological Science* 25, no. 4 (2014): 934–942.

126 **The researchers concluded:** Waters et al., "Stress Contagion," 939.

127 **Infants are likely to develop an insecure-avoidant style:** Ogden et al., *Trauma and the Body*, 48–50; van der Kolk, *The Body Keeps the Score*, 116; Siegel, *The Developing Mind*, 283. Mary Ainsworth developed the Strange Situation experiment to provide an empirical methodology to determine an infant's or young child's attachment style. The experiment goes through eight "episodes," with each episode featuring a different combination of mother, child, and stranger. From another room, observers code the child's behavior every fifteen seconds, scoring the attachment style based on the coded behavior. Based on her findings, Ainsworth developed three attachment styles—secure, insecure-avoidant, and insecure-anxious. See M. D. S. Ainsworth and S. M. Bell, "Attachment, Exploration, and

Separation: Illustrated by the Behavior of One-Year-Olds in a Strange Situation," *Child Development* 41, no. 1 (1970): 49–67. A fourth attachment style, insecure-disorganized, was observed by Mary Main and her colleagues.

127 **avoidant children and adults often experience a large incongruence:** van der Kolk, *The Body Keeps the Score*, 116; Siegel, *The Developing Mind*, 287–290; Ogden et al., *Trauma and the Body*, 48–50.

128 **their default programming generally tends toward strategies associated with . . . the dorsal PSNS:** Ogden et al., *Trauma and the Body*, 48–50; Siegel, *The Developing Mind*, 287–290; van der Kolk, *The Body Keeps the Score*, 116; Levine and Heller, *Attached*, chap. 6.

129 **If they have a partner, avoidant adults tend to be reluctant to seek support:** Levine and Heller, *Attached*, chap. 6; J. N. Fish et al., "Characteristics of Those Who Participate in Infidelity: The Role of Adult Attachment and Differentiation in Extradyadic Experiences," *American Journal of Family Therapy* 40, no. 3 (2012): 214–229; E. S. Allen and D. H. Baucom, "Adult Attachment and Patterns of Extradyadic Involvement," *Family Process* 43, no. 4 (2004): 467–488; E. Selcuk, V. Zayas, and C. Hazan, "Beyond Satisfaction: The Role of Attachment in Marital Functioning," *Journal of Family Theory and Review* 2, no. 4 (2010): 258–279.

129 **About one quarter of all adults have the insecure-avoidant attachment style:** Levine and Heller, *Attached*, 8; Meng et al., "Associations between Adult Attachment Style and Mental Health Care Utilization"; Mickelson et al., "Adult Attachment in a Nationally Representative Sample."

129 **studies suggest that this attachment style is more common among older age groups:** For review, see Magai et al., "Attachment Styles."

129 **Infants will develop [the insecure-anxious] attachment style if their mother interacts with them unpredictably:** Siegel, *The Developing Mind*, 283–284; Ogden et al., *Trauma and the Body*, 50–51.

130 **anxious children may have more awareness of their internal states and more congruence:** Ogden et al., *Trauma and the Body*, 50–51; Siegel, *The Developing Mind*, 283–284; van der Kolk, *The Body Keeps the Score*, 116.

130 **This preoccupation can manifest in many different ways:** Siegel, *The Developing Mind*, 290–292; Ogden et al., *Trauma and the Body*, 50–51; Levine and Heller, *Attached*, chap. 5; S. Reynolds, H. R. Searight, and S. Ratwik, "Adult Attachment Styles and Rumination in the Context of Intimate Relationships," *North American Journal of Psychology* 16, no. 3 (2014): 495–506; O. Gillath et al., "Attachment-Style Differences in the Ability to Suppress Negative Thoughts: Exploring the Neural Correlates," *NeuroImage* 28 (2005): 835–847; S. K. K. Nielsen et al., "Adult Attachment Style and Anxiety—The Mediating Role of Emotion Regulation," *Journal of Affective Disorders* 218 (2017): 253–259.

130 **People with an anxious attachment style are more vigilant to changes in others' emotions:** R. C. Fraley et al., "Adult Attachment and the Perception of Emotional Expressions: Probing the Hyperactivating Strategies Underlying Anxious Attachment," *Journal of Personality* 74, no. 4 (2006): 1163–1190; P. Vrtička, D. Sander, and P. Vuilleumier, "Influence of Adult Attachment Style on the Perception of Social and Non-Social Emotional Scenes," *Journal of Social and Personal Relationships* 29, no. 4 (2012): 530–544; Selcuk et al., "Beyond Satisfaction"; Levine and Heller, *Attached*, chap. 5.

130 **anxiously attached adults tend to prefer intense and enmeshed relationships:** Siegel, *The Developing Mind*, 290–292; Selcuk et al., "Beyond Satisfaction"; Gillath et al., "Attachment-Style Differences"; Levine and Heller, *Attached*, Chapter 5.

131 **more likely than securely attached adults to engage in infidelity:** Allen and Baucom, "Adult Attachment"; Fish et al., "Characteristics of Those Who Participate in Infidelity"; J. Davila and T. N. Bradbury, "Attachment Insecurity and the Distinction between Unhappy Spouses Who Do and Do Not Divorce," *Journal of Family Psychology* 15, no. 3 (2001): 371–393; Selcuk et al., "Beyond Satisfaction."

131 **6 to 22 percent of all adults have an insecure-anxious attachment style:** Levine and Heller, *Attached*, 8; Meng et al., "Associations between Adult Attachment Style and Mental Health Care Utilization"; Mickelson et al., "Adult Attachment in a Nationally Representative Sample."

131 **Most adult attachment research doesn't measure and include this attachment style:** When it is measured, it is called "anxious-avoidant" or "fearful-avoidant." See Bartholomew and Shaver, "Methods of Assessing Adult Attachment."

131 **a meta-analysis found that 15 percent of infants and children:** van Ijzendoorn et al., "Disorganized Attachment in Early Childhood."

131 **Infants will develop an insecure-disorganized style:** van Ijzendoorn et al., "Disorganized Attachment in Early Childhood"; Siegel, *The Developing Mind*, 284; Ogden et al., *Trauma and the Body*, 51–54; E. Hesse and M. Main, "Frightened, Threatening, and Dissociative Parental Behavior in Low-Risk Samples: Description, Discussion, and Interpretations," *Development and Psychopathology* 18, no. 2 (2006): 309–343; van der Kolk, *The Body Keeps the Score*, 117–120.

132 **disorganized attachment "can be viewed as a second-generation effect":** Hesse and Main, "Frightened, Threatening, and Dissociative Parental Behavior," 310.

132 **These children rarely experience interactive repair:** Hesse and Main, "Frightened, Threatening, and Dissociative Parental Behavior," 310; Ogden et al., *Trauma and the Body*, 51–54; Siegel, *The Developing Mind*, 284; van der Kolk, *The Body Keeps the Score*, 117–120; van Ijzendoorn et al., "Disorganized Attachment in Early Childhood."

132 **may also engage in unusual behaviors to self-soothe:** van Ijzendoorn et al., "Disorganized Attachment in Early Childhood," 226.

133 **Disorganized children often have to "grow up fast":** van der Kolk, *The Body Keeps the Score*, 117–120; J. L. Borelli et al., "Links between Disorganized Attachment Classification and Clinical Symptoms in School-Aged Children," *Journal of Child and Family Studies* 19, no. 3 (2010): 243–256.

133 **likely to develop the faultiest neuroception:** Porges, *The Polyvagal Theory*, 17–19; Ogden et al., *Trauma and the Body*, 51–54; van der Kolk, *The Body Keeps the Score*, 121–122.

133 **3 to 5 percent of all adults have a disorganized attachment style:** van der Kolk, *The Body Keeps the Score*, 116; Levine and Heller, *Attached*, 9.

134 **insecurely attached adults are significantly more likely to use destructive and coercive behaviors:** Gillath et al., "Attachment-Style Differences"; Allen and Baucom, "Adult Attachment"; Davila and Bradbury, "Attachment Insecurity"; Selcuk et al., "Beyond Satisfaction"; Fish et al., "Characteristics of Those Who Participate in Infidelity."

134 **Insecurely attached adults are also significantly more likely to experience sleep disturbances:** L. A. McWilliams and S. J. Bailey, "Associations between Adult Attachment Ratings and Health Conditions: Evidence from the National Comorbidity Survey Replication," *Health Psychology* 29, no. 4 (2010): 446–453; G. C. Adams and L. A. McWilliams, "Relationships between Adult Attachment Style Ratings and Sleep Disturbances in a Nationally Representative Sample," *Journal of Psychosomatic Research* 79, no. 1 (2015): 37–42; L. A. McWilliams, "Adult Attachment Insecurity Is Positively Associated with Medically Unexplained Chronic Pain," *European Journal of Pain* 21, no. 8 (2017): 1378–1383; L. A. McWilliams, B. J. Cox, and M. W. Enns, "Impact of Adult Attachment Styles on Pain and Disability Associated with Arthritis in a Nationally Representative Sample," *Clinical Journal of Pain* 16, no. 4 (2000): 360–364.

135 **insecurely attached adults are at greater risk of developing mental illnesses:** K. N. Levy, "The Implications of Attachment Theory and Research for Understanding Borderline Personality Disorder," *Development and Psychopathology* 17, no. 4 (2005): 959–986; S. Woodhouse, S. Ayers, and A. P. Field, "The Relationship between Adult Attachment Style and Post-Traumatic Stress Symptoms: A Meta-Analysis," *Journal of Anxiety Disorders* 35 (2015): 103–117; J. Feeney et al., "Attachment Insecurity, Depression,

and the Transition to Parenthood," *Personal Relationships* 10, no. 4 (2003): 475–493; M. J. Bakermans-Kranenburg and M. H. van Ijzendoorn, "The First 10,000 Adult Attachment Interviews: Distributions of Adult Attachment Representations in Clinical and Non-Clinical Groups," *Attachment and Human Development* 11, no. 3 (2009): 223–263; A. Schindler et al., "Heroin as an Attachment Substitute? Differences in Attachment Representations between Opioid, Ecstasy and Cannabis Abusers," *Attachment and Human Development* 11, no. 3 (2009): 307–330; A. Schindler and S. Broning, "A Review on Attachment and Adolescent Substance Abuse: Empirical Evidence and Implications for Prevention and Treatment," *Substance Abuse* 36, no. 3 (2015): 304–313; A. Bifulco et al., "Adult Attachment Style. I: Its Relationship to Clinical Depression," *Social Psychiatry and Psychiatric Epidemiology* 37, no. 2 (2002): 50–59; A. Buchheim, B. Strauss, and H. Kachele, "The Differential Relevance of Attachment Classification for Psychological Disorders," *Psychotherapy and Psychosomatic Medical Psychology* 52, no. 3–4 (2002): 128–133; F. Declercq and J. Willemsen, "Distress and Post-Traumatic Stress Disorders in High Risk Professionals: Adult Attachment Style and the Dimensions of Anxiety and Avoidance," *Clinical Psychology and Psychotherapy* 13, no. 4 (2006): 256–263; Nielsen et al., "Adult Attachment Style and Anxiety"; A. Marganska, M. Gallagher, and R. Miranda, "Adult Attachment, Emotion Dysregulation, and Symptoms of Depression and Generalized Anxiety Disorder," *American Journal of Orthopsychiatry* 83, no. 1 (2013): 131–141; H. Unterrainer et al., "Addiction as an Attachment Disorder: White Matter Impairment Is Linked to Increased Negative Affective States in Poly-Drug Use," *Frontiers in Human Neuroscience* 11 (2017): 208; A. M. Ponizovsky et al., "Attachment Insecurity and Psychological Resources Associated with Adjustment Disorders," *American Journal of Orthopsychiatry* 81, no. 2 (2011): 265–276; A. Bifulco et al., "Adult Attachment Style. II: Its Relationship to Psychosocial Depressive-Vulnerability," *Social Psychiatry and Psychiatric Epidemiology* 37, no. 2 (2002): 60–67.

CHAPTER 7

139 **Parents with insecure attachment styles and narrow(ed) windows are more likely to create the environmental conditions:** Bruce E. Wexler, *Brain and Culture: Neurobiology, Ideology, and Social Change* (Cambridge, MA: MIT Press, 2006), 100; Pat Ogden, Kekuni Minton, and Clare Pain, *Trauma and the Body: A Sensorimotor Approach to Psychotherapy* (New York: Norton, 2006), chap. 6; Daniel J. Siegel, *The Developing Mind: How Relationships and the Brain Interact to Shape Who We Are* (New York: Guilford, 1999), chap. 8; Robert M. Sapolsky, *Behave: The Biology of Humans at Our Best and Worst* (New York: Penguin, 2017), chap. 7, esp. 221–222.

139 **Parents play a critical role in helping children:** Siegel, *The Developing Mind*, chap. 8; Wexler, *Brain and Culture*, chap. 3; K. Chase Stovall-McClough and M. Cloitre, "Unresolved Attachment, PTSD, and Dissociation in Women with Childhood Abuse Histories," *Journal of Consulting and Clinical Psychology* 74, no. 2 (2006): 219–228; Ogden et al., *Trauma and the Body*, chap. 3; M. S. Scheeringa and C. H. Zeanah, "A Relational Perspective on PTSD in Early Childhood," *Journal of Traumatic Stress* 14, no. 4 (2001): 799–815; G. N. Neigh, C. F. Gillespie, and C. B. Nemeroff, "The Neurobiological Toll of Child Abuse and Neglect," *Trauma, Violence and Abuse* 10, no. 4 (2009): 389–410.

139 **For instance, mothers with an insecure attachment style:** P. A. Brennan et al., "Maternal Depression and Infant Cortisol: Influences of Timing, Comorbidity and Treatment," *Journal of Child Psychology and Psychiatry* 49, no. 10 (2008): 1099–1107; R. Yehuda et al., "Transgenerational Effects of Posttraumatic Stress Disorder in Babies of Mothers Exposed to the World Trade Center Attacks during Pregnancy," *Journal of Clinical Endocrinology and Metabolism* 90 (2005): 4115–4118; Robert C. Scaer, *The Trauma Spectrum: Hidden Wounds and Human Resiliency* (New York: Norton, 2005), 106–07; I. S. Yim et al., "Biological and Psychosocial Predictors of Postpartum Depression: Systematic Review and Call for Integration," *Annual Review of Clinical Psychology* 11 (2015): 99–137.

Carefully controlled studies with other mammals also show that exposing a pregnant mother-to-be to a stressful or abusive environment is akin to exposing her fetus directly, in terms of the neurobiological effects on that baby when it's born. For review, see Neigh et al., "The Neurobiological Toll of Child Abuse and Neglect."

140 **One recent study with U.S. National Guard families:** J. Snyder et al., "Parent–Child Relationship Quality and Family Transmission of Parent Posttraumatic Stress Disorder Symptoms and Child Externalizing and Internalizing Symptoms Following Fathers' Exposure to Combat Trauma," *Development and Psychopathology* 28, no. 4, pt. 1 (2016): 947–969.

140 **Other research echoes these transgenerational stress contagion effects:** For a review, see Neigh et al., "The Neurobiological Toll of Child Abuse and Neglect."

141 **having a parent who experienced either multiple military deployments or deployment-related PTSD:** A. Chandra et al., "Children on the Homefront: The Experience of Children from Military Families," *Pediatrics* 125 (2010): 16–25; J. Douglas Bremner, *Does Stress Damage the Brain? Understanding Trauma-Related Disorders from a Mind-Body Perspective* (New York: Norton, 2005), 152; A. C. Davidson and D. J. Mellor, "The Adjustment of Children of Australian Vietnam Veterans: Is There Evidence for the Transgenerational Transmission of the Effects of War-Related Trauma?" *Australian and New Zealand Journal of Psychiatry* 35, no. 3 (2001): 345–351; F. A. Al-Turkait and J. U. Ohaeri, "Psychopathological Status, Behavior Problems, and Family Adjustment of Kuwaiti Children Whose Fathers Were Involved in the First Gulf War," *Child and Adolescent Psychiatry and Mental Health* 2, no. 1 (2008): 1–12; P. Lester et al., "The Long War and Parental Combat Deployment: Effects on Military Children and at-Home Spouses," *Journal of the American Academy of Child and Adolescent Psychiatry* 49 (2010): 310–320.

142 **several studies about other kinds of family hardship corroborate their cascading effects:** K. J. Kim et al., "Reciprocal Influences between Stressful Life Events and Adolescent Internalizing and Externalizing Problems," *Child Development* 74, no. 1 (2003): 127–143; R. D. Conger et al., "A Family Process Model of Economic Hardship and Adjustment of Early Adolescent Boys," *Child Development* 63, no. 3 (1992): 526–541; M. Cui and R. D. Conger, "Parenting Behavior as Mediator and Moderator of the Association between Marital Problems and Adolescent Maladjustment," *Journal of Research on Adolescence* 18, no. 2 (2008): 261–284; M. Cui et al., "Intergenerational Transmission of Relationship Aggression: A Prospective Longitudinal Study," *Journal of Family Psychology* 24, no. 6 (2010): 688–697; T. J. Schofield et al., "Harsh Parenting, Physical Health, and the Protective Role of Positive Parent-Adolescent Relationships," *Social Science and Medicine* 157 (2016): 18–26; R. D. Conger and K. J. Conger, "Understanding the Processes through Which Economic Hardship Influences Families and Children," *Handbook of Families and Poverty* 5 (2008): 64–78; K. A. S. Wickrama et al., "Family Antecedents and Consequences of Trajectories of Depressive Symptoms from Adolescence to Young Adulthood: A Life Course Investigation," *Journal of Health and Social Behavior* 49, no. 4 (2008): 468–483.

143 **It asked twenty-five thousand Kaiser patients in the San Diego area:** Jane Ellen Stevens, "The Adverse Childhood Experiences Study—The Largest, Most Important Public Health Study You Never Heard of—Began in an Obesity Clinic," acestoohigh .com/2012/10/03/the-adverse-childhood-experiences-study-the-largest-most-important -public-health-study-you-never-heard-of-began-in-an-obesity-clinic/.

143 **only 36 percent reported no ACEs:** V. J. Felitti et al., "Relationship of Childhood Abuse and Household Dysfunction to Many of the Leading Causes of Death in Adults: The Adverse Childhood Experiences (ACE) Study," *American Journal of Preventive Medicine* 14, no. 4 (1998): 245–258.

144 **One study tracked a cohort of minority children:** I'm using three or more here, because the Chicago study did not specify four or more. J. P. Mersky, J. Topitzes, and

A. J. Reynolds, "Impacts of Adverse Childhood Experiences on Health, Mental Health, and Substance Use in Early Adulthood: A Cohort Study of an Urban, Minority Sample in the U.S.," *Child Abuse and Neglect* 37, no. 11 (2013): 917–925. See also N. J. Burke et al., "The Impact of Adverse Childhood Experiences on an Urban Pediatric Population," *Child Abuse and Neglect* 35, no. 6 (2011): 408–413; M. E. Jimenez et al., "Adverse Childhood Experiences and ADHD Diagnosis at Age 9 Years in a National Urban Sample," *Academic Pediatrics* 17, no. 4 (2017): 356–361.

144 **U.S. military service-members during the all-volunteer force (AVF) era:** J. R. Blosnich et al., "Disparities in Adverse Childhood Experiences among Individuals with a History of Military Service," *JAMA Psychiatry* 71, no. 9 (2014): 1041–1048.

144 **One recent study examined more than sixty thousand Americans:** The eleven categories measured in this study are household mental illness, parental separation or divorce, household drug use, household alcohol abuse, household physical abuse, incarcerated household member, exposure to domestic violence, emotional abuse, touched sexually, made to touch another sexually, forced to have sex.

145 **consistent with earlier research suggesting that military populations may experience more ACEs:** A. C. Iverson et al., "Influence of Childhood Adversity on Health among Male UK Military Personnel," *British Journal of Psychiatry* 191 (2007): 506–511; U. A. Kelly et al., "More Than Military Sexual Trauma: Interpersonal Violence, PTSD, and Mental Health in Women Veterans," *Research in Nursing and Health* 34, no. 6 (2011): 457–467; T. Woodruff, R. Kelty, and D. R. Segal, "Propensity to Serve and Motivation to Enlist among American Combat Soldiers," *Armed Forces and Society* 32, no. 3 (2006): 353–366; J. R. Schultz et al., "Child Sexual Abuse and Adulthood Sexual Assault among Military Veteran and Civilian Women," *Military Medicine* 171, no. 8 (2006): 723–728; L. Trent et al., "Alcohol Abuse among U.S. Navy Recruits Who Were Maltreated in Childhood," *Alcohol and Alcoholism* 42, no. 4 (2007): 370–375; Kathy Roth-Douquet and Frank Schaffer, *AWOL: The Unexcused Absence of America's Upper Classes from Military Service—And How It Hurts Our Country* (New York: HarperCollins, 2006).

145 **other high-stress professions haven't received as much systematic study:** See, for example, I. H. Stanley et al., "Career Prevalence and Correlates of Suicidal Thoughts and Behaviors among Firefighters," *Journal of Affective Disorders* 187 (2015): 163–171.

145 **adults—especially men—tend to underreport their ACE experiences:** J. Hardt and M. Rutter, "Validity of Adult Retrospective Reports of Adverse Childhood Experiences: Review of the Evidence," *Journal of Child Psychology and Psychiatry* 45, no. 2 (2004): 260–273.

146 **research shows that early-life chronic stress leads to two structural changes in the developing brain:** R. J. Davidson, D. C. Jackson, and N. H. Kalin, "Emotion, Plasticity, Context, and Regulation: Perspectives from Affective Neuroscience," *Psychological Bulletin* 126, no. 6 (2000): 890–909; J. L. Hanson et al., "Structural Variations in Prefrontal Cortex Mediate the Relationship between Early Childhood Stress and Spatial Working Memory," *Journal of Neuroscience* 32, no. 23 (2012): 7917–7925; R. J. Davidson and B. S. McEwen, "Social Influences on Neuroplasticity: Stress and Interventions to Promote Well-Being," *Nature Neuroscience* 15, no. 5 (2012): 689–695; Sapolsky, *Behave*, 194–201; S. J. Lupien et al., "Effects of Stress throughout the Lifespan on the Brain, Behaviour and Cognition," *Nature Reviews Neuroscience* 10, no. 6 (2009): 434–445; V. G. Carrion, C. F. Weems, and A. L. Reiss, "Stress Predicts Brain Changes in Children: A Pilot Longitudinal Study on Youth Stress, Posttraumatic Stress Disorder, and the Hippocampus," *Pediatrics* 119, no. 3 (2007): 509–516; F. L. Woon and D. W. Hedges, "Hippocampal and Amygdala Volumes in Children and Adults with Childhood Maltreatment-Related Posttraumatic Stress Disorder: A Meta-Analysis," *Hippocampus* 18, no. 8 (2008): 729–736.

147 **hyperreactive amygdalae linked with childhood adversity increase the risk of developing anxiety disorders:** Sapolsky, *Behave*, 194–201.

147 **If someone first experiences an anxiety disorder, depression, or another mood disorder during childhood or puberty:** M. M. Weissman et al., "Depressed Adolescents Grown Up," *Journal of the American Medical Association* 281, no. 18 (1999): 1707–1713; M. M. Weissman et al., "Children with Prepubertal-Onset Major Depressive Disorder and Anxiety Grown Up," *Archives of General Psychiatry* 56, no. 9 (1999): 794–801.

147 **children who witness frequent conflict between their parents:** A. C. Schermerhorn, "Associations of Child Emotion Recognition with Interparental Conflict and Shy Child Temperament Traits," *Journal of Social and Personal Relationships* (2018): doi:10.1177/0265407518762606.

147 **Each time kindling occurs:** Scaer, *The Trauma Spectrum,* 62–64; Pat Ogden and Janina Fisher, *Sensorimotor Psychotherapy: Interventions for Trauma and Attachment* (New York: Norton, 2015), 181.

147 **chronic stress and trauma during childhood compromises the development of our ventral PSNS circuit:** Stephen W. Porges, *The Polyvagal Theory: Neurophysiological Foundations of Emotions, Attachment, Communication, and Self-Regulation* (New York: Norton, 2011), chap. 16.

148 **childhood adversity impairs and dysregulates our endocrine (hormone) system:** Neigh et al., "The Neurobiological Toll of Child Abuse and Neglect."; J. J. Mann and D. M. Currier, "Stress, Genetics and Epigenetic Effects on the Neurobiology of Suicidal Behavior and Depression," *European Psychiatry* 25, no. 5 (2010): 268–271.

148 **childhood adversity also adversely affects the immune system:** G. E. Miller, E. Chen, and K. J. Parker, "Psychological Stress in Childhood and Susceptibility to the Chronic Diseases of Aging: Moving toward a Model of Behavioral and Biological Mechanisms," *Psychological Bulletin* 137, no. 6 (2011): 959–97; G. E. Miller and E. Chen, "Harsh Family Climate in Early Life Presages the Emergence of a Proinflammatory Phenotype in Adolescence," *Psychological Science* 21, no. 6 (2010): 848–856; G. Arango Duque and A. Descoteaux, "Macrophage Cytokines: Involvement in Immunity and Infectious Diseases," *Frontiers in Immunology* 5, no. 491 (2014): 1–12; N. Slopen et al., "Early Origins of Inflammation: An Examination of Prenatal and Childhood Social Adversity in a Prospective Cohort Study," *Psychoneuroendocrinology* 51 (2015): 403–413; C. P. Fagundes, R. Glaser, and J. K. Kiecolt-Glaser, "Stressful Early Life Experiences and Immune Dysregulation across the Lifespan," *Brain, Behavior, and Immunity* 27, no. 1 (2013): 8–12; R. Nusslock and G. E. Miller, "Early-Life Adversity and Physical and Emotional Health across the Lifespan: A Neuroimmune Network Hypothesis," *Biological Psychiatry* 80, no. 1 (2016): 23–32; Gary Kaplan and Donna Beech, *Total Recovery: Solving the Mystery of Chronic Pain and Depression* (New York: Rodale, 2014), chaps. 4, 7.

148 **chronic inflammation can manifest in many different ways:** V. H. Perry and C. Holmes, "Microglial Priming in Neurodegenerative Disease," *Nature Reviews Neurology* 10, no. 4 (2014): 217–224; V. H. Perry and J. Teeling, "Microglia and Macrophages of the Central Nervous System: The Contribution of Microglia Priming and Systemic Inflammation to Chronic Neurodegeneration," *Seminars in Immunopathology* 35, no. 5 (2013): 601–612; Kaplan and Beech, *Total Recovery,* chap. 4; C. M. Eklund, "Proinflammatory Cytokines in CRP Baseline Regulation," *Advances in Clinical Chemistry* 48 (2009): 111–136; A. A. Appleton et al., "Divergent Associations of Adaptive and Maladaptive Emotion Regulation Strategies with Inflammation," *Health Psychology* 32, no. 7 (2013): 748–756; Bruce S. McEwen and Elizabeth Norton Lasley, *The End of Stress as We Know It* (Washington, D.C.: Joseph Henry, 2002), chap. 6.

149 **Cortisol *overproduction* has been linked:** McEwen and Lasley, *The End of Stress as We Know It,* 64, chap. 6.

149 **The tendency toward addiction after childhood adversity likely comes from three neurobiological adaptations:** Sapolsky, *Behave,* 196–197; J. T. Yorgason et al., "Social Isolation Rearing Increases Dopamine Uptake and Psychostimulant Potency in the

Striatum," *Neuropharmacology* 101 (2016): 471–479; L. M. Oswald et al., "History of Childhood Adversity Is Positively Associated with Ventral Striatal Dopamine Responses to Amphetamine," *Psychopharmacology* 231, no. 12 (2014): 2417–2433; A. N. Karkhanis et al., "Social Isolation Rearing Increases Nucleus Accumbens Dopamine and Norepinephrine Responses to Acute Ethanol in Adulthood," *Alcoholism: Clinical and Experimental Research* 38, no. 11 (2014): 2770–2779; T. R. Butler et al., "Adolescent Social Isolation as a Model of Heightened Vulnerability to Comorbid Alcoholism and Anxiety Disorders," *Alcoholism: Clinical and Experimental Research* 40, no. 6 (2016): 1202–1214; A. N. Karkhanis et al., "Early-Life Social Isolation Stress Increases Kappa Opioid Receptor Responsiveness and Downregulates the Dopamine System," *Neuropsychopharmacology* 41, no. 9 (2016): 2263–2274.

149 as Gabor Maté, a physician who treats and researches addictions, explains: Gabor Maté, *In the Realm of Hungry Ghosts: Close Encounters with Addiction* (Berkeley, CA: North Atlantic, 2010), 171, chaps. 13, 15.

149 the brain's dopamine system becomes "lazy": Maté, *In the Realm of Hungry Ghosts*, 42.

150 Childhood dysregulation of the dopamine system and the HPA axis may also contribute to depression: Sapolsky, *Behave*, 197.

150 In animal studies, learned helplessness gets conditioned: Scaer, *The Trauma Spectrum*, 56; Bessel A. van der Kolk, *The Body Keeps the Score: Brain, Mind, and Body in the Healing of Trauma* (New York: Penguin, 2015), 29–31; Christopher Peterson, Steven F. Maier, and Martin E. P. Seligman, *Learned Helplessness: A Theory for the Age of Personal Control* (New York: Oxford University Press, 1993); B. A. van der Kolk et al., "Inescapable Shock, Neurotransmitters, and Addiction to Trauma: Toward a Psychobiology of Post Traumatic Stress," *Biological Psychiatry* 20, no. 3 (1985): 314–325.

150 Learned helplessness may contribute to depression in adulthood: Scaer, *The Trauma Spectrum*, 56–57; Sapolsky, *Behave*, 197; C. Anacker, K. J. O'Donnell, and M. J. Meaney, "Early Life Adversity and the Epigenetic Programming of Hypothalamic-Pituitary-Adrenal Function," *Dialogues in Clinical Neuroscience* 16, no. 3 (2014): 321–333; Peterson et al., *Learned Helplessness*.

150 Learned helplessness can also contribute to conditioning a default survival strategy: Scaer, *The Trauma Spectrum*, 56–57; B. A. van der Kolk, "Clinical Implications of Neuroscience Research in PTSD," *Annals of the New York Academy of Sciences* 1071, no. 1 (2006): 277–293.

151 A . . . consequence of childhood adversity is dysregulation of the endogenous opioid system . . . and trauma reenactment: Scaer, *The Trauma Spectrum*, 88; B. A. van der Kolk, "The Compulsion to Repeat the Trauma," *Psychiatric Clinics of North America* 12, no. 2 (1989): 389–411; Peter A. Levine, *Waking the Tiger: Healing Trauma* (Berkeley, CA: North Atlantic, 1997), 173.

151 we're more likely to simply reenact the implicit memory: Scaer, *The Trauma Spectrum*, 88–95; van der Kolk, *The Body Keeps the Score*, 31–33; Ogden and Fisher, *Sensorimotor Psychotherapy*, chap. 21; Levine, *Waking the Tiger*, chap. 13; P. Payne and M. A. Crane-Godreau, "The Preparatory Set: A Novel Approach to Understanding Stress, Trauma, and the Bodymind Therapies," *Frontiers in Human Neuroscience* 9, no. 178 (2015): doi:10.3389/fnhum.2015.00178; P. Payne, P. A. Levine, and M. A. Crane-Godreau, "Somatic Experiencing: Using Interoception and Proprioception as Core Elements of Trauma Therapy," *Frontiers in Psychology* 6, no. 93 (2015): doi:10.3389/fpsyg.2015.00093.

152 many traumatized people seek out high-arousal activities: van der Kolk, *The Body Keeps the Score*, 31; van der Kolk, "The Compulsion to Repeat the Trauma."

152 these activities may trigger an endorphin rush: van der Kolk, "The Compulsion to Repeat the Trauma"; Scaer, *The Trauma Spectrum*, 89–92; T. Woodman, N. Cazenave, and C. Le Scanff, "Skydiving as Emotion Regulation: The Rise and Fall of Anxiety Is Moderated by Alexithymia," *Journal of Sport and Exercise Psychology* 30, no. 3 (2008): 424–433.

152 **A similar pattern exists in abusive relationships:** van der Kolk, "The Compulsion to Repeat the Trauma," 400. See also D. G. Dutton and S. L. Painter, "Traumatic Bonding: The Development of Emotional Attachments in Battered Women and Other Relationships of Intermittent Abuse," *Victimology: An International Journal* 6, no. 1–4 (1981): 139–155; Lenore E. A. Walker, *The Battered Woman Syndrome*, 4th ed. (New York: Springer, 2017).

153 **For school-aged children, ACE exposure has been linked:** Jimenez et al., "Adverse Childhood Experiences"; J. C. Spilsbury et al., "Profiles of Behavioral Problems in Children Who Witness Domestic Violence," *Violence and Victims* 23, no. 1 (2008): 3–17; Burke et al., "The Impact of Adverse Childhood Experiences"; van der Kolk, *The Body Keeps the Score*, chap. 9.

153 **For adults, ACE exposure has been linked with a wide range of health problems:** Mersky et al., "Impacts of Adverse Childhood Experiences"; E. A. Greenfield, "Child Abuse as a Life-Course Social Determinant of Adult Health," *Maturitas* 66, no. 1 (2010): 51–55; K. W. Springer et al., "Long-Term Physical and Mental Health Consequences of Childhood Physical Abuse: Results from a Large Population-Based Sample of Men and Women," *Child Abuse and Neglect* 31, no. 5 (2007): 517–530; K. W. Springer, "Childhood Physical Abuse and Midlife Physical Health: Testing a Multi-Pathway Life Course Model," *Social Science and Medicine* 69, no. 1 (2009): 138–146; D. P. Chapman et al., "Adverse Childhood Experiences and the Risk of Depressive Disorders in Adulthood," *Journal of Affective Disorders* 82, no. 2 (2004): 217–225; Felitti et al., "Relationship of Childhood Abuse and Household Dysfunction"; V. J. Felitti, "Adverse Childhood Experiences and Adult Health," *Academic Pediatrics* 9, no. 3 (2009): 131–132; D. W. Brown et al., "Adverse Childhood Experiences and the Risk of Premature Mortality," *American Journal of Preventive Medicine* 37, no. 5 (2009): 389–396; R. Bruffaerts et al., "Childhood Adversities as Risk Factors for Onset and Persistence of Suicidal Behaviour," *British Journal of Psychiatry* 197, no. 1 (2010): 20–27; R. C. Kessler et al., "Childhood Adversities and Adult Psychopathology in the WHO World Mental Health Surveys," *British Journal of Psychiatry* 197, no. 5 (2010): 378–385; S. R. Dube et al., "Childhood Abuse, Household Dysfunction, and the Risk of Attempted Suicide throughout the Life Span: Findings from the Adverse Childhood Experiences Study," *Journal of the American Medical Association* 286, no. 24 (2001): 3089–3096; S. R. Dube et al., "Exposure to Abuse, Neglect, and Household Dysfunction among Adults Who Witnessed Intimate Partner Violence as Children: Implications for Health and Social Services," *Violence and Victims* 17, no. 1 (2002): 3–17; S. R. Dube et al., "Long-Term Consequences of Childhood Sexual Abuse by Gender of Victim," *American Journal of Preventive Medicine* 28, no. 5 (2005): 430–438; S. R. Dube et al., "Childhood Abuse, Neglect, and Household Dysfunction and the Risk of Illicit Drug Use: The Adverse Childhood Experiences Study," *Pediatrics* 111, no. 3 (2003): 564–572; B. S. Brodsky and B. Stanley, "Adverse Childhood Experiences and Suicidal Behavior," *Psychiatric Clinics of North America* 31, no. 2 (2008): 223–235; R. F. Anda et al., "Adverse Childhood Experiences and Smoking during Adolescence and Adulthood," *Journal of the American Medical Association* 282, no. 17 (1999): 1652–1658; R. F. Anda et al., "The Enduring Effects of Abuse and Related Adverse Experiences in Childhood," *European Archives of Psychiatry and Clinical Neuroscience* 256, no. 3 (2006): 174–186; van der Kolk, *The Body Keeps the Score*, chap. 9; Sapolsky, *Behave*, 194–197; Mann and Currier, "Stress, Genetics and Epigenetic Effects"; Neigh et al., "The Neurobiological Toll of Child Abuse and Neglect."

154 **people with ACE histories end up experiencing abuse and violence again—or abusing others—as adolescents and adults:** van der Kolk, *The Body Keeps the Score*, 146–147, 68; L. M. Renner and K. S. Slack, "Intimate Partner Violence and Child Maltreatment: Understanding Intra- and Intergenerational Connections," *Child Abuse and Neglect* 30, no. 6 (2006): 599–617; S. Desai et al., "Childhood Victimization and Subsequent Adult Revictimization Assessed in a Nationally Representative Sample of

Women and Men," *Violence and Victims* 17, no. 6 (2002): 639–653; C. S. Widom, S. J. Czaja, and M. A. Dutton, "Childhood Victimization and Lifetime Revictimization," *Child Abuse and Neglect* 32, no. 8 (2008): 785–796; Sapolsky, *Behave*, 194–201; A. B. Amstadter et al., "Predictors of Physical Assault Victimization: Findings from the National Survey of Adolescents," *Addictive Behaviors* 36, no. 8 (2011): 814–820; A. M. Begle et al., "Longitudinal Pathways of Victimization, Substance Use, and Delinquency: Findings from the National Survey of Adolescents," *Addictive Behaviors* 36, no. 7 (2011): 682–689; R. Acierno et al., "Risk Factors for Rape, Physical Assault, and Posttraumatic Stress Disorder in Women," *Journal of Anxiety Disorders* 13, no. 6 (1999): 541–563; J. R. Cougle, H. Resnick, and D. G. Kilpatrick, "Does Prior Exposure to Interpersonal Violence Increase Risk of PTSD Following Subsequent Exposure?," *Behaviour Research and Therapy* 47, no. 12 (2009): 1012–1017; S. L. Buka et al., "Youth Exposure to Violence: Prevalence, Risks and Consequences," *American Journal of Orthopsychiatry* 71, no. 3 (2001): 298–310; J. B. Bingenheimer, R. T. Brennan, and F. J. Earls, "Firearm Violence Exposure and Serious Violent Behavior," *Science* 308, no. 5726 (2005): 1323–1326; R. T. Leeb, L. E. Barker, and T. W. Strine, "The Effect of Childhood Physical and Sexual Abuse on Adolescent Weapon Carrying," *Journal of Adolescent Health* 40, no. 6 (2007): 551–558; R. Spano, C. Rivera, and J. M. Bolland, "Are Chronic Exposure to Violence and Chronic Violent Behavior Closely Related Developmental Processes during Adolescence?," *Criminal Justice and Behavior* 37, no. 10 (2010): 1160–1179; Cui et al., "Intergenerational Transmission of Relationship Aggression"; Audra Burch, "Linking Childhood Trauma to Prison's Revolving Door," *New York Times*, October 16, 2017.

159 **you could also take the ACE survey:** You can find it at acestudy.org/the-ace-score .html.

CHAPTER 8

163 **Shock trauma's effects are magnified:** M. Laurie Leitch and Elaine Miller-Karas, *Veterans' Resiliency Model, Level 1 Training Manual* (Claremont, CA: Trauma Resource Institute, 2009), 25.

164 **The width of the window we bring to a crisis depends on our allostatic load:** Bruce S. McEwen and Elizabeth Norton Lasley, *The End of Stress as We Know It* (Washington, D.C.: Joseph Henry, 2002), 6–10, 29–33, 55–66; Robert M. Sapolsky, *Why Zebras Don't Get Ulcers*, 3rd ed. (New York: Holt, 2004), 9–16; Pat Ogden, Kekuni Minton, and Clare Pain, *Trauma and the Body: A Sensorimotor Approach to Psychotherapy* (New York: Norton, 2006), 33–40.

165 **Israeli leaders dismissed several warnings:** Charles Duhigg, *Smarter Faster Better: The Transformative Power of Real Productivity* (New York: Random House, 2016), 103–106, 109–115.

166 **As the researchers note, Dayan's behavior:** U. Bar-Joseph and R. McDermott, "Personal Functioning under Stress: Accountability and Social Support of Israeli Leaders in the Yom Kippur War," *Journal of Conflict Resolution* 52, no. 1 (2008): 144–170.

166 **Dayan made poor decisions during the war:** Bar-Joseph and McDermott, "Personal Functioning under Stress," 156.

166 **Dayan's social engagement skills also showed strain:** Bar-Joseph and McDermott, "Personal Functioning under Stress," 159.

167 **Elazar "did not show any dramatic or notable signs of distress":** Bar-Joseph and McDermott, "Personal Functioning under Stress," 164–65.

167 **Elazar was also able to offer and receive social support:** Bar-Joseph and McDermott, "Personal Functioning under Stress," 165.

170 **soldiers joined the U.S. Army with significantly higher rates:** R. C. Kessler et al., "Thirty-Day Prevalence of DSM-IV Mental Disorders among Nondeployed Soldiers in the U.S. Army: Results from the Army Study to Assess Risk and Resilience in Servicemembers (Army Starrs)," *JAMA Psychiatry* 71, no. 5 (2014): 504–513.

170 the risk of being raped as an adult is six times as high: Bessel A. van der Kolk, *The Body Keeps the Score: Brain, Mind, and Body in the Healing of Trauma* (New York: Penguin, 2015), 146–147; S. Desai et al., "Childhood Victimization and Subsequent Adult Revictimization Assessed in a Nationally Representative Sample of Women and Men," *Violence and Victims* 17, no. 6 (2002): 639–653; R. Spano, C. Rivera, and J. M. Bolland, "Are Chronic Exposure to Violence and Chronic Violent Behavior Closely Related Developmental Processes during Adolescence?," *Criminal Justice and Behavior* 37, no. 10 (2010): 1160–1179; R. T. Leeb, L. E. Barker, and T. W. Strine, "The Effect of Childhood Physical and Sexual Abuse on Adolescent Weapon Carrying," *Journal of Adolescent Health* 40, no. 6 (2007): 551–558; R. Acierno et al., "Risk Factors for Rape, Physical Assault, and Posttraumatic Stress Disorder in Women," *Journal of Anxiety Disorders* 13, no. 6 (1999): 541–563.

171 peritraumatic dissociation . . . has been shown to be the single biggest predictor of developing PTSD: E. Ozer et al., "Predictors of Posttraumatic Stress Disorder and Symptoms in Adults: A Meta-Analysis," *Psychological Bulletin* 129, no. 1 (2003): 52–73. See also C. W. Hoge et al., "Association of Posttraumatic Stress Disorder with Somatic Symptoms, Health Care Visits, and Absenteeism among Iraq War Veterans," *American Journal of Psychiatry* 164, no. 1 (2007): 150–153; J. Douglas Bremner, *Does Stress Damage the Brain? Understanding Trauma-Related Disorders from a Mind-Body Perspective* (New York: Norton, 2005), chap. 1; Robert C. Scaer, *The Trauma Spectrum: Hidden Wounds and Human Resiliency* (New York: Norton, 2005), chap. 9.

171 Some people who've experienced an MVA . . . develop a group of symptoms: Scaer, *The Trauma Spectrum*, 228–229.

171 a strong link between the people who eventually developed whiplash syndrome and their prior developmental and/or relational trauma: Scaer, *The Trauma Spectrum*, 228.

172 several studies with American, Canadian, and British troops and veterans: G. J. G. Asmundson, K. D. Wright, and M. B. Stein, "Pain and PTSD Symptoms in Female Veterans," *European Journal of Pain* 8, no. 4 (2004): 345–350; C. R. Brewin, B. Andrews, and J. D. Valentine, "Meta-Analysis of Risk Factors for Posttraumatic Stress Disorder in Trauma-Exposed Adults," *Journal of Consulting and Clinical Psychology* 68, no. 5 (2000): 345–350; J. D. Bremner et al., "Childhood Physical Abuse and Combat-Related Posttraumatic Stress Disorder in Vietnam Veterans," *American Journal of Psychiatry* 150, no. 2 (1993): 235–239; K. G. Lapp et al., "Lifetime Sexual and Physical Victimization among Male Veterans with Combat-Related Post-Traumatic Stress Disorder," *Military Medicine* 170, no. 9 (2005): 787–790; A. C. Iverson et al., "Influence of Childhood Adversity on Health among Male UK Military Personnel," *British Journal of Psychiatry* 191 (2007): 506–511; C. A. LeardMann, B. Smith, and M. A. Ryan, "Do Adverse Childhood Experiences Increase the Risk of Postdeployment Posttraumatic Stress Disorder in U.S. Marines?," *BMC Public Health* 10, no. 437 (2010): 1–8; O. A. Cabrera et al., "Childhood Adversity and Combat as Predictors of Depression and Post-Traumatic Stress in Deployed Troops," *American Journal of Preventive Medicine* 33, no. 2 (2007): 77–82; A. M. Fritch et al., "The Impact of Childhood Abuse and Combat-Related Trauma on Postdeployment Adjustment," *Journal of Traumatic Stress* 23, no. 2 (2010): 248–254; E. A. Dedert et al., "Association of Trauma Exposure with Psychiatric Morbidity in Military Veterans Who Have Served since September 11, 2001," *Journal of Psychiatric Research* 43, no. 9 (2009): 830–836; G. A. Gahm et al., "Relative Impact of Adverse Events and Screened Symptoms of Posttraumatic Stress Disorder and Depression among Active Duty Soldiers Seeking Mental Health Care," *Journal of Clinical Psychology* 63, no. 3 (2007): 199–211; L. Trent et al., "Alcohol Abuse among U.S. Navy Recruits Who Were Maltreated in Childhood," *Alcohol and Alcoholism* 42, no. 4 (2007): 370–375; T. C. Smith et al., "Prior Assault and Posttraumatic Stress Disorder after Combat Deployment," *Epidemiology* 19, no. 3 (2008): 505–512; C. P. Clancy et al., "Lifetime

Trauma Exposure in Veterans with Military-Related Posttraumatic Stress Disorder: Association with Current Symptomatology," *Journal of Clinical Psychiatry* 67, no. 9 (2006): 1346–1353; J. Sareen et al., "Adverse Childhood Experiences in Relation to Mood and Anxiety Disorders in a Population-Based Sample of Active Military Personnel," *Psychological Medicine* 43, no. 01 (2013): 73–84.

172 **Several studies with paramedics and police show a similar pattern:** N. Pole et al., "Associations between Childhood Trauma and Emotion-Modulated Psychophysiological Responses to Startling Sounds: A Study of Police Cadets," *Journal of Abnormal Psychology* 116, no. 2 (2007): 352–361; R. G. Maunder et al., "Symptoms and Responses to Critical Incidents in Paramedics Who Have Experienced Childhood Abuse and Neglect," *Emergency Medicine Journal* 29, no. 3 (2012): 222–227; V. M. Follette, M. M. Polusny, and K. Milbeck, "Mental Health and Law Enforcement Professionals: Trauma History, Psychological Symptoms, and Impact of Providing Services to Child Sexual Abuse Survivors," *Professional Psychology: Research and Practice* 25, no. 3 (1994): 275–282; C. Otte et al., "Association between Childhood Trauma and Catecholamine Response to Psychological Stress in Police Academy Recruits," *Biological Psychiatry* 57, no. 1 (2005): 27–32; L. M. Rouse et al., "Law Enforcement Suicide Discerning Etiology through Psychological Autopsy," *Police Quarterly* 18, no. 1 (2015): 79–108.

173 **a year after Hurricane Harvey flooded Houston:** Manny Fernandez, "1 Year Later, Relief Stalls for Poorest in Houston," *New York Times*, September 3, 2018.

CHAPTER 9

179 **Most Americans aren't getting enough sleep:** National Sleep Foundation, "Sleep Health Index 2014—Highlights," sleepfoundation.org/sleep-health-index-2014 -highlights; P. M. Krueger and E. M. Friedman, "Sleep Duration in the United States: A Cross-Sectional Population-Based Study," *American Journal of Epidemiology* 169, no. 9 (2009): 1052–1063.

179 **According to the Department of Labor's annual time use surveys:** L. Hale, "Who Has Time to Sleep?," *Journal of Public Health* 27, no. 2 (2005): 205–211; M. Basner et al., "American Time Use Survey: Sleep Time and Its Relationship to Waking Activities," *Sleep* 30, no. 9 (2007): 1085–1095; Maggie Jones, "How Little Sleep Can You Get Away With," *New York Times Magazine*, April 11, 2011.

179 **One recent study with five thousand U.S. and Canadian police:** S. M. W. Rajaratnam et al., "Sleep Disorders, Health, and Safety in Police Officers," *Journal of the American Medical Association* 306, no. 23 (2011): 2567–2578; S. M. James et al., "Shift Work: Disrupted Circadian Rhythms and Sleep—Implications for Health and Well-Being," *Current Sleep Medicine Reports* 3, no. 2 (2017): 104–112; M. Price, "The Risks of Night Work," *Monitor on Psychology* 42, no. 1 (2011): 38; B. Vila, G. B. Morrison, and D. J. Kenney, "Improving Shift Schedule and Work-Hour Policies and Practices to Increase Police Officer Performance, Health, and Safety," *Police Quarterly* 5, no. 1 (2002): 4–24; T. C. Neylan et al., "Critical Incident Exposure and Sleep Quality in Police Officers," *Psychosomatic Medicine* 64, no. 2 (2002): 345–352; S. Garbarino et al., "Sleep Disorders and Daytime Sleepiness in State Police Shiftworkers," *Archives of Environmental Health* 57, no. 2 (2002): 167–173; D. X. Swenson, D. Waseleski, and R. Hartl, "Shift Work and Correctional Officers: Effects and Strategies for Adjustment," *Journal of Correctional Health Care* 14, no. 4 (2008): 299–310; L. K. Barger et al., "Neurobehavioral, Health, and Safety Consequences Associated with Shift Work in Safety-Sensitive Professions," *Current Neurology and Neuroscience Reports* 9, no. 2 (2009): 155–164; L. K. Barger et al., "Impact of Extended-Duration Shifts on Medical Errors, Adverse Events, and Attentional Failures," *PLOS Medicine* 3, no. 12 (2006): e487.

180 **Several studies show that service-members, both during deployment and after returning home:** Office of the Surgeon Multi-National Force–Iraq, Office of the Command Surgeon, and Office of the Surgeon General United States Army Medical

Command, *Mental Health Advisory Team (MHAT)-V Report—Operation Iraqi Freedom 06-08: Iraq, and Operation Enduring Freedom 8: Afghanistan* (Washington, D.C.: U.S. Army Medical Command, 2008); A. L. Peterson et al., "Sleep Disturbance during Military Deployment," *Military Medicine* 173, no. 3 (2008): 230–235; A. D. Seelig et al., "Sleep Patterns before, during, and after Deployment to Iraq and Afghanistan," *Sleep* 33, no. 12 (2010): 1615–1622.

180 **72 percent of three thousand soldiers in a U.S. Army brigade combat team:** D. D. Luxton et al., "Prevalence and Impact of Short Sleep Duration in Redeployed OIF Soldiers," *Sleep* 34, no. 9 (2011): 1189–1195. See also V. Mysliwiec et al., "Sleep Disorders and Associated Medical Comorbidities in Active Duty Military Personnel," *Sleep* 36, no. 2 (2013): 167–174.

180 **Differing responses to caffeine sensitivity:** H. P. A. van Dongen and G. Belenky, "Individual Differences in Vulnerability to Sleep Loss in the Work Environment," *Industrial Health* 47, no. 5 (2009): 518–526.

180 **Two different experiments subjected about one hundred healthy volunteers to differing lengths of sleep deprivation:** G. Belenky et al., "Patterns of Performance Degradation and Restoration during Sleep Restriction and Subsequent Recovery: A Sleep Dose-Response Study," *Journal of Sleep Research* 12, no. 1 (2003): 1–12; H. P. A. Van Dongen et al., "The Cumulative Cost of Additional Wakefulness: Dose-Response Effects on Neurobehavioral Functions and Sleep Physiology from Chronic Sleep Restriction and Total Sleep Deprivation," *Sleep* 26, no. 2 (2003): 117–126.

181 **Other research equates the cognitive impairment after twenty-four hours awake with a blood alcohol concentration of 0.1 percent:** National Sleep Foundation, "Drowsy Driving Prevention Week: Facts and Stats," drowsydriving.org/about/facts-and -stats/.

181 **Several studies have found this same result:** van Dongen and Belenky, "Individual Differences in Vulnerability to Sleep Loss"; R. Leproult et al., "Individual Differences in Subjective and Objective Alertness during Sleep Deprivation Are Stable and Unrelated," *American Journal of Physiology—Regulatory, Integrative and Comparative Physiology* 284, no. 2 (2003): R280–R290; H. P. A. Van Dongen et al., "Systematic Interindividual Differences in Neurobehavioral Impairment from Sleep Loss: Evidence of Trait-Like Differential Vulnerability," *Sleep* 27, no. 3 (2004): 423–433; Q. Mu et al., "Decreased Brain Activation during a Working Memory Task at Rested Baseline Is Associated with Vulnerability to Sleep Deprivation," *Sleep* 28, no. 4 (2005): 433–448.

181 **Sleep deprivation impairs executive functioning:** A. Gohar et al., "Working Memory Capacity Is Decreased in Sleep-Deprived Internal Medicine Residents," *Journal of Clinical Sleep Medicine* 5, no. 3 (2009): 191–197; J. Lim and D. F. Dinges, "A Meta-Analysis of the Impact of Short-Term Sleep Deprivation on Cognitive Variables," *Psychological Bulletin* 136 (2010): 375–389; J. C. Lo et al., "Effects of Partial and Acute Total Sleep Deprivation on Performance across Cognitive Domains, Individuals and Circadian Phase," *PLOS One* 7, no. 9 (2012): e45987; Q. Mu et al., "Decreased Cortical Response to Verbal Working Memory Following Sleep Deprivation," *Sleep* 28, no. 1 (2005): 55–67; M. E. Smith, L. K. McEvoy, and A. Gevins, "The Impact of Moderate Sleep Loss on Neurophysiologic Signals during Working-Memory Task Performance," *Sleep* 25, no. 7 (2002): 784–794; Barger et al., "Neurobehavioral, Health, and Safety Consequences"; H. Lee, L. Kim, and K. Suh, "Cognitive Deterioration and Changes of P300 during Total Sleep Deprivation," *Psychiatry and Clinical Neurosciences* 57, no. 5 (2003): 490–496.

181 **60 percent of adult drivers in the United States said they'd driven while drowsy:** National Sleep Foundation, "Drowsy Driving Prevention Week: Facts and Stats," drowsydriving.org/about/facts-and-stats/.

181 **U.S. drowsy drivers were responsible for one in five crashes:** Brian C. Tefft, *Prevalence of Motor Vehicle Crashes Involving Drowsy Drivers, United States, 2009–2013* (Washington, D.C.: AAA Foundation for Traffic Safety, 2014); Centers for Disease

Control and Prevention, "Drowsy Driving: Asleep at the Wheel," www.cdc.gov/features
/dsdrowsydriving/index.html.

182 **the two 2017 deadly collisions in the Pacific Ocean:** Dave Phillips and Eric Schmitt,
"Strains on Crews and Vessels Set the Stage for Navy Crashes," *New York Times*, August
28, 2017; Helene Cooper, "Navy Leaders Admit Fleet Is Pulled Thin," *New York Times*,
September 20, 2017.

182 **Numerous rail and truck crashes . . . have also been linked with sleep
deprivation:** Bill Chappell, "Regulators Pull Plan to Test Truckers, Train Operators for
Sleep Apnea," NPR, August 8, 2017.

182 **The Exxon Valdez disaster:** Rubin Naiman, "Seven Ways Inadequate Sleep Negatively
Impacts Health," interview, National Institute for the Clinical Application of Behavioral
Medicine (July 25, 2011), www.pacificariptide.com/files/interview-from-the-national
-institute-for-the-clinical-application-of-behavioral-medicine.pdf; A. Williamson et al.,
"The Link between Fatigue and Safety," *Accident Analysis and Prevention* 43, no. 2 (2011):
498–515.

182 **Fragmented sleep . . . may be at least as detrimental:** A. I. Luik et al., "Associations
of the 24-Hour Activity Rhythm and Sleep with Cognition: A Population-Based Study of
Middle-Aged and Elderly Persons," *Sleep Medicine* 16, no. 7 (2015): 850–855; A. I. Luik
et al., "24-Hour Activity Rhythm and Sleep Disturbances in Depression and Anxiety:
A Population-Based Study of Middle-Aged and Older Persons," *Depression and Anxiety* 32,
no. 9 (2015): 684–692; A. I. Luik et al., "Stability and Fragmentation of the Activity
Rhythm across the Sleep-Wake Cycle: The Importance of Age, Lifestyle, and Mental
Health," *Chronobiology International* 30, no. 10 (2013): 1223–1230.

182 **sleep deprivation may also adversely affect our social engagement . . . [and] ability
to make ethical decisions:** S. Yoo et al., "The Human Emotional Brain without
Sleep—a Prefrontal Amygdala Disconnect," *Current Biology* 17, no. 20 (2007): R877–
R878; W. D. S. Killgore et al., "Sleep Deprivation Impairs Recognition of Specific
Emotions," *Neurobiology of Sleep and Circadian Rhythms* 3, Suppl. C (2017): 10–16; B. L.
Reidy et al., "Decreased Sleep Duration Is Associated with Increased fMRI Responses to
Emotional Faces in Children," *Neuropsychologia* 84, Suppl. C (2016): 54–62; L. Beattie
et al., "Social Interactions, Emotion and Sleep: A Systematic Review and Research
Agenda," *Sleep Medicine Reviews* 24, Suppl. C (2015): 83–100; C. M. Barnes et al., "Lack of
Sleep and Unethical Conduct," *Organizational Behavior and Human Decision Processes* 115,
no. 2 (2011): 169–180; C. M. Barnes, B. C. Gunia, and D. T. Wagner, "Sleep and Moral
Awareness," *Journal of Sleep Research* 24, no. 2 (2015): 181–188; W. D. S. Killgore et al.,
"The Effects of 53 Hours of Sleep Deprivation on Moral Judgment," *Sleep* 30, no.
3 (2007): 345–352; O. K. Olsen, S. Pallesen, and E. Jarle, "The Impact of Partial
Sleep Deprivation on Moral Reasoning in Military Officers," *Sleep* 33, no. 8 (2010):
1086–1090.

183 **Chronic sleep deprivation is linked with three changes that make sleep-deprived
brains jumpier:** Yoo et al., "The Human Emotional Brain without Sleep"; E. J. van
Someren et al., "Disrupted Sleep: From Molecules to Cognition," *Journal of Neuroscience* 35,
no. 41 (2015): 13889–13895; Reidy et al., "Decreased Sleep Duration." Other recent
experimental research measuring the electrical activity in the cortex (the thinking brain)
shows that the longer we stay awake, the jumpier, more excitable, and hypersensitive
the sleep-deprived brain gets—responding to external stimuli with progressively stronger
spikes of activity. This progressive buildup of excitability only gets rebalanced when we
sleep. This finding helps to explain why our risk for hallucinations and seizures increases
the longer we stay awake, and why depressive symptoms dissipate during sleep
deprivation. R. Huber et al., "Human Cortical Excitability Increases with Time Awake,"
Cerebral Cortex 23, no. 2 (2012): 1–7.

183 **40 percent of police officers had at least one sleep disorder:** Rajaratnam et al., "Sleep
Disorders, Health, and Safety in Police Officers."

183 **firefighters with insomnia or nightmares:** M. A. Hom et al., "The Association between Sleep Disturbances and Depression among Firefighters: Emotion Dysregulation as an Explanatory Factor," *Journal of Clinical Sleep Medicine* 12, no. 2 (2016): 235–245.

183 **insomnia is the single strongest predictor of clinical depression:** Naiman, "Seven Ways Inadequate Sleep Negatively Impacts Health."

183 **chronic sleep deprivation and shift work can lead to detrimental epigenetic changes:** van Someren et al., "Disrupted Sleep"; James et al., "Shift Work"; K. Ackermann et al., "Effect of Sleep Deprivation on Rhythms of Clock Gene Expression and Melatonin in Humans," *Chronobiology International* 30, no. 7 (2013): 901–909.

183 **sleep-deprived people are more likely to have low HRV:** James et al., "Shift Work"; M. Glos et al., "Cardiac Autonomic Modulation and Sleepiness: Physiological Consequences of Sleep Deprivation Due to 40h of Prolonged Wakefulness," *Physiology and Behavior* 125, Suppl. C (2014): 45–53; D. Grimaldi et al., "Adverse Impact of Sleep Restriction and Circadian Misalignment on Autonomic Function in Healthy Young Adults," *Hypertension* 68, no. 1 (2016): 243–250.

183 **Sleep deprivation especially affects the stress hormone cortisol:** James et al., "Shift Work."

184 **Our circadian rhythm affects our mood:** Price, "The Risks of Night Work"; van Dongen and Belenky, "Individual Differences in Vulnerability to Sleep Loss"; James et al., "Shift Work"; Barger et al., "Neurobehavioral, Health, and Safety Consequences"; D. Dawson and K. Reid, "Fatigue, Alcohol and Performance Impairment," *Nature* 388, no. 6639 (1997): 235.

184 **increased cortisol levels interact with . . . hormones that control appetite:** van Someren et al., "Disrupted Sleep"; A. V. Nedeltcheva and F. Scheer, "Metabolic Effects of Sleep Disruption, Links to Obesity and Diabetes," *Current Opinion in Endocrinology, Diabetes, and Obesity* 21, no. 4 (2014): 293–298; James et al., "Shift Work."

184 **70 percent of Americans were overweight or obese:** Centers for Disease Control and Prevention, "Obesity and Overweight," www.cdc.gov/faststats.overwt.htm.

184 **One study tracked healthy volunteers' sleep duration and "sleep efficiency":** S. Cohen et al., "Sleep Habits and Susceptibility to the Common Cold," *Archives of Internal Medicine* 169, no. 1 (2009): 62–67.

185 **Other studies have tracked how chronic sleep deprivation harms our ability to create antibodies:** P. A. Bryant, J. Trinder, and N. Curtis, "Sick and Tired: Does Sleep Have a Vital Role in the Immune System?," *Nature Reviews Immunology* 4, no. 6 (2004): 457–467; K. Spiegel, J. F. Sheridan, and E. van Cauter, "Effect of Sleep Deprivation on Response to Immunizaton," *Journal of the American Medical Association* 288, no. 12 (2002): 1471–1472.

185 **Not getting enough sleep also increases our risk for chronic inflammation:** J. M. Mullington et al., "Sleep Loss and Inflammation," *Best Practice and Research: Clinical Endocrinology and Metabolism* 24, no. 5 (2010): 775–784; H. K. Meier-Ewert et al., "Effect of Sleep Loss on C-Reactive Protein, an Inflammatory Marker of Cardiovascular Risk," *Journal of the American College of Cardiology* 43, no. 4 (2004): 678–683; R. Leproult, U. Holmback, and E. van Canter, "Circadian Misalignment Augments Markers of Insulin Resistance and Inflammation, Independently of Sleep Loss," *Diabetes* 63, no. 6 (2014): 1860–1869; S. Floam et al., "Sleep Characteristics as Predictor Variables of Stress Systems Markers in Insomnia Disorder," *Journal of Sleep Research* 24, no. 3 (2015): 296–304.

185 **Circadian misalignment can also suppress melatonin production:** James et al., "Shift Work."

185 **in numerous polls since 1989:** See, for instance, Gallup, "In U.S., 55% of Workers Get Sense of Identity from Their Job," news.gallup.com/poll/175400/workers-sense-identity-job.aspx.

185 **Vacations are not considered a non-negotiable social right:** Juliet Schore, *The Overworked American: The Unexpected Decline of Leisure* (New York: Basic, 1993); Rebecca

Ray, Milla Sanes, and John Schmitt, *No-Vacation Nation Revisited* (Washington, D.C.: Center for Economic and Policy Research, 2013); G. Richards, "Vacations and the Quality of Life: Patterns and Structures," *Journal of Business Research* 44 (1999): 189–198.

185 **vacation benefits are not distributed equally:** Ray et al., *No-Vacation Nation Revisited.*

186 **In 1995, one third of U.S. workers took less than half of their vacation time:** Richards, "Vacations and the Quality of Life."

186 **54 percent still didn't use all their paid vacation:** Project Time Off, "The State of the American Vacation" (2017), www.projecttimeoff.com/state-american-vacation-2017; Harvard Chan School of Public Health, "The Workplace and Health" (2016), www.npr .org/documents/2016/jul/HarvardWorkplaceandHealthPollReport.pdf.

186 **Three perceived barriers to taking time off stood out . . . "work martyrs" were less likely:** Project Time Off, "The State of the American Vacation," 7–9; Harvard Chan School of Public Health, "The Workplace and Health," 5.

186 **taking time off can alleviate burnout:** Richards, "Vacations and the Quality of Life."

186 **American workaholics also put in long hours:** Harvard Chan School of Public Health, "The Workplace and Health."

187 **In another survey, millennials . . . were far more likely:** Daniel Victor, "If You're Sick, Stay Away from Work. If You Can't, Here Is What Doctors Advise," *New York Times*, November 13, 2017.

187 **some of this workaholism is likely not by choice:** Alissa Quart, *Squeezed: Why Our Families Can't Afford America* (New York: HarperCollins, 2018); David Love, "The Real National Emergency is Not at the Border," CNN (March 11, 2019), www.cnn.com /2019/03/11/opinions/national-emergency-is-economic-inequality-and-greed-love/index .html.

188 **Managing mortality concerns depletes our executive functioning capacity:** M. T. Gailliot, B. J. Schmeichel, and R. F. Baumeister, "Self-Regulatory Processes Defend against the Threat of Death: Effects of Self-Control Depletion and Trait Self-Control on Thoughts of Fear and Dying," *Journal of Personality and Social Psychology* 91, no. 4 (2006): 49–62.

188 **Several psychological stressors also may lead our survival brain to neurocept danger:** Robert C. Scaer, *The Trauma Spectrum: Hidden Wounds and Human Resiliency* (New York: Norton, 2005), 132–133.

188 **studies with the British civil service:** M. G. Marmot et al., "Health Inequalities among British Civil Servants: The Whitehall II Study," *Lancet* 337, no. 8754 (1991): 1387–1393.

188 **people in low-status roles are more vigilant:** See, for example, P. K. Smith et al., "Lacking Power Impairs Executive Functions," *Psychological Science* 19, no. 5 (2008): 441–447.

189 **perhaps the most stressful combination of work characteristics:** Jon Kabat-Zinn, *Full Catastrophe Living (Revised Edition): Using the Wisdom of Your Body and Mind to Face Stress, Pain, and Illness* (New York: Bantam, 2013), chap. 30.

189 **95 percent of workers said working privately . . . was important:** Geoffrey James, "9 Reasons That Open-Space Offices Are Insanely Stupid," *Inc.*, February 25, 2016; J. Kim and R. de Dear, "Workspace Satisfaction: The Privacy-Communication Trade-Off in Open-Plan Offices," *Journal of Environmental Psychology* 36 (2013): 18–26; E. S. Bernstein and S. Turban, "The Impact of the 'Open' Workspace on Human Collaboration," *Philosophical Transactions of the Royal Society of London. Series B, Biological Sciences* 373 (2018): doi:10.1098/rstb.2017.0239.

190 **Emotional labor is common with jobs that require face-to-face or voice-to-voice contact with the public:** Arlie Russell Hochschild, *The Managed Heart: Commercialization of Human Feeling* (Berkeley: University of California Press, 2012).

190 **one study found that female managers were more likely:** Kristin Wong, "This Gender Gap Can't Be Stressed Enough," *New York Times*, November 26, 2018.

190 **emotional labor increases psychological strain and depletes mental resources:**
U. R. Hülsheger and A. F. Schewe, "On the Costs and Benefits of Emotional Labor:
A Meta-Analysis of Three Decades of Research," *Journal of Occupational Health Psychology*
16, no. 3 (2011): 361–389; M. J. Zyphur et al., "Self-Regulation and Performance in
High-Fidelity Simulations: An Extension of Ego-Depletion Research," *Human Performance*
20, no. 2 (2007): 103–118; L. S. Goldberg and A. A. Grandey, "Display Rules versus
Display Autonomy: Emotion Regulation, Emotional Exhaustion, and Task Performance
in a Call Center Simulation," *Journal of Occupational Health Psychology* 12 (2007): 301–318;
M. Muraven, D. M. Tice, and R. F. Baumeister, "Self-Control as a Limited Resource:
Regulatory Depletion Patterns," *Journal of Personality and Social Psychology* 74, no. 3
(1998): 774–789; D. Holman, D. Martinez-Iñigo, and P. Totterdell, "Emotional Labour,
Well-Being and Performance," in *The Oxford Handbook of Organizational Well-Being*, edited
by C. L. Cooper and S. Cartwright (Oxford, UK: Oxford University Press, 2008),
331–355; D. Martínez-Iñigo et al., "Emotional Labour and Emotional Exhaustion:
Interpersonal and Intrapersonal Mechanisms," *Work and Stress* 21 (2007): 30–47.

191 **empirical research shows that women consistently report higher stress levels than
men:** Wong, "This Gender Gap Can't Be Stressed Enough."

191 **partners in a romantic relationship form one neurobiological unit:** B. J. Baker
et al., "Marital Support, Spousal Contact, and the Course of Mild Hypertension," *Journal
of Psychosomatic Research* 55, no. 3 (2003): 229–233; J. A. Coan, H. S. Schaefer, and R. J.
Davidson, "Lending a Hand: Social Regulation of the Neural Response to Threat,"
Psychological Science 17, no. 12 (2006): 1032–1039; Amir Levine and Rachel S. F. Heller,
Attached: The New Science of Adult Attachment and How It Can Help You Find and Keep Love
(New York: Tarcher/Penguin, 2010), chap. 2.

191 **One study asked middle-aged women about their marriages:** W. M. Troxel et al.,
"Marital Quality and Occurrence of the Metabolic Syndrome in Women," *Archives of
Internal Medicine* 165, no. 9 (2005): 1022–1027.

192 **nearly half of Americans:** Arthur C. Brooks, "How Loneliness Is Tearing America
Apart," *New York Times*, November 24, 2018.

192 **people who are chronically lacking in social contacts:** D. Umberson and
J. Karas Montez, "Social Relationships and Health: A Flashpoint for Health Policy,"
Journal of Health and Social Behavior 51, no. 1 Suppl. (2010): S54–S66; Jane E. Brody,
"Social Interaction Is Critical for Mental and Physical Health," *New York Times*,
June 13, 2017.

192 **Kabat-Zinn calls the stress reaction cycle:** Kabat-Zinn, *Full Catastrophe Living*,
chap. 19.

195 **for every hour of interrupted sleep the previous night, people wasted 8.4 minutes:**
Gretchen Rubin, *Better Than Before* (New York: Crown, 2015), 52.

195 **being "too tired" is the most common reason that people give for procrastination:**
Piers Steel, *The Procrastination Equation: How to Stop Putting Things Off and Start Getting
Stuff Done* (New York: Harper, 2010), 147.

CHAPTER 10

199 **In turn, the mind-body system gets focused on the immediate needs of survival:**
Bruce S. McEwen and Elizabeth Norton Lasley, *The End of Stress as We Know It*
(Washington, D.C.: Joseph Henry, 2002), 73.

199 **anticipatory stress about pain is linked:** Gary Kaplan and Donna Beech, *Total
Recovery: Solving the Mystery of Chronic Pain and Depression* (New York: Rodale, 2014), 180;
Brian Resnick, "100 Million Americans Have Chronic Pain. Very Few Use One of the
Best Tools to Treat It," *Vox*, August 16, 2018, www.vox.com/science-and-health/2018/5
/17/17276452/chronic-pain-treatment-psychology-cbt-mindfulness-evidence.

204 **stress-spectrum illnesses, both physical and mental, share a common origin:**
J. Douglas Bremner, *Does Stress Damage the Brain? Understanding Trauma-Related Disorders*

from a Mind-Body Perspective (New York: Norton, 2005), 34–37; B. A. van der Kolk et al., "Disorders of Extreme Stress: The Empirical Foundation of a Complex Adaptation to Trauma," *Journal of Traumatic Stress* 18, no. 5 (2005): 389–399; Robert C. Scaer, *The Trauma Spectrum: Hidden Wounds and Human Resiliency* (New York: Norton, 2005), chap. 9.

205 **trauma clinicians say that about two thirds of the population tend toward "stuck on high"**: When pressed, several clinicians cite a brain imaging study of seven women who'd been diagnosed with PTSD after sexual and/or physical abuse. R. A. Lanius et al., "Brain Activation during Script-Driven Imagery Induced Dissociative Responses in PTSD: A Functional Magnetic Resonance Imaging Investigation," *Biological Psychiatry* 52, no. 4 (2002): 305–311. Researchers used script-driven imagery of each woman's traumatic event to elicit traumatic stress activation—and 70 percent of these women had stuck-on-high responses, while 30 percent had stuck-on-low responses. Given its narrow focus on women diagnosed with PTSD, however, this study is clearly not generalizable to both genders or to the downstream effects of all three pathways to building allostatic load.

205 **The HPA axis tends to produce elevated levels of adrenaline . . . the mind-body system is hyperactively trying to cope**: Robert M. Sapolsky, *Why Zebras Don't Get Ulcers*, 3rd ed. (New York: Holt, 2004), 319–320.

206 **symptoms associated with "stuck on high" include**: Pat Ogden, Kekuni Minton, and Clare Pain, *Trauma and the Body: A Sensorimotor Approach to Psychotherapy* (New York: Norton, 2006), 33–34; Peter A. Levine, *Waking the Tiger: Healing Trauma* (Berkeley, CA: North Atlantic, 1997), chap. 11; Scaer, *The Trauma Spectrum*, chap. 9.

206 **"stuck on low" behavioral responses fall along the freeze spectrum**: Levine, *Waking the Tiger*, 136–137.

206 **the HPA axis produces too much cortisol and other glucocorticoids . . . "stuck on low" is associated with underactive coping**: Sapolsky, *Why Zebras Don't Get Ulcers*, 318–319; Ogden et al., *Trauma and the Body*, 34–35; Scaer, *The Trauma Spectrum*, chap. 9.

207 **The third pattern oscillates between "stuck on high" and "stuck on low"**: Scaer, *The Trauma Spectrum*, 212–216; Ogden et al., *Trauma and the Body*, 34.

207 **Scaer suggests that this third pattern is especially common with the passage of time**: Scaer, *The Trauma Spectrum*, 215.

208 **One classic indicator of this particular trajectory of dysregulation is persistent low levels of cortisol**: Scaer, *The Trauma Spectrum*, 215.

209 **I have grouped the symptoms into five categories**: Symptoms in these charts were drawn from Elizabeth A. Stanley and John M. Schaldach, *Mindfulness-Based Mind Fitness Training (MMFT) Course Manual*, 2nd ed. (Alexandria, VA: Mind Fitness Training Institute, 2011), 84–88; Levine, *Waking the Tiger*, chap. 11; McEwen and Lasley, *The End of Stress as We Know It*, 64; Diane Poole Heller and Laurence S. Heller, *Crash Course: A Self-Healing Guide to Auto Accident Trauma and Recovery* (Berkeley, CA: North Atlantic, 2001), 47–54.

214 **Nature's most striking example of allostatic load comes from salmon**: McEwen and Lasley, *The End of Stress as We Know It*, 10, 31.

CHAPTER 11

224 **we experience greater stress arousal when we perceive something to be novel**: S. J. Lupien, "Brains under Stress," *Canadian Journal of Psychiatry* 54, no. 1 (2009): 4–5; Bruce S. McEwen and Elizabeth Norton Lasley, *The End of Stress as We Know It* (Washington, D.C.: Joseph Henry, 2002), 32.

225 **SIT is expected to improve mission performance**: Jennifer Kavanagh, *Stress and Performance: A Review of the Literature and Its Applicability to the Military* (Arlington, VA: RAND, 2005); Kelsey L. Larsen and Elizabeth A. Stanley, "Conclusion: The Way Forward," in *Bulletproofing the Psyche: Preventing Mental Health Problems in Our Military and Veterans*, edited by Kate Hendricks Thomas and David Albright (Santa Barbara, CA: Praeger, 2018), 233–253; J. E. Driskell and J. H. Johnston, "Stress Exposure Training," in

Making Decisions under Stress: Implications for Individual and Team Training, edited by J. A. Cannon-Bowers and E. Salas (Chicago: American Psychological Association, 1998), 191–217; T. Saunders et al., "The Effect of Stress Inoculation Training on Anxiety and Performance," *Journal of Occupational Health Psychology* 1, no. 2 (1996): 170–186; R. A. Dienstbier, "Arousal and Physiological Toughness: Implications for Mental and Physical Health," *Psychological Review* 96, no. 1 (1989): 84–100; E. A. Stanley, "Neuroplasticity, Mind Fitness, and Military Effectiveness," in *Bio-Inspired Innovation and National Security*, edited by R. E. Armstrong et al. (Washington, D.C.: National Defense University Press, 2010), 257–279.

225 **in a study of civilian firefighters:** M. R. Baumann, C. L. Gohm, and B. L. Bonner, "Phased Training for High-Reliability Occupations: Live-Fire Exercises for Civilian Firefighters," *Human Factors* 53, no. 5 (2011): 548–557.

226 **a common belief is that SIT should be as stressful:** R. W. Cone, "The Changing National Training Center," *Military Review* 86, no. 3 (2006): 70.

227 **SIT is linked with cognitive degradation, anxiety, mood disturbances:** Kavanagh, *Stress and Performance*; H. R. Lieberman et al., "Severe Decrements in Cognition Function and Mood Induced by Sleep Loss, Heat, Dehydration, and Undernutrition during Simulated Combat," *Biological Psychiatry* 57, no. 4 (2005): 422–429; H. R. Lieberman et al., "Effects of Caffeine, Sleep Loss, and Stress on Cognitive Performance and Mood during U.S. Navy Seal Training," *Psychopharmacology* 164, no. 3 (2002): 250–261; C. A. Morgan et al., "Stress-Induced Deficits in Working Memory and Visuo-Constructive Abilities in Special Operations Soldiers," *Biological Psychiatry* 60, no. 7 (2006): 722–729; C. A. Morgan et al., "Relationships among Plasma Dehydroepiandrosterone Sulfate and Cortisol Levels, Symptoms of Dissociation, and Objective Performance in Humans Exposed to Acute Stress," *Archives of General Psychiatry* 61, no. 8 (2004): 819–825; C. A. Morgan et al., "Neuropeptide-Y, Cortisol, and Subjective Distress in Humans Exposed to Acute Stress: Replication and Extension of Previous Report," *Biological Psychiatry* 52, no. 2 (2002): 136–142; C. A. Morgan et al., "Relationship among Plasma Cortisol, Catecholamines, Neuropeptide Y, and Human Performance during Exposure to Uncontrollable Stress," *Psychosomatic Medicine* 63, no. 3 (2001): 412–422; A. P. Jha et al., "Minds 'at Attention': Mindfulness Training Curbs Attentional Lapses in Military Cohorts," *PLOS One* 10, no. 2 (2015): e0116889; A. P. Jha, A. B. Morrison, S. C. Parker, and E. A. Stanley, "Practice Is Protective: Mindfulness Training Promotes Cognitive Resilience in High-Stress Cohorts," *Mindfulness* 8, no. 1 (2017): 46–58; A. P. Jha et al., "Examining the Protective Effects of Mindfulness Training on Working Memory Capacity and Affective Experience," *Emotion* 10, no. 1 (2010): 54–64; A. P. Jha et al., "Short-Form Mindfulness Training Protects against Working-Memory Degradation over High-Demand Intervals," *Journal of Cognitive Enhancement* 1, no. 2 (2017): 154–171; E. A. Stanley et al., "Mindfulness-Based Mind Fitness Training: A Case Study of a High-Stress Predeployment Military Cohort," *Cognitive and Behavioral Practice* 18, no. 4 (2011): 566–576.

228 **skills training is usually highly specific:** C. S. Green and D. Bavelier, "Exercising Your Brain: A Review of Human Brain Plasticity and Training-Induced Learning," *Psychology and Aging* 23, no. 4 (2008): 692–701; H. A. Slagter, R. J. Davidson, and A. Lutz, "Mental Training as a Tool in the Neuroscientific Study of Brain and Cognitive Plasticity," *Frontiers in Human Neuroscience* 5, no. 1 (2011): 1–12; P. R. Roelfsema, A. van Ooyen, and T. Watanabe, "Perceptual Learning Rules Based on Reinforcers and Attention," *Trends in Cognitive Sciences* 14, no. 2 (2010): 64–71; C. Basak et al., "Can Training in a Real-Time Strategy Video Game Attenuate Cognitive Decline in Older Adults?," *Psychology and Aging* 23, no. 4 (2008): 765–777; K. Ball et al., "Effects of Cognitive Training Interventions with Older Adults: A Randomized Controlled Trial," *Journal of the American Medical Association* 288, no. 18 (2002): 2271–2281; J. C. Allaire and M. Marsiske, "Intraindividual Variability May Not Always Indicate Vulnerability in Elders' Cognitive

Performance," *Psychology and Aging* 20, no. 3 (2005): 390–401; A. F. Kramer and S. L. Willis, "Cognitive Plasticity and Aging," *Psychology of Learning and Motivation* 43 (2003): 267–302.

228 **The evolutionary purpose behind domain-general learning:** K. MacDonald, "Domain-General Mechanisms: What They Are, How They Evolved, and How They Interact with Modular, Domain-Specific Mechanisms to Enable Cohesive Human Groups," *Behavioral and Brain Sciences* 37, no. 4 (2014): 430–431.

228 **Domain-general learning paradigms are typically more complex:** Green and Bavelier, "Exercising Your Brain," 693; J. M. Chien and W. Schneider, "Neuroimaging Studies of Practice-Related Change: fMRI and Meta-Analytic Evidence of a Domain-General Control Network for Learning," *Cognitive Brain Research* 25 (2005): 607–623.

229 **action video-gamers have shown a variety of improvements:** For review, see Green and Bavelier, "Exercising Your Brain."

229 **domain-general learning effects from musical training:** A. B. Graziano, M. Peterson, and G. L. Shaw, "Enhanced Learning of Proportional Math through Music Training and Spatial-Temporal Training," *Neurological Research* 21, no. 2 (1999): 139–152; Y. C. Ho, M. C. Cheung, and A. S. Chan, "Music Training Improves Verbal but Not Visual Memory: Cross-Sectional and Longitudinal Explorations in Children," *Neuropsychology* 17, no. 3 (2003): 439–450; F. H. Rauscher et al., "Music Training Causes Long-Term Enhancement of Preschool Children's Spatial-Temporal Reasoning," *Neurological Research* 19, no. 1 (1997): 2–8; E. G. Schellenberg, "Music Lessons Enhance IQ," *Psychological Science* 15, no. 8 (2004): 511–514.

229 **expert players of several sports:** E. Kioumourtzoglou et al., "Differences in Several Perceptual Abilities between Experts and Novices in Basketball, Volleyball and Water-Polo," *Perceptual and Motor Skills* 86, no. 3, pt. 1 (1998): 899–912; Green and Bavelier, "Exercising Your Brain."

229 **aerobic exercise has been linked with a range of cognitive improvements:** For reviews, see C. H. Hillman, K. I. Erickson, and A. F. Kramer, "Be Smart, Exercise Your Heart: Exercise Effects on Brain and Cognition," *National Review of Neuroscience* 9 (2008): 58–65; A. F. Kramer and K. I. Erickson, "Capitalizing on Cortical Plasticity: Influence of Physical Activity on Cognition and Brain Function," *Trends in Cognitive Sciences* 11, no. 8 (2007): 342–348; K. F. Hsiao, "Can We Combine Learning with Augmented Reality Physical Activity?," *Journal of Cybertherapy and Rehabilitation* 3, no. 1 (2010): 51–62.

229 **"brain training" regimens have deliberately parsed out different cognitive processes for training:** Green and Bavelier, "Exercising Your Brain"; Slagter et al., "Mental Training as a Tool."

230 **"brain training" regimens highlight a trade-off inherent in how we learn:** M. Ahissar and S. Hochstein, "The Reverse Hierarchy Theory of Visual Perceptual Learning," *Trends in Cognitive Sciences* 8, no. 10 (2004): 457–464.

230 **The first form of mental training that does provide domain-general learning is visualization of a physical skill:** Slagter et al., "Mental Training as a Tool."

230 **two forms of mindfulness meditation:** Slagter et al., "Mental Training as a Tool."

231 **Different domain-general learning effects are available after different amounts of FA and OM practice:** Slagter et al., "Mental Training as a Tool."

231 **the longer a cabbie drove a taxi:** E. A. Maguire et al., "Navigation-Related Structural Change in the Hippocampi of Taxi Drivers," *Proceedings of the National Academy of Sciences* 97 (2000): 4398–4403; E. A. Maguire et al., "Navigation Expertise and the Human Hippocampus: A Structural Brain Imaging Analysis," *Hippocampus* 13, no. 2 (2003): 250–259.

231 **hippocampal changes have not been observed in London bus drivers:** E. A. Maguire, K. Woollet, and H. J. Spiers, "London Taxi Drivers and Bus Drivers: A Structural MRI and Neuropsychological Analysis," *Hippocampus* 16 (2006): 1091–1101.

231 **these training regimens involve embodied learning . . . [and] moderate stress:** Green and Bavelier, "Exercising Your Brain."

232 **The original domain-general training regimen is actually instrumental parenting:** Bruce E. Wexler, *Brain and Culture: Neurobiology, Ideology, and Social Change* (Cambridge, MA: MIT Press, 2006), chap. 3.

235 **the Warrior embodies *a path of service*:** Carol S. Pearson, *Awakening the Heroes Within* (San Francisco: HarperSanFrancisco, 1991), chap. 8.

236 **In perhaps the oldest Western description of the profession of arms:** E. Hamilton, and H. Cairns, *The Collected Dialogues of Plato* (Princeton, NJ: Princeton University Press, 1987), 627.

236 **The best-known warrior body practices include the martial arts:** Winston L. King, *Zen and the Way of the Sword: Arming the Samurai Psyche* (Oxford, UK: Oxford University Press, 1993), 65.

237 **Stoic warriors in ancient Greece and Rome engaged in daily contemplative practices:** Nancy Sherman, *Stoic Warriors: The Ancient Philosophy Behind the Military Mind* (New York: Oxford University Press, 2005), 67, 117.

237 **Japanese samurai were trained in zazen . . . and koans:** King, *Zen and the Way of the Sword*, 160–161, 67; Jeffrey K. Mann, *When Buddhists Attack: The Curious Relationship between Zen and the Martial Arts* (Rutland, VT: Tuttle, 2012), 78–79.

237 **In *The Book of Five Rings*:** Miyamoto Musashi, *The Book of Five Rings*, translated by B. J. Brown, Y. Kashiwagi, and W. H. Barrett (New York: Bantam, 1982), 57–58.

237 **recent laboratory research on skill acquisition:** A. K. Ericsson, "Deliberate Practice and Acquisition of Expert Performance: A General Overview," *Academic Emergency Medicine* 15, no. 11 (2008): 988–994; K. Yarrow, P. Brown, and J. W. Krakauer, "Inside the Brain of an Elite Athlete: The Neural Processes That Support High Achievement in Sports," *Nature Reviews Neuroscience* 10, no. 8 (2009): 585–596; U. Debarnot et al., "Experts Bodies, Experts Minds: How Physical and Mental Training Shape the Brain," *Frontiers in Human Neuroscience* 8 (2014): doi:10.3389/fnhum.2014.00280.

238 **we develop character traits through repetition:** Aristotle, *Nicomachean Ethics*, ii, 4.

238 **As samurai Musashi put it:** Musashi, *The Book of Five Rings*, 12. See also Richard Strozzi-Heckler, *In Search of the Warrior Spirit: Teaching Awareness Disciplines to the Green Berets, Third Edition* (Berkeley, CA: North Atlantic Books, 2003), 33–34.

238 **"One's ultimate aim is to do all in one's power to shoot straight":** Quoted in Sherman, *Stoic Warriors*, 33.

238 **Sun Tzu explained it this way:** Sun Tzu, *The Art of War: Denma Translation* (Boston: Shambala, 2002), 16.

240 **"These are the victories of the military lineage":** Sun Tzu, *The Art of War*, 5.

240 **the Tibetan word for warrior, pawo:** Chogyam Trungpa, *Shambala: The Sacred Path of the Warrior* (Boston: Shambala, 2007), 9.

241 **the Spartan king Agesilaus was reputedly once asked:** Steven Pressfield, *The Warrior Ethos* (Los Angeles: Black Irish Entertainment, 2011), 13.

241 **As Epictetus explains, *a warrior's job is to find his agency*:** Sherman, *Stoic Warriors*, 10.

CHAPTER 12

245 **interoception, which includes the ability to recognize bodily sensations:** D. C. Johnson et al., "Modifying Resilience Mechanisms in At-Risk Individuals: A Controlled Study of Mindfulness Training in Marines Preparing for Deployment," *American Journal of Psychiatry* 171, no. 8 (2014): 844–853.

246 **Neuroscientists argue that the insula and the ACC together provide top-down control:** H. D. Critchley et al., "Human Cingulate Cortex and Autonomic Control: Converging Neuroimaging and Clinical Evidence," *Brain: A Journal of Neurology* 126, no. 10 (2003): 2139–2152; Stephen W. Porges, *The Polyvagal Theory: Neurophysiological Foundations of Emotions, Attachment, Communication, and Self-Regulation* (New York: Norton,

2011), 76–79; S. N. Garfinkel and H. D. Critchley, "Interoception, Emotion and Brain: New Insights Link Internal Physiology to Social Behaviour," *Social Cognitive and Affective Neuroscience* 8, no. 3 (2013): 231–234; H. D. Critchley et al., "Neural Systems Supporting Interoceptive Awareness," *Nature Neuroscience* 7, no. 2 (2004): 189–195; B. A. van der Kolk, "Clinical Implications of Neuroscience Research in PTSD," *Annals of the New York Academy of Sciences* 1071, no. 1 (2006): 277–293; Richard J. Davidson and Sharon Begley, *The Emotional Life of Your Brain: How Its Unique Patterns Affect the Way You Think, Feel, and Live—And How You Can Change Them* (New York: Plume, 2012), 78–81.

246 **regions of the insula and ACC also serve as the brain's pain distress network:** Specifically, the dorsal ACC and anterior insula. Matthew D. Lieberman, *Social: Why Our Brains Are Wired to Connect* (New York: Crown, 2013), chap. 3.

246 **we can improve the functioning of this regulatory loop by cultivating interoceptive awareness:** P. Payne, P. A. Levine, and M. A. Crane-Godreau, "Somatic Experiencing: Using Interoception and Proprioception as Core Elements of Trauma Therapy," *Frontiers in Psychology* 6, no. 93 (2015): doi:10.3389/fpsyg.2015.00093; van der Kolk, "Clinical Implications of Neuroscience Research in PTSD"; A. Feder, E. J. Nestler, and D. S. Charney, "Psychobiology and Molecular Genetics of Resilience," *Nature Reviews Neuroscience* 10 (2009): 446–457; Porges, *The Polyvagal Theory*.

247 **In brain imaging studies . . . military and civilian "elite performers":** M. P. Paulus et al., "Differential Brain Activation to Angry Faces by Elite Warfighters: Neural Processing Evidence for Enhanced Threat Detection," *PLOS One* 5, no. 4 (2010): e10096; M. P. Paulus et al., "Subjecting Elite Athletes to Inspiratory Breathing Load Reveals Behavioral and Neural Signatures of Optimal Performers in Extreme Environments," *PLOS One* 7, no. 2 (2012): e29394; A. N. Simmons et al., "Altered Insula Activation in Anticipation of Changing Emotional States: Neural Mechanisms Underlying Cognitive Flexibility in Special Operations Forces Personnel," *Neuroreport* 23, no. 4 (2012): 234–239; N. J. Thom et al., "Detecting Emotion in Others: Increased Insula and Decreased Medial Prefrontal Cortex Activation during Emotion Processing in Elite Adventure Racers," *Social Cognitive and Affective Neuroscience* 9, no. 2 (2014): 225–231.

247 **infants and children who experience difficulty developing their ventral PSNS circuit:** Porges, *The Polyvagal Theory*, chap. 5.

247 **Compromised interoceptive functioning plays a critical role among adults with depression, anxiety disorders:** M. B. Stein et al., "Increased Amygdala and Insula Activation during Emotion Processing in Anxiety-Prone Subjects," *American Journal of Psychiatry* 164, no. 2 (2007): 318–327; M. P. Paulus and J. L. Stewart, "Interoception and Drug Addiction," *Neuropharmacology* 76 (2014): 342–350; K. Domschke et al., "Interoceptive Sensitivity in Anxiety and Anxiety Disorders: An Overview and Integration of Neuro-biological Findings," *Clinical Psychology Review* 30, no. 1 (2010): 1–11; J. A. Avery et al., "Major Depressive Disorder Is Associated with Abnormal Interoceptive Activity and Functional Connectivity in the Insula," *Biological Psychiatry* 76, no. 3 (2014): 258–266; M. P. Paulus and M. B. Stein, "An Insular View of Anxiety," *Biological Psychiatry* 60, no. 4 (2006): 383–387; M. P. Paulus and M. B. Stein, "Interoception in Anxiety and Depression," *Brain Structure and Function* 214, no. 5–6 (2010): 451–463; Davidson and Begley, *The Emotional Life of Your Brain*, 81; van der Kolk, "Clinical Implications of Neuroscience Research in PTSD"; Pat Ogden, Kekuni Minton, and Clare Pain, *Trauma and the Body: A Sensorimotor Approach to Psychotherapy* (New York: Norton, 2006), 7; Robert C. Scaer, *The Trauma Spectrum: Hidden Wounds and Human Resiliency* (New York: Norton, 2005), 62–64.

248 **MMFT Marines showed significant changes in interoceptive functioning:** Johnson et al., "Modifying Resilience Mechanisms in At-Risk Individuals"; L. Haase et al., "Mindfulness-Based Training Attenuates Insula Response to an Aversive Interoceptive Challenge," *Social Cognitive and Affective Neuroscience* 11, no. 1 (2016): 182–190.

248 **As a recent *New York Times* article put it:** David Gelles, "How to Meditate," *New York Times*, December 25, 2016. Italics added.

250 a recent study evaluating . . . mindfulness training with more than three hundred middle and high school students: C. Johnson et al., "Effectiveness of a School-Based Mindfulness Program for Transdiagnostic Prevention in Young Adolescents," *Behaviour Research and Therapy* 81 (2016): 1–11.

250 To counter this bias, her team interviewed experienced practitioners/teachers: J. R. Lindahl et al., "The Varieties of Contemplative Experience: A Mixed-Methods Study of Meditation-Related Challenges in Western Buddhists," *PLOS One* 12, no. 5 (2017): e0176239. See also W. B. Britton, *Current Opinion in Psychology* 28 (2019): 159–165; P. L. Dobkin, J. A. Irving, and S. Amar, "For Whom May Participation in a Mindfulness-Based Stress Reduction Program Be Contraindicated?," *Mindfulness* 3, no. 1 (2011): 44–50; V. Follette, K. Palm, and A. Pearson, "Mindfulness and Trauma: Implications for Treatment," *Journal of Rational-Emotive and Cognitive-Behavior Therapy* 24 (2006): 45–61; C. Strauss et al., "Mindfulness-Based Interventions for People Diagnosed with a Current Episode of an Anxiety or Depressive Disorder: A Meta-Analysis of Randomised Controlled Trials," *PLOS One* 9, no. 4 (2014): e96110; L. C. Waelde, "Dissociation and Meditation," *Journal of Trauma and Dissociation* 5, no. 2 (2004): 147–162; E. Shonin, W. van Gordon, and M. D. Griffiths, "Are There Risks Associated with Using Mindfulness in the Treatment of Psychopathology?," *Clinical Practice* 11, no. 4 (2014): 398–392; A. W. Hanley et al., "Mind the Gaps: Are Conclusions about Mindfulness Entirely Conclusive?," *Journal of Counseling and Development* 94, no. 1 (2016): 103–113; S. F. Santorelli, "Mindfulness-Based Stress Reduction (MBSR): Standards of Practice" (Worcester, MA: UMass Medical School Center for Mindfulness in Medicine, Health Care, and Society, 2014).

250 Center for Mindfulness . . . states that MBSR is not advised during active PTSD or other mental illness: UMass Medical School Center for Mindfulness in Medicine, Health Care, and Society, "FAQ—Stress Reduction," www.umassmed.edu/cfm/stress -reduction/faqs; Santorelli, "Mindfulness-Based Stress Reduction (MBSR)."

251 six out of ten Marines were actively experiencing multiple symptoms of dysregulation: E. A. Stanley et al., "Mindfulness-Based Mind Fitness Training: A Case Study of a High-Stress Predeployment Military Cohort," *Cognitive and Behavioral Practice* 18, no. 4 (2011): 566–576.

253 the integration of information between our thinking brain and our survival brain can become disconnected: Bessel A. van der Kolk, *The Body Keeps the Score: Brain, Mind and Body in the Healing of Trauma* (New York: Viking, 2014); Pat Ogden and Janina Fisher, *Sensorimotor Psychotherapy: Interventions for Trauma and Attachment* (New York: Norton, 2015); Payne et al., "Somatic Experiencing"; Ogden et al., *Trauma and the Body.*

253 awareness of breathing, which they consider to be "relatively neutral sensory stimuli": J. D. Teasdale and M. Chaskalson (Kulananda), "How Does Mindfulness Transform Suffering? II: The Transformation of Dukkha," *Contemporary Buddhism* 12, no. 1 (2011): 103–112.

255 body-based trauma therapies for reregulating the nervous system and survival brain after trauma: Ogden and Fisher, *Sensorimotor Psychotherapy*; Peter A. Levine, *Waking the Tiger: Healing Trauma* (Berkeley, CA: North Atlantic, 1997); M. L. Leitch, "Somatic Experiencing Treatment with Tsunami Survivors in Thailand: Broadening the Scope of Early Intervention," *Traumatology* 13, no. 3 (2007): 11–20; Ogden et al., *Trauma and the Body*; M. L. Leitch, J. Vanslyke, and M. Allen, "Somatic Experiencing Treatment with Social Service Workers Following Hurricanes Katrina and Rita," *Social Work* 54, no. 1 (2009): 9–18; Payne et al., "Somatic Experiencing."

257 I strongly recommend you seek out a trained professional: You can find a certified practitioner for Somatic Experiencing at sepractitioner.membergrove.com. You can find a certified practitioner for sensorimotor psychotherapy at www.sensorimotor psychotherapy.org/referral.html.

257 **received variants of the eight-week MMFT course:** We have tested five different variants of the eight-week course in research with predeployment troops: a twenty-four-hour version, a twenty-hour version, a sixteen-hour version, and two different eight-hour versions, which focused on either (1) in-class practice of and discussion about the exercises or (2) the intellectual content related to the neurobiology of stress and resilience. From the empirical findings, arguably the best variant is the twenty-hour version.

258 **MMFT participants showed improved cognitive performance:** A. P. Jha et al., "Examining the Protective Effects of Mindfulness Training on Working Memory Capacity and Affective Experience," *Emotion* 10, no. 1 (2010): 54–64; Stanley et al., "Mindfulness-Based Mind Fitness Training"; A. P. Jha et al., "Minds 'at Attention': Mindfulness Training Curbs Attentional Lapses in Military Cohorts," *PLOS One* 10, no. 2 (2015): e0116889; A. P. Jha, A. B. Morrison, S. C. Parker, and E. A. Stanley, "Practice Is Protective: Mindfulness Training Promotes Cognitive Resilience in High-Stress Cohorts," *Mindfulness* 8, no. 1 (2017): 46–58; A. P. Jha et al., "Short-Form Mindfulness Training Protects against Working-Memory Degradation over High-Demand Intervals," *Journal of Cognitive Enhancement* 1, no. 2 (2017): 154–171.

258 **MMFT participants showed significantly more efficient stress arousal . . . followed by a more complete recovery:** Johnson et al., "Modifying Resilience Mechanisms in At-Risk Individuals." Indeed, the MMFT Marines' stress arousal pattern was similar to those found in other empirical studies with humans and mammals, where resilience has been linked with rapid activation of the stress response and its efficient termination. Feder et al., "Psychobiology and Molecular Genetics of Resilience"; E. R. De Kloet, M. Joels, and F. Holsboer, "Stress and the Brain: From Adaptation to Disease," *Nature Reviews Neuroscience* 6 (2005): 463–475.

258 **MMFT Marines showed lower concentrations of neuropeptide Y:** Johnson et al., "Modifying Resilience Mechanisms in At-Risk Individuals." NPY levels are used as resilience indicators in two ways. First, resilience is associated with higher NPY concentrations in the blood during a resting state. In addition, the ability to secrete additional NPY during stress arousal and then discharge it quickly afterward, as the MMFT Marines did, is a sign of an adaptive stress response and efficient recovery, another facet of resilience. For more about NPY as a resilience indicator, see J. Pernow et al., "Plasma Neuropeptide Y–Like Immunoreactivity and Catecholamines during Various Degrees of Sympathetic Activation in Man," *Clinical Psychology* 6 (1986): 561–578; R. Yehuda, S. Brand, and R. Yang, "Plasma Neuropeptide-Y Concentrations in Combat Exposed Veterans; Relationship to Trauma Exposure, Recovery from PTSD, and Coping," *Biological Psychiatry* 59 (2006): 660–663; C. A. Morgan et al., "Plasma Neuropeptide-Y Concentrations in Humans Exposed to Military Survival Training," *Biological Psychiatry* 47 (2000): 902–909; C. A. Morgan et al., "Relationship among Plasma Cortisol, Catecholamines, Neuropeptide-Y, and Human Performance during Exposure to Uncontrollable Stress," *Psychosomatic Medicine* 63, no. 3 (2001): 412–422; C. A. Morgan et al., "Neuropeptide-Y, Cortisol, and Subjective Distress in Humans Exposed to Acute Stress: Replication and Extension of Previous Report," *Biological Psychiatry* 52, no. 2 (2002): 136–142; Feder et al., "Psychobiology and Molecular Genetics of Resilience"; Z. Zhou et al., "Genetic Variation in Human NPY Expression Affects Stress Response and Emotion," *Nature* 452, no. 7190 (2008): 997–1001; T. J. Sajdyk, A. Shekhar, and D. R. Gehlert, "Interactions between NPY and CRF in the Amygdala to Regulate Emotionality," *Neuropeptides* 38 (2004): 225–234.

259 **MMFT Marines also experienced significant improvements in sleep quality . . . [and] insulin-like growth factor (IGF-1):** S. R. Sterlace et al., "Hormone Regulation under Stress: Recent Evidence from Warfighters on the Effectiveness of Mindfulness-Based Mind Fitness Training in Building Stress Resilience," poster presentation at the Society for Neuroscience Annual Meeting, New Orleans, LA (October 2012).

261 **Skepticism is a natural and healthy response to claims about the benefits of anything:** Elizabeth A. Stanley and John M. Schaldach, *Mindfulness-Based Mind Fitness Training (MMFT) Course Manual*, 2nd ed. (Alexandria, VA: Mind Fitness Training Institute, 2011), 199–209.

262 **physically overtraining can lead to physical injury and exhaustion:** R. Meeusen et al., "Prevention, Diagnosis and Treatment of the Overtraining Syndrome," *European Journal of Sports Science* 6 (2006): 1–14.

CHAPTER 13

271 **Paying attention to and engaging with the external environment:** B. A. van der Kolk, "Clinical Implications of Neuroscience Research in PTSD," *Annals of the New York Academy of Sciences* 1071, no. 1 (2006): 277–293; Robert C. Scaer, *The Trauma Spectrum: Hidden Wounds and Human Resiliency* (New York: Norton, 2005), 49–50; Pat Ogden, Kekuni Minton, and Clare Pain, *Trauma and the Body: A Sensorimotor Approach to Psychotherapy* (New York: Norton, 2006), 17; Stephen W. Porges, *The Polyvagal Theory: Neurophysiological Foundations of Emotions, Attachment, Communication, and Self-Regulation* (New York: Norton, 2011), 76–79.

271 **physical symptoms of discharging stress activation:** Peter A. Levine, *Waking the Tiger: Healing Trauma* (Berkeley, CA: North Atlantic, 1997); Scaer, *The Trauma Spectrum*, 44–45, 48–49.

271 **our encounter actually conferred survival benefits:** Scaer, *The Trauma Spectrum*, 54–55.

272 **Rat pups who received such attentive licking:** Bruce E. Wexler, *Brain and Culture: Neurobiology, Ideology, and Social Change* (Cambridge, MA: MIT Press, 2006), 91; A. Feder, E. J. Nestler, and D. S. Charney, "Psychobiology and Molecular Genetics of Resilience," *Nature Reviews Neuroscience* 10 (2009): 446–457; A. S. Fleming et al., "Mothering Begets Mothering," *Pharmacology Biochemistry and Behavior* 73, no. 1 (2002): 61–75; D. Francis et al., "Nongenomic Transmission across Generations of Maternal Behavior and Stress Responses in the Rat," *Science* 286, no. 5442 (1999): 1155–1158; A. Gonzalez et al., "Intergenerational Effects of Complete Maternal Deprivation and Replacement Stimulation on Maternal Behavior and Emotionality in Female Rats," *Developmental Psychobiology* 38, no. 1 (2001): 11–32; I. C. G. Weaver et al., "Epigenetic Programming by Maternal Behavior," *Nature Neuroscience* 7, no. 8 (2004): 847–854; R. M. Sapolsky, "Mothering Style and Methylation," *Nature Neuroscience* 7, no. 8 (2004): 791–792.

272 **the "safe home base" of secure attachment allows humans to wire a wide window:** Daniel J. Siegel, *The Developing Mind: How Relationships and the Brain Interact to Shape Who We Are* (New York: Guilford, 1999), 278–283; Bessel A. van der Kolk, *The Body Keeps the Score: Brain, Mind, and Body in the Healing of Trauma* (New York: Penguin, 2015), 110–114; Ogden et al., *Trauma and the Body*, 41–48.

274 **experiments with chicks, conducted many decades ago:** Studies cited in Scaer, *The Trauma Spectrum*, 54–55.

275 **experiencing challenge and recovering completely afterward conveys survival advantages:** Feder et al., "Psychobiology and Molecular Genetics of Resilience"; D. S. Charney, "Psychobiological Mechanisms of Resilience and Vulnerability: Implications for Successful Adaptation to Extreme Stress," *American Journal of Psychiatry* 161 (2004): 95–216; E. R. De Kloet, M. Joels, and F. Holsboer, "Stress and the Brain: From Adaptation to Disease," *Nature Reviews Neuroscience* 6 (2005): 463–475.

276 **trauma extinction actually involves forming a new implicit memory:** Bruce S. McEwen and Elizabeth Norton Lasley, *The End of Stress as We Know It* (Washington, D.C.: Joseph Henry, 2002), 37–38; Joseph LeDoux, *The Emotional Brain: The Mysterious Underpinnings of Emotional Life* (New York: Touchstone, 1998); M. R. Delgado, A. Olsson, and E. A. Phelps, "Extending Animal Models of Fear Conditioning to Humans," *Biological Psychiatry* 73 (2006): 39–48; Feder et al., "Psychobiology and Molecular Genetics of Resilience"; M. R. Delgado et al., "Neural Circuitry Underlying the

Regulation of Conditioned Fear and Its Relation to Extinction," *Neuron* 59, no. 5 (2008): 829–838; M. R. Milad et al., "Thickness of Ventromedial Prefrontal Cortex in Humans Is Correlated with Extinction Memory," *Proceedings of the National Academy of Sciences* 102, no. 30 (2005): 10706–10711; D. Schiller et al., "From Fear to Safety and Back: Reversal of Fear in the Human Brain," *Journal of Neuroscience* 28 (2008): 11517–11525.

276 **This "learning by doing" . . . teaches the survival brain:** van der Kolk, "Clinical Implications of Neuroscience Research in PTSD"; J. LeDoux and J. M. Gorman, "A Call to Action: Overcoming Anxiety through Active Coping," *American Journal of Psychiatry* 12 (2001): 1953–1955; Ogden et al., *Trauma and the Body*, 40.

277 **Table 13.1 gives a complete list of the signs of discharge:** Levine, *Waking the Tiger*; Scaer, *The Trauma Spectrum*, 44–45; Elizabeth A. Stanley and John M. Schaldach, *Mindfulness-Based Mind Fitness Training (MMFT) Course Manual*, 2nd ed. (Alexandria, VA, Mind Fitness Training Institute, 2011), 96–97.

280 **Being inside our window allows us to integrate top-down (thinking brain) and bottom-up (survival brain) processing:** Pat Ogden and Janina Fisher, *Sensorimotor Psychotherapy: Interventions for Trauma and Attachment* (New York: Norton, 2015), 182–83; Ogden et al., *Trauma and the Body*, 27–29, 40; P. Payne, P. A. Levine, and M. A. Crane-Godreau, "Somatic Experiencing: Using Interoception and Proprioception as Core Elements of Trauma Therapy," *Frontiers in Psychology* 6, no. 93 (2015): doi.org/10.3389 /fpsyg.2015.00093; Siegel, *The Developing Mind*, 253; van der Kolk, "Clinical Implications of Neuroscience Research in PTSD."

280 **the clinician provides surrogate interoceptive awareness to facilitate their client's recovery:** Ogden and Fisher, *Sensorimotor Psychotherapy*, chap. 26, esp. 544.

281 **metaphorically stand with one foot in the traumatic past . . . and one foot in the present:** Ogden et al., *Trauma and the Body*, 40; Ogden and Fisher, *Sensorimotor Psychotherapy*, chap. 26; van der Kolk, "Clinical Implications of Neuroscience Research in PTSD."

281 **the survival brain learns that it's no longer helpless:** Ogden and Fisher, *Sensorimotor Psychotherapy*, 545–546; van der Kolk, "Clinical Implications of Neuroscience Research in PTSD."

284 **Resilience is an active process:** Feder et al., "Psychobiology and Molecular Genetics of Resilience."

CHAPTER 14

296 **Whenever we struggle against reality:** The ideas in this section are particularly influenced by what I've learned from several wonderful teachers, especially Adyashanti and Rodney Smith. I am deeply grateful for their teaching.

303 **Stress arousal points to a primary emotion . . . we usually also experience one of the categorical (or secondary) emotions:** Daniel J. Siegel, *The Developing Mind: How Relationships and the Brain Interact to Shape Who We Are* (New York: Guilford, 1999), chap. 4; Bruce S. McEwen and Elizabeth Norton Lasley, *The End of Stress as We Know It* (Washington, D.C.: Joseph Henry, 2002), 36–37; R. J. Davidson, D. C. Jackson, and N. H. Kalin, "Emotion, Plasticity, Context, and Regulation: Perspectives from Affective Neuroscience," *Psychological Bulletin* 126, no. 6 (2000): 890–909.

303 **Which categorical emotion gets triggered in any situation:** Pat Ogden, Kekuni Minton, and Clare Pain, *Trauma and the Body: A Sensorimotor Approach to Psychotherapy* (New York: Norton, 2006), 11–14; Siegel, *The Developing Mind*, chap. 4; E. Halperin and R. Pliskin, "Emotions and Emotion Regulation in Intractable Conflict: Studying Emotional Processes within a Unique Context," *Political Psychology* 36, no. S1 (2015): 119–150.

305 **Damasio's research into somatic markers:** Antonio Damasio, *Descartes' Error: Emotion, Reason, and the Human Brain* (New York: Penguin, 1994), chaps. 8, 9.

305 **emotions filter our perceptions, bias our information search, and influence our risk-taking tendencies and behavior:** J. N. Druckman and R. McDermott, "Emotion

and the Framing of Risky Choice," *Political Behavior* 30, no. 3 (2008): 297–321; T. Brader and G. E. Marcus, "Emotion and Political Psychology," in *The Oxford Handbook of Political Psychology*, 2nd ed., edited by L. Huddy, D. O. Sears, and J. S. Levy (Oxford, UK: Oxford University Press, 2013), 165–204; Halperin and Pliskin, "Emotions and Emotion Regulation in Intractable Conflict"; J. Renshon, J. J. Lee, and D. Tingley, "Emotions and the Micro-Foundations of Commitment Problems," *International Organization* 71, no. S1 (2017): S189–S218; E. A. Stanley, "War Duration and the Micro-Dynamics of Decision-Making under Stress," *Polity* 50, no. 2 (2018): 178–200; Karl E. Weick, *Sensemaking in Organizations* (New York: Sage, 1995), 91–105.

306 **emotion suppression may actually be adaptive in the short term:** G. A. Bonanno, "Loss, Trauma, and Human Resilience: Have We Underestimated the Human Capacity to Thrive after Extremely Aversive Events?," *American Psychologist* 59, no. 1 (2004): 20–28.

306 **Habitual emotion suppression is linked with many allostatic load-building effects:** E. A. Stanley and K. L. Larsen, "Emotion Dysregulation and Military Suicidality since 2001: A Review of the Literature," *Political Psychology* 40, no. 1 (2019): 147–163; S. Lam et al., "Emotion Regulation and Cortisol Reactivity to a Social-Evaluative Speech Task," *Psychoneuroendocrinology* 34, no. 9 (2009): 1355–1362; J. J. Gross and R. W. Levenson, "Emotional Suppression: Physiology, Self-Report, and Expressive Behavior," *Journal of Personality and Social Psychology* 64, no. 6 (1993): 970–986; J. J. Gross and R. W. Levenson, "Hiding Feelings: The Acute Effects of Inhibiting Negative and Positive Emotion," *Journal of Abnormal Psychology* 106, no. 1 (1997): 95–103; A. A. Appleton et al., "Divergent Associations of Adaptive and Maladaptive Emotion Regulation Strategies with Inflammation," *Health Psychology* 32, no. 7 (2013): 748–756; I. B. Mauss and J. J. Gross, "Emotion Suppression and Cardiovascular Disease: Is Hiding Feelings Bad for Your Heart?," in *Emotional Expression and Health: Advances in Theory, Assessment and Clinical Applications*, edited by I. Nyklicek, L. Temoshok, and A. Vingerhoets (New York: Routledge, 2004), 60–80; D. DeSteno, J. J. Gross, and L. Kubzansky, "Affective Science and Health: The Importance of Emotion and Emotion Regulation," *Health Psychology* 32, no. 5 (2013): 474–486; K. B. Koh et al., "The Relation between Anger Expression, Depression, and Somatic Symptoms in Depressive Disorders and Somatoform Disorders," *Journal of Clinical Psychiatry* 66, no. 4 (2005): 485–491; W. D. S. Killgore et al., "The Effects of Prior Combat Experience on the Expression of Somatic and Affective Symptoms in Deploying Soldiers," *Journal of Psychosomatic Research* 60, no. 4 (2006): 379–385; C. G. Beevers et al., "Depression and the Ironic Effects of Thought Suppression: Therapeutic Strategies for Improving Mental Control," *Clinical Psychology: Science and Practice* 6, no. 2 (1999): 133–148; J. G. Beck et al., "Rebound Effects Following Deliberate Thought Suppression: Does PTSD Make a Difference?," *Behavior Therapy* 37, no. 2 (2006): 170–180; M. T. Feldner et al., "Anxiety Sensitivity–Physical Concerns as a Moderator of the Emotional Consequences of Emotion Suppression during Biological Challenge: An Experimental Test Using Individual Growth Curve Analysis," *Behaviour Research and Therapy* 44, no. 2 (2006): 249–272; M. A. Hom et al., "The Association between Sleep Disturbances and Depression among Firefighters: Emotion Dysregulation as an Explanatory Factor," *Journal of Clinical Sleep Medicine* 12, no. 2 (2016): 235–245; H. Braswell and H. I. Kushner, "Suicide, Social Integration, and Masculinity in the U.S. Military," *Social Science and Medicine* 74, no. 4 (2012): 530–536; L. Campbell-Sills et al., "Effects of Suppression and Acceptance on Emotional Responses of Individuals with Anxiety and Mood Disorders," *Behaviour Research and Therapy* 44, no. 9 (2006): 1251–1263; E. B. Elbogen et al., "Risk Factors for Concurrent Suicidal Ideation and Violent Impulses in Military Veterans," *Psychological Assessment* 30, no. 4 (2017): 425–435; N. K. Y. Tang and C. Crane, "Suicidality in Chronic Pain: A Review of the Prevalence, Risk Factors and Psychological Links," *Psychological Medicine* 36, no. 5 (2006): 575–586.

306 **habitual emotion suppression is also linked with substance abuse, overeating:** Braswell and Kushner, "Suicide, Social Integration, and Masculinity in the U.S.

Military"; Stanley and Larsen, "Emotion Dysregulation and Military Suicidality since 2001"; E. B. Elbogen et al., "Violent Behaviour and Post-Traumatic Stress Disorder in U.S. Iraq and Afghanistan Veterans," *British Journal of Psychiatry* 204, no. 5 (2014): 368–375; Elbogen et al., "Risk Factors for Concurrent Suicidal Ideation"; G. Green et al., "Exploring the Ambiguities of Masculinity in Accounts of Emotional Distress in the Military among Young Ex-Servicemen," *Social Science and Medicine* 71, no. 8 (2010): 1480–1488; Y. I. Nillni et al., "Deployment Stressors and Physical Health among OEF/ OIF Veterans: The Role of PTSD," *Health Psychology* 33, no. 11 (2014): 1281–1287; R. P. Auerbach, J. R. Z. Abela, and M. R. Ho, "Responding to Symptoms of Depression and Anxiety: Emotion Regulation, Neuroticism, and Engagement in Risky Behaviors," *Behaviour Research and Therapy* 45, no. 9 (2007): 2182–2191; S. Fischer, K. G. Anderson, and G. T. Smith, "Coping with Distress by Eating or Drinking: Role of Trait Urgency and Expectancies," *Psychology of Addictive Behaviors* 18, no. 3 (2004): 269–274; A. L. Teten et al., "Intimate Partner Aggression Perpetrated and Sustained by Male Afghanistan, Iraq, and Vietnam Veterans with and without Posttraumatic Stress Disorder," *Journal of Interpersonal Violence* 25, no. 9 (2010): 1612–1630; J. E. McCarroll et al., "Deployment and the Probability of Spousal Aggression by U.S. Army Soldiers," *Military Medicine* 175, no. 5 (2010): 352–356; Q. M. Biggs et al., "Acute Stress Disorder, Depression, and Tobacco Use in Disaster Workers Following 9/11," *American Journal of Orthopsychiatry* 80, no. 4 (2010): 586–592; R. M. Bray and L. L. Hourani, "Substance Use Trends among Active Duty Military Personnel: Findings from the United States Department of Defense Health Related Behavior Surveys, 1980–2005," *Addiction* 102, no. 7 (2007): 1092–1101; I. G. Jacobson et al., "Alcohol Use and Alcohol-Related Problems before and after Military Combat Deployment," *Journal of the American Medical Association* 300, no. 6 (2008): 663–675.

306 **Emotional arousal degrades our impulse control and executive functioning:** See Chapter 5. See also Stanley, "War Duration"; Renshon et al., "Emotions and the Micro-Foundations of Commitment Problems"; G. Lowenstein and J. S. Lerner, "The Role of Affect in Decision Making," in *Handbook of Affective Science*, edited by R. J. Davidson, K. R. Scherer, and H. H. Goldsmith (Oxford, UK: Oxford University Press, 2003), 619–642.

308 **at high arousal levels, we tend to focus narrowly on information we perceive to be psychologically central:** Weick, *Sensemaking in Organizations*, 91–105; Jennifer Kavanagh, *Stress and Performance: A Review of the Literature and Its Applicability to the Military* (Arlington, VA: RAND, 2005), 17–19, 32–33; Stanley, "War Duration"; Elizabeth A. Stanley, *Paths to Peace: Domestic Coalition Shifts, War Termination and the Korean War* (Stanford, CA: Stanford University Press, 2009), chap. 2; Daniel Kahneman, *Thinking, Fast and Slow* (New York: Macmillan, 2011), chaps. 3, 4; M. J. Dugas, P. Gosselin, and R. Ladouceur, "Intolerance of Uncertainty and Information Processing: Evidence of Biased Recall and Interpretations," *Cognitive Therapy and Research* 29, no. 1 (2005): 57–70; M. J. Dugas, P. Gosselin, and R. Ladouceur, "Intolerance of Uncertainty and Worry: Investigating Specificity in a Non-Clinical Sample," *Cognitive Therapy and Research* 25, no. 5 (2001): 551–558.

308 **negative information is more likely to capture our attention:** R. F. Baumeister et al., "Bad Is Stronger Than Good," *Review of General Psychology* 5, no. 4 (2001): 323–370; P. Rozin and E. B. Royzman, "Negativity Bias, Negativity Dominance, and Contagion," *Personality and Social Psychology Review* 5, no. 4 (2001): 296–320.

308 **"many good events can overcome the psychological effects of a single bad one":** Baumeister et al., "Bad Is Stronger Than Good," 323. As the authors note, "Given the large number of patterns in which bad outweighs good . . . the lack of exception suggests how basic and powerful is the greater power of bad" (362).

310 **empirical studies have shown the benefits of "expressive writing":** For reviews, see J. W. Pennebaker and C. K. Chung, "Expressive Writing: Connections to Physical and Mental

Health," in *The Oxford Handbook of Health Psychology*, edited by H. S. Friedman (Oxford, UK: Oxford University Press, 2011), 417–437; J. Frattaroli, "Experimental Disclosure and Its Moderators: A Meta-Analysis," *Psychological Bulletin* 132, no. 6 (2006): 823–865.

311 **Emotional and physical pain use the same neural circuits:** Matthew D. Lieberman, *Social: Why Our Brains Are Wired to Connect* (New York: Crown, 2013), chap. 3.

311 **catastrophizing is linked with greater pain intensity:** Gary Kaplan and Donna Beech, *Total Recovery: Solving the Mystery of Chronic Pain and Depression* (New York: Rodale, 2014), 180; Brian Resnick, "100 Million Americans Have Chronic Pain. Very Few Use One of the Best Tools to Treat It," *Vox*, August 16, 2018, www.vox.com/science-and-health/2018/5/17 /17276452/chronic-pain-treatment-psychology-cbt-mindfulness-evidence.

311 **we activate the same *pain distress network*:** Lieberman, *Social*, chap. 3.

311 **people working in high-stress environments may be prone to expressing emotional distress via somatization:** Killgore et al., "The Effects of Prior Combat Experience"; C. W. Hoge et al., "Combat Duty in Iraq and Afghanistan, Mental Health Problems, and Barriers to Care," *New England Journal of Medicine* 351, no. 1 (2004): 13–22; C. W. Hoge et al., "Mental Health Problems, Use of Mental Health Services, and Attrition from Military Service After Returning from Deployment to Iraq or Afghanistan," *Journal of the American Medical Association* 295, no. 9 (2006): 1023–1032; C. W. Hoge et al., "Association of Posttraumatic Stress Disorder with Somatic Symptoms, Health Care Visits, and Absenteeism among Iraq War Veterans," *American Journal of Psychiatry* 164 (2007): 150–153; T. M. Greene-Shortridge, T. W. Britt, and C. A. Castro, "The Stigma of Mental Health Problems in the Military," *Military Medicine* 172, no. 2 (2007): 157–161; I. H. Stanley, M. A. Hom, and T. E. Joiner, "A Systematic Review of Suicidal Thoughts and Behaviors among Police Officers, Firefighters, EMTs, and Paramedics," *Clinical Psychology Review* 44 (2016): 25–44; Stanley and Larsen, "Emotion Dysregulation and Military Suicidality since 2001"; C. J. Bryan et al., "Understanding and Preventing Military Suicide," *Archives of Suicide Research* 16, no. 2 (2012): 95–110; Green et al., "Exploring the Ambiguities of Masculinity."

312 **study about the 82nd Airborne Division soldiers:** Killgore et al., "The Effects of Prior Combat Experience"

312 **around 85 percent of people with lower back pain have nothing diagnosable wrong:** Resnick, "100 Million Americans Have Chronic Pain."

312 **about half of people with major depression also experience chronic pain:** B. A. Arnow et al., "Comorbid Depression, Chronic Pain, and Disability in Primary Care," *Psychosomatic Medicine* 68, no. 2 (2006): 262–268; W. M. Compton et al., "Changes in the Prevalence of Major Depression and Comorbid Substance Use Disorders in the United States between 1991–1992 and 2001–2002," *American Journal of Psychiatry* 163, no. 12 (2006): 2141–2147; Kaplan and Beech, *Total Recovery*, 72–76, 92–97.

312 **chronic pain and drug misuse significantly correlate with suicidal ideation and violent impulses:** Tang and Crane, "Suicidality in Chronic Pain"; Elbogen et al., "Risk Factors for Concurrent Suicidal Ideation."

312 **chronic use of almost any analgesic . . . increases hypersensitization of pain receptors . . . [and] gut permeability:** Kaplan and Beech, *Total Recovery*, 78–79, 144–145.

313 **about 20 percent of back surgery patients still have chronic pain after their expensive procedures:** Resnick, "100 Million Americans Have Chronic Pain"; Kaplan and Beech, *Total Recovery*, 78–79, 56, 144–145, 89.

CHAPTER 15

323 **boundaries are relational:** Pat Ogden and Janina Fisher, *Sensorimotor Psychotherapy: Interventions for Trauma and Attachment* (New York: Norton, 2015), chap. 19.

327 **the fatigue and discomfort we feel at first-stage muscle failure don't come from the working muscles:** T. D. Noakes, A. St. Clair Gibson, and E. V. Lambert, "From Catastrophe to Complexity: A Novel Model of Integrative Central Neural Regulation of

Effort and Fatigue during Exercise in Humans: Summary and Conclusions," *British Journal of Sports Medicine* 39, no. 2 (2005): 120–124.

327 **related research about the placebo effect:** O. Atasoy, "Your Thoughts Can Release Abilities beyond Normal Limits," *Scientific American*, November 26, 2013, www.scienti ficamerican.com/article/your-thoughts-can-release-abilities-beyond-normal-limits/#; U. W. Weger and S. Loughnan, "Mobilizing Unused Resources: Using the Placebo Concept to Enhance Cognitive Performance," *Quarterly Journal of Experimental Psychology* 66, no. 1 (2013): 23–28.

329 **Limiting beliefs are persuasive aspects of:** Marc Lesser, *Less: Accompling More by Doing Less* (Novato, CA: New World Library, 2009), chap. 7.

335 **observers have commented that resistance is often fear-based:** Tara Mohr, *Playing Big: Find Your Voice, Your Mission, Your Message* (New York: Gotham, 2014); Eric Maisel, *Fearless Creating: A Step-by-Step Guide to Starting and Completing Your Work of Art* (New York: Tarcher/Putnam, 1995); Steven Pressfield, *The War of Art: Break through the Blocks and Win Your Inner Creative Battles* (New York: Black Irish Entertainment, 2002).

335 **Steven Pressfield argues that resistance is proportional to love:** Pressfield, *The War of Art*, 37–38.

CHAPTER 16

341 **In our technocentric culture, we've taken this anticipation approach to the extreme:** Elizabeth A. Stanley, *Techno-Blinders: How the U.S. Techno-Centric Security System Is Endangering National Security* (unpublished manuscript, 2018).

341 **With enough measurement, data, calculation, and/or analysis, we assume certainty is possible:** Jakob Arnoldi, *Risk* (Cambridge, UK: Polity, 2009), chap. 5; Joost Van Loon, *Risk and Technological Culture: Towards a Sociology of Virulence* (London: Routledge, 2002), 186–187, 94; Ulrich Beck, *Risk Society—Towards a New Modernity* (London: Sage, 1992).

341 **the massive rush today toward "big data" and "data-mining" methodologies:** Viktor Mayer-Schoenberger and Kenneth Cukier, *Big Data: A Revolution That Will Transform How We Live, Work, and Think* (New York: Houghton Mifflin Harcourt, 2013), chap. 5.

342 **The narrower our window, the more intolerant of uncertainty we'll be:** Karl E. Weick, *Sensemaking in Organizations* (New York: Sage, 1995), 47; M. J. Dugas et al., "Intolerance of Uncertainty and Worry: Investigating Specificity in a Non-Clinical Sample," *Cognitive Therapy and Research* 25, no. 5 (2001): 551–558.

344 **Planning 2.0 provides a new structured outlet:** Although I practiced Planning 2.0 before coaching for the National Center of Faculty Development and Diversity, the faculty writing boot camp for which I coach teaches a similar technique. Both my Georgetown students and the professors I've coached have told me how empowered they feel using such planning techniques.

354 **with complex decisions, people are generally happier when they choose the intuitive answer that arrives from unconscious processing:** A. Dijksterhuis and L. F. Nordgren, "A Theory of Unconscious Thought," *Perspectives on Psychological Science* 1, no. 2 (2006): 95–109.

CHAPTER 17

355 **habits get stored in the basal ganglia:** Judson Brewer, *The Craving Mind: From Cigarettes to Smartphones to Love—Why We Get Hooked and How We Can Break Bad Habits* (New Haven, CT: Yale University Press, 2017), 1–7; Charles Duhigg, *The Power of Habit: Why We Do What We Do in Life and Business* (New York: Random House, 2012), 13–21.

355 **a study of individuals who worked out at least three times a week:** K. A. Finlay, D. Trafimow, and A. Villarreal, "Predicting Exercise and Health Behavioral Intentions: Attitudes, Subjective Norms, and Other Behavioral Determinants," *Journal of Applied Social Psychology* 32, no. 2 (2002): 342–356.

356 **studies in which people kept behavioral diaries:** W. Wood, J. M. Quinn, and D. A. Kashy, "Habits in Everyday Life: Thought, Emotion, and Action," *Journal of Personality and Social Psychology* 83 (2002): 1281–1297; J. M. Quinn and W. Wood, "Habits across the Lifespan," unpublished paper (2006).

356 **intentions account for only about 30 percent of our behavior:** M. S. Hagger, N. L. D. Chatzisarantis, and S. Biddle, "A Meta-Analytic Review of the Theories of Reasoned Action and Planned Behavior: Predictive Validity and the Contribution of Additional Variables," *Journal of Sport and Exercise Psychology* 24 (2002): 3–32; N. L. D. Chatzisarantis and M. S. Hagger, "Mindfulness and the Intention-Behavior Relationship within the Theory of Planned Behavior," *Personality and Social Psychology Bulletin* 33, no. 5 (2007): 663–676.

356 **Once the neurobiological structures in our brain and nervous system get established:** Bruce E. Wexler, *Brain and Culture: Neurobiology, Ideology, and Social Change* (Cambridge, MA: MIT Press, 2006), chap. 4.

357 **the three most stressful life events:** D. K. Harmon, M. Masuda, and T. H. Holmes, "The Social Readjustment Rating Scale: A Cross-Cultural Study of Western Europeans and Americans," *Journal of Psychosomatic Research* 14 (1970): 391–400.

357 **It takes time for the new reality to become familiar:** Wexler, *Brain and Culture*, 173.

358 **you start New Year's Day with a long list of resolutions:** David DeSteno, "How to Keep Your Resolutions," *New York Times*, December 31, 2017.

366 **Seventy percent of our immune system:** H. J. Wu and E. Wu, "The Role of Gut Microbiota in Immune Homeostasis and Autoimmunity," *Gut Microbes* 3, no. 1 (2012): 4–14; Elizabeth Lipski, *Digestive Wellness: Strengthen the Immune System and Prevent Disease through Healthy Digestion*, 4th ed. (New York: McGraw-Hill, 2012); Gary Kaplan and Donna Beech, *Total Recovery: Solving the Mystery of Chronic Pain and Depression* (New York: Rodale, 2014), 120–121.

366 **The microorganisms in the microbiome have epigenetic powers:** Kaplan and Beech, *Total Recovery*, 117–123; Wu and Wu, "The Role of Gut Microbiota"; Dale E. Bredesen, *The End of Alzheimer's: The First Program to Prevent and Reverse Cognitive Decline* (New York: Avery, 2017).

366 **Our intestinal tract also has 100 million nerves:** Lipski, *Digestive Wellness*; Kaplan and Beech, *Total Recovery*, 117.

367 **things that contribute to microbiome imbalances:** Lipski, *Digestive Wellness*; Kaplan and Beech, *Total Recovery*, chap. 5; Moises Valasquez-Manoff, "The Germs That Love Diet Soda," *New York Times*, April 8, 2018.

368 **fruits and vegetables with high pesticide residues:** The 2018 Dirty Dozen—which tend to have the greatest pesticide residues—are strawberries, spinach, nectarines, apples, grapes, peaches, cherries, pears, tomatoes, celery, potatoes, sweet bell peppers, and hot peppers. In contrast, the 2018 Clean Fifteen—which tend to have the least pesticide residues and thus are safer as non-organic produce—are avocados, sweet corn, pineapples, cabbage, onions, sweet frozen peas, papayas, asparagus, mangos, eggplant, honeydew melon, kiwi, cantaloupe, cauliflower, and broccoli. See the Environmental Working Group's "2018 Shopper's Guide to Pesticides in Produce," April 10, 2018, www.ewg.org /release/out-now-ewg-s-2018-shopper-s-guide-pesticides-produce#.

369 **Although caffeine has been shown to improve memory:** J. D. Lane et al., "Caffeine Affects Cardiovascular and Neuroendocrine Activation at Work and Home," *Psychosomatic Medicine* 64, no. 4 (2002): 595–603; D. Borota et al., "Post-Study Caffeine Administration Enhances Memory Consolidation in Humans," *Nature Neuroscience* 17 (2014): 201–203; S. E. Meredith et al., "Caffeine Use Disorder: A Comprehensive Review and Research Agenda," *Journal of Caffeine Research* 3, no. 3 (2013): 114–130; Deane Alban, "All about Caffeine Addiction and Withdrawal and How to Quit," Be Brain Fit blog, June 2, 2018, bebrainfit.com/caffeine-addiction-withdrawal.

370 **Many sleep medications actually backfire:** Bredesen, *The End of Alzheimer's*, 192–193.

370 **About 5 percent of the population has sleep apnea:** Kaplan and Beech, *Total Recovery*, 182–183.

371 **These foods late at night can really whack your circadian rhythm and cortisol cycle:** Anahad O'Conner, "The Pitfalls of Late-Night Snacking," *New York Times*, July 24, 2018.

371 **exercise . . . has been linked empirically with better sleep:** National Sleep Foundation, *2013 Sleep in America Poll: Exercise and Sleep* (Arlington, VA: National Sleep Foundation, 2013).

372 **Melatonin gets suppressed by light:** B. Wood et al., "Light Level and Duration of Exposure Determine the Impact of Self-Luminous Tablets on Melatonin Suppression," *Applied Ergonomics* 44, no. 2 (2013): 237–240.

373 **several specific ways exercise helps us widen our windows:** H. Eyre, E. Papps, and B. Baune, "Treating Depression and Depression-Like Behavior with Physical Activity: An Immune Perspective," *Frontiers in Psychiatry* 4, no. 3 (2013): 1–27; National Sleep Foundation, *2013 Sleep in America Poll*; R. A. Kohman et al., "Exercise Reduces Activation of Microglia Isolated from Hippocampus and Brain of Aged Mice," *Journal of Neuroinflammation* 10, no. 1 (2013): 885; Gretchen Reynolds, "Exercise May Starve a Cold," *New York Times*, December 22, 2015; Gretchen Reynolds, "When Exercise Takes a Vacation," *New York Times*, August 7, 2018.

374 **Research shows that a lack of social connections:** D. Umberson and J. Karas Montez, "Social Relationships and Health: A Flashpoint for Health Policy," *Journal of Health and Social Behavior* 51, no. 1 Suppl. (2010): S54–S66; Ruth Whippman, "Happiness Is Other People," *New York Times*, October 29, 2017; Jane E. Brody, "Social Interaction Is Critical for Mental and Physical Health," *New York Times*, June 13, 2017.

374 **the most consistent predictor of happiness:** Matthew D. Lieberman, *Social: Why Our Brains Are Wired to Connect* (New York: Crown, 2013), 247; Whippman, "Happiness Is Other People."

374 **a significant decline in American social connectivity:** L. Bruni and L. Stanca, "Watching Alone: Relational Goods, Television and Happiness," *Journal of Economic Behavior and Organization* 65, no. 3 (2008): 506–528; Lieberman, *Social*, 246–256.

375 **The more time teenagers spend looking at screens:** Jeanne M. Twenge, "Have Smartphones Destroyed a Generation?" *Atlantic*, September 2017.

375 **Social isolation . . . has grown dramatically:** Arthur C. Brooks, "How Loneliness Is Tearing America Apart," *New York Times*, November 24, 2018; Susan Scutti, "Loneliness Peaks at Three Key Ages, Study Finds—But Wisdom May Help," CNN, December 20, 2018; M. McPherson, L. Smith-Lovin, and M. E. Brashears, "Social Isolation in America: Changes in Core Discussion Networks over Two Decades," *American Sociological Review* 71, no. 3 (2006): 353–375.

375 **Nearly half of all meals:** Whippman, "Happiness Is Other People."

375 **one recent study showed that most lonely people were married:** C. M. Perissinotto and K. E. Covinsky, "Living Alone, Socially Isolated or Lonely—What Are We Measuring?," *Journal of General Internal Medicine* 29, no. 11 (2014): 1429–1431; Brooks, "How Loneliness Is Tearing America Apart."

376 **loneliness increases stress hormones:** Perissinotto and Covinsky, "Living Alone, Socially Isolated or Lonely"; Umberson and Karas Montez, "Social Relationships and Health"; Jane E. Brody, "How Loneliness Takes a Toll on Our Health," *New York Times*, December 12, 2017; Brody, "Social Interaction Is Critical for Mental and Physical Health."

CHAPTER 18

379 **I went to a different periodontist:** A grateful shout-out to Dr. Peter Farzin!

381 **Researchers who study emotion contagion:** Elaine Hatfield, John T. Cacioppo, and Richard L. Rapson, *Emotional Contagion* (Cambridge, UK: Cambridge University Press, 1993); Daniel Goleman, *Emotional Intelligence* (New York: Bantam, 1995), chap. 7; Bruce E. Wexler, *Brain and Culture: Neurobiology, Ideology, and Social Change* (Cambridge, MA:

MIT Press, 2006), chap. 3; Daniel J. Siegel, *The Developing Mind: How Relationships and the Brain Interact to Shape Who We Are* (New York: Guilford 1999), chaps. 4, 8.

381 **when we experience distress from social pain:** Matthew D. Lieberman, *Social: Why Our Brains Are Wired to Connect* (New York: Crown, 2013), chap. 3.

382 **motivated to stay connected to others through endorphins and oxytocin:** Lieberman, *Social*, 92–95.

382 **our brains have a system of mirror neurons:** Marco Iacoboni, *Mirroring People: The Science of Empathy and How We Connect with Others* (New York: Picador, 2009), chaps. 3, 4.

382 **The mirror system may also be implicated in imitative violence:** For reviews, see L. R. Huesmann and L. D. Taylor, "The Role of Media Violence in Violent Behavior," *Annual Review of Public Health* 27 (2006): 393–415; Iacoboni, *Mirroring People*, 204–210; Robert M. Sapolsky, *Behave: The Biology of Humans at Our Best and Worst* (New York: Penguin, 2017), 197–198.

383 **humans have evolved some additional structures:** Lieberman, *Social*, chaps. 5, 6, 9.

383 **in one experiment that tried to get students to use sunscreen:** E. B. Falk et al., "Predicting Persuasion-Induced Behavior Change from the Brain," *Journal of Neuroscience* 30, no. (2010): 8421–8424; Lieberman, *Social*, chap. 8.

385 **If the person we're speaking with is dysregulated:** Stephen W. Porges, *The Polyvagal Theory: Neurophysiological Foundations of Emotions, Attachment, Communication, and Self-Regulation* (New York: Norton, 2011), 58–59.

388 **leaders can affect how their subordinates will interpret:** P. Bartone, "Resilience under Military Operational Stress: Can Leaders Influence Hardiness?," *Military Psychology* 18, Suppl. (2006): 131–148; T. W. Britt et al., "How Leaders Can Influence the Impact That Stressors Have on Soldiers," *Military Medicine* 169, no. 7 (2004): 541–545.

392 **Highlighting adaptive capacity in music:** F. J. Barrett, "Creativity and Improvisation in Jazz and Organizations: Implications for Organizational Learning," *Organization Science* 9, no. 5 (1998): 605–622.

394 **64 percent of Americans in a recent Pew poll:** David Brooks, "The Siege Mentality Problem," *New York Times*, November 13, 2017.

394 **When a group feels threatened, it's easier for them to dehumanize:** M. Cikara, M. M. Botvinick, and S. T. Fiske, "Us versus Them: Social Identity Shapes Neural Responses to Intergroup Competition and Harm," *Psychological Science* 22, no. 3 (2011): 306–313; L. W. Chang, A. R. Krosch, and M. Cikara, "Effects of Intergroup Threat on Mind, Brain, and Behavior," *Current Opinion in Psychology* 11 (2016): 69–73.

395 **a Pew poll in August 2018 found that 78 percent of Americans:** Michael Scherer and Robert Costa, "'Rock Bottom': Supreme Court Fight Reveals a Country on the Brink," *Washington Post*, October 6, 2018.

396 **A 2018 Gallup poll shows that most Americans believe climate change won't affect them personally . . . a recent United Nations report projects that the worst effects will start occurring by 2040:** Coral Davenport, "Major Climate Report Describes a Strong Risk of Crisis as Early as 2040," *New York Times*, October 7, 2018; Megan Brenan and Lydia Saad, "Global Warming Concern Steady Despite Some Partisan Shifts," Gallup, March 28, 2018.

396 **65 percent of new vehicle sales:** Robert Ferris, "The Steadily Disappearing American Car," CNBC, April 6, 2018.

396 **we've mostly on relied on our all-volunteer force:** Mark Thompson, "Here's Why the U.S. Military Is a Family Business," *Time*, March 10, 2016; Kathy Roth-Douquet and Frank Schaffer, *AWOL: The Unexcused Absence of America's Upper Classes from Military Service—And How It Hurts Our Country* (New York: HarperCollins, 2006); Sarah Hautzinger and Jean Scandlyn, *Beyond Post-Traumatic Stress: Homefront Struggles with the Wars on Terror* (Walnut Creek, CA: Left Coast, 2014).

396 **two thirds of American civilians in 2011 said the disproportionate burden shouldered by U.S. troops:** Pew Research Center, "War and Sacrifice in the Post-9/11

Era" (2011); Chris Marvin, "Americans Are Viewing Veterans All Wrong," U.S. Department of Veterans Affairs blog, November 17, 2014, www.blogs.va.gov/VAntage /16011/americans-are-viewing-veterans-all-wrong; Benedict Carey, "After Thriving in Combat Tours, Veterans Are Struggling at Home," *New York Times*, May 30, 2016.

397 **these conflicts have cost more than $2 trillion, financed on the nation's credit card:** Sarah Kreps, *Taxing Wars: The American Way of War Finance and the Decline of Democracy* (Oxford, UK: Oxford University Press, 2018).

397 **the large allostatic loads that many Americans outside the military carry:** Centers for Disease Control and Prevention, "Obesity and Overweight," http://www.cdc.gov /faststats.overwt.htm; Gary Kaplan and Donna Beech, *Total Recovery: Solving the Mystery of Chronic Pain and Depression* (New York: Rodale, 2014), 180–182; Dan Keating and Lenny Bernstein, "U.S. Suicide Rate Has Risen Sharply in the 21st Century," *Washington Post*, April 22, 2016; Gina Kolata, "New High Blood Pressure Norm to Affect Millions," *New York Times*, November 14, 2017; R. C. Kessler, K. R. Merikangas, and P. S. Wang, "Prevalence, Comorbidity, and Service Utilization for Mood Disorders in the United States at the Beginning of the Twenty-First Century," *Annual Review of Clinical Psychology* 3 (2007): 137–158; Jen Doll, "How to Combat Your Anxiety, One Step at a Time," *New York Times*, December 21, 2017; Adeel Hassan, "Deaths from Drugs and Suicide Reach a Record in U.S.," *New York Times*, March 7, 2019; Amanda Erickson, "Opioid Abuse in the U.S. Is So Bad It's Lowering Life Expectancy. Why Hasn't the Epidemic Hit Other Countries?," *Washington Post*, December 28, 2017; Max Fisher and Josh Keller, "Only One Thing Explains Mass Shootings in the United States," *New York Times*, November 8, 2017.

398 **in one recent study, video gamers playing a nonviolent immersive game:** U. W. Weger and S. Loughnan, "Virtually Numbed: Immersive Video Gaming Alters Real-Life Experience," *Psychonomic Bulletin and Review* 21, no. 2 (2014): 562–565.

399 **We could address income inequality:** Alissa Quart, *Squeezed: Why Our Families Can't Afford America* (New York: HarperCollins, 2018); Nelson D. Schwartz, "Workers Needed, but Drug Testing Thins Pool," *New York Times*, July 25, 2017.

399 **The national debt now tops $21 trillion:** Harriet Torry, "U.S. Household Debt Continues to Climb in 3rd Quarter," *Wall Street Journal*, November 16, 2018; Kathryn Watson, "Under Trump's Watch, National Debt Tops $21 Trillion for First Time Ever," CBS News, March 17, 2018.